Advances in Renewable and Sustainable Engineering from the 5th International Conference on Sustainable Engineering Techniques

Advances in Renewable and Sustainable Engineering from the 5th International Conference on Sustainable Engineering Techniques

Guest Editors

Iqbal M. Mujtaba
Wadhah Amer Hatem
Mudhar Al-Obaidi

MDPI

Basel • Beijing • Wuhan • Barcelona • Belgrade • Novi Sad • Cluj • Manchester

Guest Editors
Iqbal M. Mujtaba
Faculty of Management,
Sciences and Engineering
University of Bradford
Bradford
UK

Wadhah Amer Hatem
Technical Institute of Baquba
Middle Technical University
Baghdad
Iraq

Mudhar Al-Obaidi
Technical Instructor Training
Institute
Middle Technical University
Baghdad
Iraq

Editorial Office
MDPI AG
Grosspeteranlage 5
4052 Basel, Switzerland

This is a reprint of the Special Issue, published open access by the journal *Processes* (ISSN 2227-9717), freely accessible at: https://www.mdpi.com/journal/processes/special_issues/1766Z37QEN.

For citation purposes, cite each article independently as indicated on the article page online and as indicated below:

Lastname, A.A.; Lastname, B.B. Article Title. *Journal Name* **Year**, *Volume Number*, Page Range.

ISBN 978-3-7258-4967-3 (Hbk)
ISBN 978-3-7258-4968-0 (PDF)
https://doi.org/10.3390/books978-3-7258-4968-0

Contents

processes

MDPI

Review

Sustainable Drilling Fluids: A Review of Nano-Additives for Improved Performance and Reduced Environmental Impact

Maaly Salah Asad [1], Mohammed Thamer Jaafar [1], Farhan Lafta Rashid [1], Hussein Togun [2,*], Musaab K. Rasheed [3], Mudhar A. Al-Obaidi [4,5], Qusay Rasheed Al-Amir [6], Hayder I. Mohammed [7] and Ioannis E. Sarris [8]

1 Petroleum Engineering Department, College of Engineering, University of Kerbala, Karbala 56001, Iraq; maaly.salah@uokerbala.edu.iq (M.S.A.); mohammed.thamer@uokerbala.edu.iq (M.T.J.); farhan.lefta@uokerbala.edu.iq (F.L.R.)
2 Department of Mechanical Engineering, College of Engineering, University of Baghdad, Baghdad 10071, Iraq
3 Institute of Technology, Middle Technical University, Baghdad 32001, Iraq; musaabk.rasheed@mtu.edu.iq
4 Technical Institute of Baquba, Middle Technical University, Baquba 32001, Iraq; dr.mudhar.alaubedy@mtu.edu.iq
5 Technical Instructor Training Institute, Middle Technical University, Baghdad 10074, Iraq
6 Mechanical Power Technical Engineering Department, College of Engineering and Technologies, Al-Mustaqbal University, Babylon 51001, Iraq; qusay.rasheed.abdal-amir@uomus.edu.iq
7 Department of Physics, College of Education, University of Garmian, Kalar 46021, Iraq; hayder.i.mohammad@garmian.edu.krd
8 Department of Mechanical Engineering, School of Engineering, University of West Attica, 12244 Athens, Greece; sarris@uniwa.gr
* Correspondence: htokan_2004@yahoo.com

Abstract: The implication of nano-additives in drilling fluids introduces a promising avenue for enhancing sustainability in the oil and gas industry. By upgrading the properties of drilling fluids, nano-additives can contribute to mitigating the drilling costs, improving wellbore stability, and minimizing the environmental impact. For example, the use of nano-additives can diminish the amount of drilling fluid required, thus reducing the volume of waste generated. Also, nano-additives can enhance the efficacy of drilling operations, leading to reduced energy consumption and greenhouse gas emissions. This review researches the potential of nano-additives in enhancing sustainable drilling practices, emphasizing the environmental advantages and economic advantages associated with their usage. Specifically, this comprehensive review will elucidate the most recent developments in drilling fluids by evaluating the impact of nano-additives. Referring to the conclusions, adding nanoparticles to drilling fluids significantly improves their characteristics. At 0.2 parts per billion (ppb), for example, the yield stress increases by 36% and the plastic viscosity increases by 17%. In addition, the inclusion of nanoparticles at a concentration of 0.6 ppb led to a significant decrease of 60% in the loss of filtrate. The measured enhancements highlight the capacity of nano-additives to augment the properties of drilling fluid, necessitating additional investigation into their prospective applications for enhancing competitiveness in the gas and oil industry. This study methodically examines the effects of these breakthroughs on scientific, commercial, and industrial sectors. It intends to provide an inclusive understanding of the possible advantages of nano-additives in drilling operations.

Keywords: drilling fluid; nano-additives; nanofluids; recent advances

1. Introduction

Addressing the possible risks associated with manufactured nanoparticles in drilling fluids (DFs) is necessary considering their widespread application across various industrial domains. Users who are exposed to this technology may have concerns about their health, safety, and the potential negative effects on the environment. Over the last 20 years, nanotechnology has also garnered great attention from the gas and oil sectors throughout the world. One potential use of nanotechnology in the gas and oil sector is the production of drilling muds containing microscopic solid particles. Minimal amounts of these particles have been shown to alleviate a number of issues, involving but not confined to increasing cutting transport, decreasing the force of friction between borehole wells and drill pipes and filtration of DF, altering rock wettability, preventing corrosion, and reinforcing borehole walls created through incompetent rocks sinking [1,2].

Nano-additives have demonstrated favorable outcomes in enhancing the characteristics of drilling fluids, namely in various sectors, such as gas and oil, cosmetics, electronics, biomedicine, and agriculture. Transferring drilling cuttings from the well's bottom to the surface for disposal is a primary use of drilling fluids. To show whether the hole is being cleaned effectively, other terms, like cutting transport efficiency and hole cleaning efficiency, are used interchangeably. The density of the drilling fluid, rate of penetration, cutting density, hole size, cutting size, flow rate, pipe rotation, cutting shape, and other characteristics may all impact the cutting transport efficiency [3–5].

DF should have three characteristics: (1) it should be easy to use, (2) economical, and (3) safe for the environment. DF serves multiple purposes, including cleaning the well by lifting cuttings from the bottom to the surface, preventing caving by supporting the formation, maintaining well tightness, lubricating and cooling the tool, transferring hydraulic power to move the drill bit, and revealing details about the drilled formation's characteristics through the process of lifting cuttings from the surface to the bottom [6].

At each step of the drilling process, one of three primary kinds of DFs—synthetic mud, water-based mud, or oil-based mud, is accessible and employed. Drilling operations often use water-based mud due to its lack of the other two types' downsides, which include increased expense, environmental impact, disposal issues, and health and safety concerns [7]. The main ingredients of water-based mud are water and other additions, such as bentonite (montmorillonite clay). Pump ability and cutting carrying capacity are both improved by their high thinning viscosities and rheological characteristics [8]. Heats greater than 250 degrees Fahrenheit may affect montmorillonite clay [9]. The instability of shale, damage to the formation, poor mud cake properties, and high loss of fluid are some of the few drawbacks of using water-based drilling fluid [10]. As such, a process known as "filtration" occurs, which is the increased loss of fluid from the DF into the formation. In this regard, a reduction in overall capacity would be the result of the deposition of solids from the DF as a result of the pressure change between the wellbore pressure and formation pressure [11].

To enhance the filtration and rheological features of water-based DFs, a number of polymeric additives have been utilized, such as polyanionic cellulose (PAC) [12,13], hydroxyethyl cellulose (HEC) [14,15], carboxymethyl cellulose [16,17], acrylamide–styrene copolymer (SBASC) [18], partially hydrolyzed polyacrylamide (PHPA) [19], polyethylene glycol [20], low- and high-density polyethylene [21], etc. Consequently, the DF contains a diversity of nanoparticles to enhance its performance [22–24].

Vryzas and Kelessidis (2017) [25] provided a comprehensive review of the state of the art on the incorporation of nanoparticles into drilling fluids with the purpose of improving their rheological and filtration properties, in addition to enhancing the shale stability and strengthening the wellbore. This review discussed and critically evaluated a number of aspects, including the kind, size, and nanoparticle form, the volumetric concentration, the inclusion of various surfactants, and the introduction of an outer magnetic field. Referring to the findings of many investigations, nanoparticles (NPs) exhibited significant potential as an addition to drilling fluid for stern drilling.

Current innovations in the use and synthesis of NPs in DF systems were provided by Cheraghian (2021) [26]. NPs were analyzed for their functions in regulating fluid loss, thermal properties, filtration characteristics, and mud cake thickness. The related literature showed that there are two main reasons to embrace nanomaterials in drilling fluid: to augment the fluid's thermal properties and to augment its mechanical–physical characteristics. In terms of thermal characteristics and HPHT settings, the results established that carbon

NPs and metal NPs are appropriate conductors. The analytical results showed that NPs might be a greener choice than other materials for drilling fluid base applications.

Tahr et al. (2023) [27] inspected the investigations' results and any gains made by utilizing the new materials in drilling fluids. A detailed review and comparison of current developments in drilling fluids was delivered by the researchers, who also sought appropriate groundbreaking materials that would increase the gas and oil industry's competitiveness. As biodegradable materials, the filtration and rheological capabilities were indicated to be significantly improved by the powders of rice husk and pomegranate peel. Titanium oxide and nano-clay also significantly altered the drilling fluid's characteristics. Therefore, these qualities may be enhanced, and the wellbore stability may be provided by combining these two unique materials—titanium and pomegranate peel.

In their assessment of nanotechnology's role in drilling engineering, Martin et al. (2023) [28] focused attention on drilling fluids and the ecological issues surrounding their post-use disposal. The nanotechnology difficulties in the gas and oil industry, the price of NP mud fluids, and the destiny and behavior of nanoparticles were detailed by researchers, who also highlighted the hazards of nanotechnology's toxicity to people, ecosystems, and the environment. Since this field is still in the early stages of industrial-scale development, more data about the dangers of nanomaterials to the human body are required. The results demonstrated that these substances might induce long-term disorders that would show up in the body years after they infiltrate human organs. In addition to their great penetration into human skin and propensity to induce new illnesses, these particles may readily disperse throughout the environment and persist in soil, water, and the air for long periods.

It is still necessary to critically evaluate the thermal properties and interpretation characteristics of these modified drilling fluids, in addition to assessing their feasibility and practicality for real-world applications, even though there have been several review studies on the application of nano-additives to enhance DF properties. Continuing from this introduction and the review studies that came before it, this study will examine the most current research published between 2019 and 2024 to evaluate the progress made in using nano-additives to promote the characteristics of DFs. Thus, it is reasonable to assume that this study will provide state-of-the-art information on drilling fluids that use nano-additive technology. More sustainable and environmentally friendly energy solutions may be possible as a result of this review, which summarizes recent research that intended to enhance the characteristics of drilling fluids. Interestingly, this review aims to consolidate diverse findings, identify knowledge gaps, and offer new insights into the application of nano-additives for enhancing drilling fluid properties. The significance of this review lies in its capacity to highlight specific counted enhancements in rheological features, such as enlarged plastic viscosity and yield stress, while also underlining the potential advantages of nano-additives for augmenting competitiveness in the gas and oil sector and improving environmental sustainability. By bridging the gap between research outcomes and practical applications, this review offers a unique viewpoint on the utilization of nano-additives in drilling fluids, pointing out to advance their implementation in the industry.

2. Nano-Additives for Improving Drilling Fluid Properties

Nowadays, NPs have their features enhanced using nanotechnology. This is attributed to the fact that the size and shape of nanomaterials, which are impacted by their interaction with mud components, affect their ability to improve fluid performance [29–32]. Drilling fluid has many uses, including stabilizing the wellbore, decreasing formation damage, and avoiding drill cuttings. Any of these tasks can be achieved by adding nanomaterials to drilling fluids [33]. Using NPs with uncommon properties, including a large surface area and strong thermal conductivity, may lessen drilling fluid difficulties. Using NPs in drilling fluids has several key advantages, including reducing mud cake and fluid loss, improving heat transmission, lubrication, and rheological characteristics, like viscosity, and eliminating hazardous compounds [26]. Multiple researchers have investigated the potential of NPs as an addition to DF.

2.1. Impact of Nanoparticles on Rheological Characteristics

Hajiabadi et al. (2019) [34] observed the rheological behaviors of inverted oil emulsions (W/O), a drilling fluid often used in the gas and oil sector, with a focus on the integration of carbon-based nanomaterials. Famous rheological models, like the model of Carreau, were developed to create and evaluate W/O fluid samples using a diversity of modified multi-walled carbon nanotube (MWCNT) additions over a temperature range. The conclusions established that the nanomaterial additions expressively boost the crucial rheological characteristics and that the behaviors of the DF samples match Carreau's model well. The differential pressure profiles obtained by injecting the prepared DFs and then damaging the core are presented in Figure 1. The base fluid exhibits a smooth trend, whereas the NDF-3 and NDF-1 samples have a sharp one. Injecting around five pore volumes of the base fluid resulted in 87 psi on the gage, but testing with nanofluids was halted after just one pore volume injection due to an abrupt and significant increase in pressure over the apparatus safety boundary.

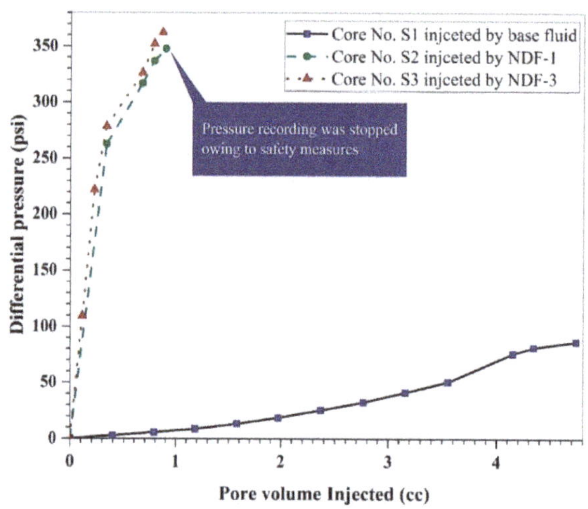

Figure 1. Differential pressure against pore volume injected via core samples (NDF-1, NDF-2, and NDF-3, with 2, 4, and 6 g of 500 mg dispersed POCNT in 80 mL of deionized water) [34] (Reproduced with permission from Journal of Petroleum Science and Engineering by Elsevier, Amsterdam, The Netherlands, 2024; Order Number: 586027021288.5).

2.2. Impact of Nanoparticles on Hole Cleaning

Traditional water-based drilling fluids typically consist of a base fluid, additives, and weighting materials to achieve desired properties, such as density and viscosity. These fluids are frequently utilized in drilling operations due to their cost-efficiency and compatibility with a wide range of geological formations. There are a number of successful implementations of nano-additives in drilling fluids in real-world drilling operations. For instance, oil and gas companies used a nano-clay additive in a deep-water drilling operation. The nano-clay has expressively enhanced the rheological properties of the drilling fluid and thermal stability, while decreasing the torque and drag and enhancing hole cleaning efficiency [35]. Nanoparticles increase the ability of colloids to interact with cuttings, resulting in more efficient cuttings transport compared to conventional water-based drilling fluids. Furthermore, the use of nano-diamonds can enhance the lubricity of the drilling fluid, reducing friction and wear on the drill bit [36]. The utilization of nanosilica additive in a high-temperature drilling operation can enhance the thermal stability of the drilling fluid, with permitting to withstand temperatures up to 250 °C. This in turn has enabled researchers to drill deeper wells and access new geothermal resources [37]. The researchers intend to introduce a full description of the advantages and performance enhancements offered by nano-enhanced DFs in a comparison against the regular formulations by highlighting these differences.

The efficiency of drilling fluids containing nano-silica in cleaning holes during directional drilling operations was considered by Boyou et al. (2019) [38]. To mimic drilling conditions in a wellbore, a system of flow loops with variable rotational speeds (0 and 150 rpm) was suggested. The cuttings removal execution of nano-enhanced DFs was compared against the traditional water-based DFs. In every experiment, the addition of silica nanoparticles improved the efficiency of cuttings transport. The results showed that several nano-enhanced water-based DFs improved cuttings' transit efficiency by 30.8% to 44% due to increased colloidal interactions with cuttings caused by nano-silica in the mud. The use of nano-silica in water-based DFs has shown potential for extended-reach drilling operations, as it improved the hole cleaning process. Figure 2 shows a comparison of the CTEs at crucial angles as a function of pipe rotation. The results demonstrate that the CTEs were improved by 18.4% and 25.4%, respectively, when the pipe was rotated with DFs containing the optimal densities and concentrations of nano-silica of 12 and 9 ppg, respectively.

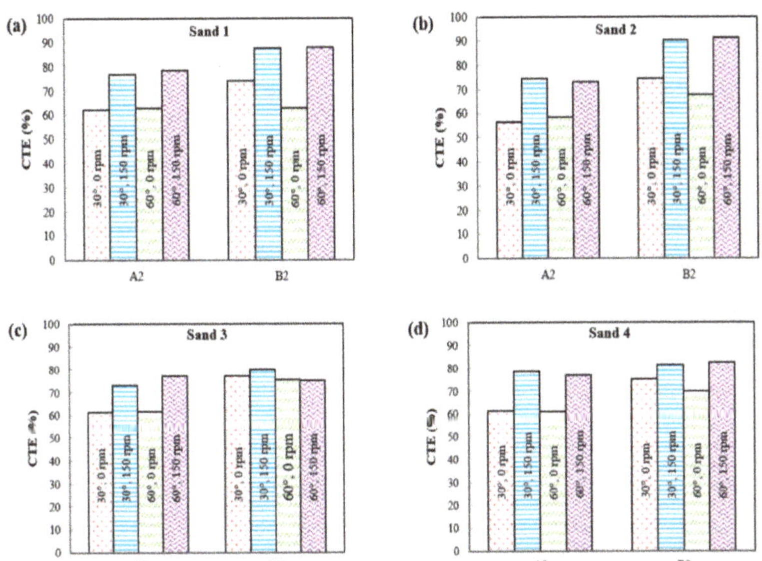

Figure 2. Mud with optimal nano-silica content and pipe rotation of 150 rpm improves CTE for cuttings sizes 1.40–1.69 mm in (**a**), 1.70–1.99 mm in (**b**), 2.00–2.79 mm in (**c**), and 2.80–4.00 mm in (**d**) at critical angles [38] (Reproduced with permission from Journal of Petroleum Science and Engineering by Elsevier, Amsterdam, The Netherlands, 2024; Order Number: 586027057598.9).

2.3. *Impact of Nanoparticles on Filtration*

The findings of drilling fluid filtration via a porous media with varying permeabilities, including silicon oxide nanoparticles, were given by Minakov et al. (2019) [39]. The examined fluid samples were first modeled after a water-based clay suspension. A dispersed phase made of aluminum nitride (AlN) particles was used. The microparticle size ranged from 1 to 10 μm, and their concentration ranged from 0.5 to 4.00 wt.%. The size of the SiO_2 nanoparticles ranged from 5 to 100 nm, and their concentration was adjusted between

0.25 and 4.00 wt.%. Findings indicate that nanoparticle size, concentration, substance, microparticle size and concentration, and ceramic filter pore diameters all have a role in the filtering of DFs, including nanoparticles. Adding nanoparticles to a micro-suspension drastically decreases its filtering capacity and changes the shape and cake thickness of those forms on the filter's surface. In addition to the the size and concentration of the nanoparticles, the DF's microparticle characteristics and the rock's pore size are the other factors that determine the additive's beneficial impact on filtration losses. The filtering loss is shown depending on the microparticle average size to filter average pore size ratio. Figure 3 shows that the least loss of filtration, in the nanoparticle's absence, corresponds to a ratio of 0.36 between the diameters of the particles and the pores.

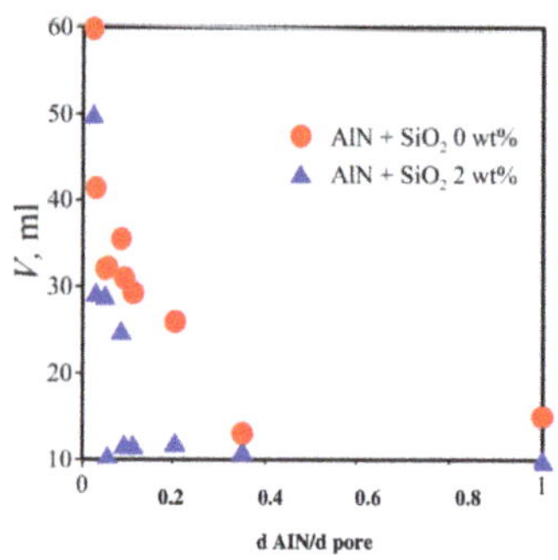

Figure 3. Loss of filtration for DF containing SiO_2 nanoparticles 5 nm in size at 2 wt.% depends on the ratio of the AlN microparticle size at 2 wt.% to the filter's average pore size, determined by mercury [39] (Reproduced with permission from Journal of Natural Gas Science and Engineering by Elsevier, Amsterdam, The Netherlands, 2024; Order Number: 586027090281.7).

While higher concentrations of nano-additives can indeed improve filtration loss control up to a certain point, excessive concentrations may lead to diminishing returns or even counterproductive effects. This phenomenon is likely due to such factors as the agglomeration of nanoparticles, increased viscosity of the drilling fluid, and potential interactions between the nano-additives and other components of the fluid. The particle size of nano-additives also plays a crucial role in their effectiveness. Smaller particles are generally more effective in bridging pores and reducing filtration due to their increased surface area and better penetration into the porous formation. However, the optimal particle size may vary depending on the specific type of nano-additive and the nature of the formation being drilled.

The critical concentration at which the benefits of nano-additives plateau or diminish can vary significantly depending on the specific type of nano-additive being used. The critical concentration for nano-additives will depend on such factors as their particle size, surface chemistry, and interactions with other components of the drilling fluid. It is essential to conduct thorough testing and optimization to determine the optimal concentration for a specific nano-additive and drilling application.

2.4. Impact of Nanoparticles on Rheology

Polyacrylamide (PAM), sodium dodecyl sulfate (SDS), and different quantities of silicon dioxide, an anionic nanoparticle, were studied in a complicated aqueous solution by Inturi et al. (2019) [40]. The impacts of various concentrations of silica NPs on the fluid's rheological characteristics are investigated at a known concentration of surfactant and polymer. We studied both hydrated and unhydrated polyacrylamide solutions at a temperature close to 90 °C to determine the impact of the polymer's thermal stability. When NPs are added to a fluid at low velocity, the viscosity of the fluid is found to be two orders of quantity lower. Nevertheless, because of the intricate structure, including the particle, surfactant and polymer, the fluid exhibited shear-thickening characteristics. The interaction of ionic charges, which causes complex structures to develop in hydrated polymers, has a significant impact on the fluid's viscosity. Figure 4 shows how the complex structure of the SP slug, including silica NPs, changes the flow behavioral characteristics.

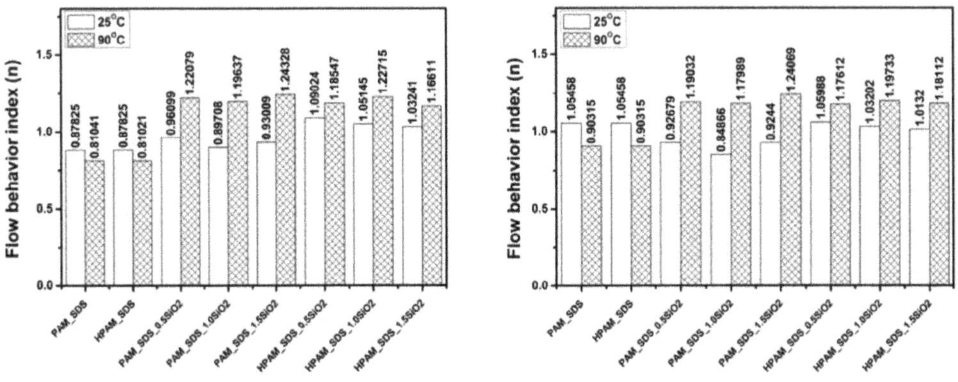

(a) **(b)**

Figure 4. (**a**) Flow index of behavior values at (**a**) 25 °C and (**b**) 90 °C [40] (Reproduced with permission from Materials Today: Proceedings by Elsevier, Amsterdam, The Netherlands, 2024; Order Number: 586027114868.6).

2.5. *Impact of Nanoparticle Type on Viscosity*

The effect of the particle size of glass spheres and polymers, barite, and calcite on the apparent viscosity of suspensions was studied by Nascimento et al. (2019) [41]. Xanthan gum (XG) and carboxymethyl cellulose (CMC) were dissolved in water and applied as viscosifying agents. The zeta potential, microstructure, and rheological properties of the suspensions were investigated. The shear stress values were greater in the suspensions containing small material particles. With CMC polymer, the viscosity was enhanced more markedly than with XG, and the solid effect took a back seat to the polymer chain feature.

In contrast to calcite suspensions, which exposed a larger electrostatic interaction with polymers, glass and barite sphere suspensions had larger zeta potentials and lower shear stress values. The ANOVA analysis was used to conduct statistical tests, and the results were assured within a 95% confidence level. Referring to the statistical tests, the shear stress values were expressively impacted by the polymer, the distribution of the particle sizes, and the chemical features of the solids.

2.6. *Impact of Copper Oxide/Polyacrylamide Nanocomposite on Thermal Properties*

The experimental results of using a copper oxide/polyacrylamide nanocomposite as an addition to water-based DF were delivered by Saboori et al. (2019) [42]. The solution polymerization method was used to generate the nanocomposite, which was then reflected by dynamic light scattering, Fourier transform infrared spectroscopy, and field emission scanning electron microscopy. The rheological, filtration, and thermal conductivity improvements for both salty and deionized water were calculated using developed methods after adding the synthesized nanocomposite to water-based DF at variable concentrations. DF viscosity was found to have significantly improved. Both the amount of fluid filtration (fluid loss) and the filter cake thickness were decreased. Filter cakes enhanced with nanocomposite addition are smoother and less porous. In terms of filtration and rheological characteristics, the nanocomposite addition performed better in deionized water-based DF; however, in salty water, it enhanced thermal conductivity more than in the deionized base DF. Compared to the water-based DF without nanocomposite, the filter cake thickness decreases by 3.7 to 69.14%, and fluid loss decreases by 2.4 to 63.27% as the nanocomposite concentrations increase, as shown in Figures 5 and 6, respectively. The enhanced performance of the nanocomposite in deionized water-based DF can be attributed to the absence of ions and impurities that may interfere with the interactions between the nanoparticles and the fluid components.

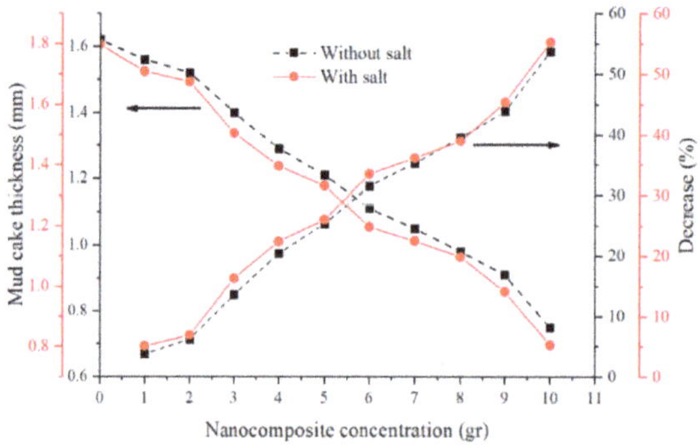

Figure 5. Nanocomposite and filter cake thickness [42] (Reproduced with permission from Powder Technology by Elsevier, Amsterdam, The Netherlands, 2024; Order Number: 586034077227.8).

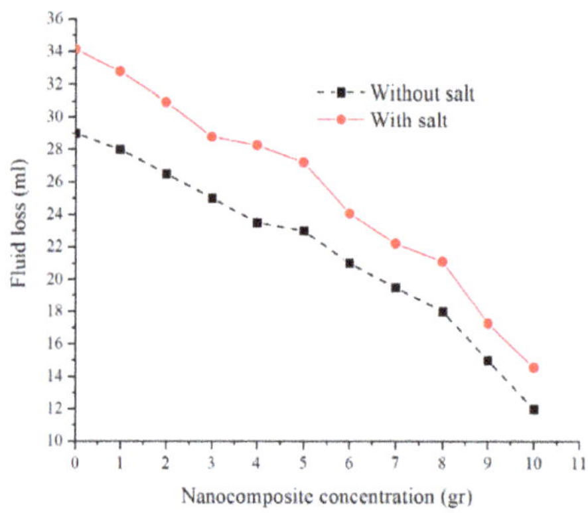

Figure 6. Nanocomposite concentration and fluid loss [42] (Reproduced with permission from Powder Technology by Elsevier, Amsterdam, The Netherlands, 2024; Order Number: 586034077227.8).

In contrast, in salty water, the presence of ions can lead to changes in the colloidal stability of the nanocomposite, affecting its rheological and filtration properties. However, the enhanced thermal conductivity in salty water may be due to the interactions between the nanoparticles and the salt ions, leading to improved heat transfer properties. By elucidating these scientific mechanisms, the authors aim to provide a deeper understanding of how the composition of the fluid medium influences the performance of nanocomposite additives in drilling fluids.

2.7. *Impact of $BiFeO_3$ Nanoparticles on Various Properties*

Along with polymeric additives, such as bentonite, Perween et al. (2019) [43] created a $BiFeO_3$ (BFO) NP using non-aqueous wet chemical procedures. The nanoparticle was then used as an addition for water-based mud. The mud's qualities were enhanced because BFO nanoparticles were compatible with clay. When nanoparticles interacted with clay and polymers, the fluid loss of apparent viscosity, API, yield point, and plastic viscosity improved. There was a 35% reduction in API filtrate loss at 20 °C prior to hot rolling and a

25% increase in AV, PV, and YP when the nanoparticle concentration was increased from 0% to 0.30 w/v%. Unlike mud devoid of nanoparticles, these characteristics were preserved even after being hot-rolled at 110 °C. The drilling fluid was made more thermally stable by adding nanoparticles and bentonite platelets.

2.8. Impact of CuO and ZnO Nanoparticles on Filtration and Rheology

At 25 °C, 50 °C, and 80 °C, as well as at 100 °C and 500 psi, Dejtaradon et al. (2019) [44] investigated the impacts of varying concentrations of CuO and ZnO NP additions on the filtration and rheological characteristics of a water-based DF. Nanofluids containing CuO and ZnO nanoparticles with concentrations ranging from 0.1 to 1 wt.% were created and then added as additives (1 vol.%) to water-based base DFs that had already been prepared. Both nanoparticle-based DFs outperformed the base fluid in terms of rheological parameters; however, ZnO performed better in general than CuO. DFs containing NP were indicated to be more stable when heated to greater temperatures. Filtration investigations achieved at high temperature and pressure (100 °C and 500 psi) and elaborated that 0.8 wt.% CuO NPs can reduce the filtration loss by 30.2%. However, 0.8 wt.% ZnO nanoparticles can reduce the filtration loss by 18.6%. The thickness of the mud cake was similarly reduced while using the CuO (27.6% reduction) and ZnO (24.6% reduction) NP fluids as compared to the base fluid. These conclusions showed that nanoparticles of zinc oxide and copper oxide can enhance the features of drilling fluids that include water and have the possibility to be employed as an addition for high-efficiency filtration loss.

2.9. Impact of Biopolymer/Nanoparticle Combination on Rheology, Filtration, and Lubrication

An innovative drilling fluid with enhanced rheological, filtration, and lubricating capacities was presented by Al-Yasiri et al. (2019) [45] using a biopolymer/nanoparticle combination to improve the interpretation of a water-based mud (WBM) formulation. Due to the complementary features of nanoparticles and the polymer, the xanthan gum–SiO_2 combination has the potential to solve operational problems in drilling operations. The innovative nanofluids were studied using thermal–gravimetric analysis, FTIR, X-ray diffraction, and SEM. The adapted drilling fluids improved upon the usual WBM formulations in many areas, comprising bit lubrication, filtrate loss, hole cleaning, and yield point. Current drilling fluids may be substantially enhanced with the help of nanoparticles if they are designed correctly. Figure 7 shows that the filtrate volume is reduced by both xanthan gum and the use of SiO_2 NPs, although the latter has a smaller percentage reduction than the former.

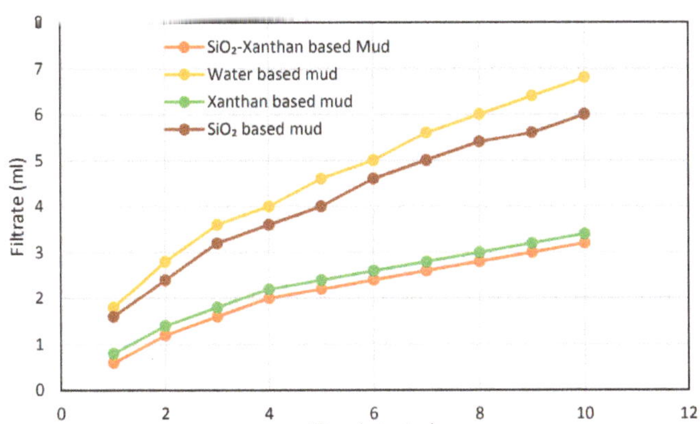

Figure 7. Sample drilling fluid filtrates [45] (Reproduced with permission from Journal of Petroleum Science and Engineering by Elsevier, Amsterdam, The Netherlands, 2024; Order Number: 586034129294.7).

The impact of laponite NPs on the DF's thermal stability was investigated by Huang et al. (2019) [46]. Radical polymerization of diallyl dimethylammonium chloride, 2-acrylamido-2-methylpropane sulfonic acid, and acrylamide monomers was used to create a terpolymer (AAD) first. At high temperatures, the onset viscosities and decomposition temperature of the AAD terpolymer were both enhanced by the inclusion of laponite nanoparticles, as shown by high-temperature, high-pressure (HTHP) rheology and thermogravimetric analysis (TGA), respectively. The next step was to mix AAD and laponite with water to create drilling fluids, which were then aged at various temperatures. This is because the ADD terpolymer's functional groups interact strongly with laponite, which causes the DFs containing laponite to have a much greater viscosity than those without laponite.

3. Optimizing Drilling Fluids with Nanomaterials

The oil and gas industry constantly pushes the boundaries of wellbore exploration, encountering increasingly complex and demanding environments. Traditional drilling fluids often have a specific barrier to meet the challenges commensurate with these severe circumstances, leading to problems, like wellbore instability, formation damage, and compromised wellbore integrity. Accordingly, nanotechnology emerges as a game-changer, offering a radical method to optimize drilling fluids.

Concerns have been expressed over the possible long-term environmental hazards associated with the usage of nanoparticle additions in drilling fluids. The hazards encompass certain factors, such as persistence, bioaccumulation, toxicity, and interactions with other pollutants. The precise characteristics of the nanoparticle, the methods by which it is released, and the surrounding environmental circumstances can impact the magnitude of these hazards.

A lifecycle assessment (LCA) is a thorough methodology used to systematically examine the environmental consequences of items and activities over their whole lifespan. The assessment of environmental concerns associated with nanoparticle additions in drilling fluids may be conducted by taking into account several aspects, including raw material extraction, nanoparticle production, drilling fluid formulation, drilling operations, and waste management. A comprehensive lifecycle assessment (LCA) enables the identification of significant environmental hotspots and the formulation of risk mitigation strategies.

Via nanoparticle integration, i.e., the integration of materials with at least one dimension in the nanometer range (billionths of a meter) into drilling fluids, the researchers revealed a new level of control and performance. These tiny particles have exceptional structures due to their high surface area and interaction with other molecules. When intentionally added to drilling fluids, they can significantly enhance many crucial features. For instance, nanoparticles can improve the rheology (flow behavior) of drilling fluids, leading to better wellbore cleaning and hole stability. In this aspect, nanoparticles can act as nano-lubricants, which enable researchers to reduce the friction between the drilling fluid and the wellbore walls. Indeed, the addition of nanoparticles can upgrade the fluid's viscosity and shear thinning characteristics, which enables researchers to improve the flow behavior and hole cleaning efficiency. Accordingly, this lessens the torque and drag on the drill string, which can lead to faster drilling rates and lower energy consumption. Also, they can act as well-organized filtration control agents, diminishing fluid loss into the formation and stopping formation damage. Specifically, the nanoparticles can act as bridge agents while plugging formation pores and mitigating fluid invasion. Furthermore, specific nanoparticles offer directed functionalities, such as shale inhibition, thermal stability improvement, and even wellbore clean-up capabilities. For instance, specific types of nanoparticles can interact with clay particles in shale formations, which in turn inhibit them from swelling and dispersing. As research in this field continues to progress, the potential of nanomaterials to revolutionize drilling fluids and unlock new prospects in wellbore exploration is enormous.

3.1. Modified Nanomaterials for Enhanced Drilling Fluid Performance

Multiwall carbon nanotubes (MWCNTs) were modified to increase their salt tolerance and thermal stability by Ma et al. (2020) [47]. An anionic polymer was functionalized onto MWCNTs utilizing the surface-initiated atom transfer radical polymerization (SI-ATRP) method. Two composites, namely MWCNTs-g-SPMA-2 and MWCNTs-g-SPMA-1, were generated by amending the monomer-to-initiator ratio. After 60 days of room temperature storage, both were systematically mixed with standard API brine and saturated brine (2 wt.% $CaCl_2$ + 8 wt.% NaCl). The modified MWCNTs also remained stable in a slightly alkaline environment and after 24 h at 170 °C. Referring to the following experiments, water-based drilling fluids using MWCNTs-g-SPMA-2 as a nano-plugging additive demonstrated greater plugging performance. To attain a plugging efficiency of 50.96%, a small quantity of 0.3 wt.% MWCNTs-g-SPMA-2 was added.

Zhong et al. (2020) [48] evaluated modified nanoparticles using FTIR, distribution of particle size, TGA, TEM, and zeta potential. The modified nanoparticles were positively charged and dispersed better in water than clean nano-silica particles. The modified nanoparticle's shale stabilizing performance was assessed using linear swelling and shale cutting hot-rolling dispersion as pressure transmission and spontaneous imbibition tests, and chemical inhibitive tests as physical plugging tests. The modified nanoparticles demonstrated physical plugging and chemical inhibition, whereas nano-silica revealed very minimal physical plugging. The grafted hyperbranched polymer significantly adsorbed onto the clay surface via hydrogen bonding and electrostatic contact, inhibiting shale swelling and hydration. The negative charge of nano-silica particles in the dispersion kept its zeta potential constant regardless of concentration, as seen in Figure 8.

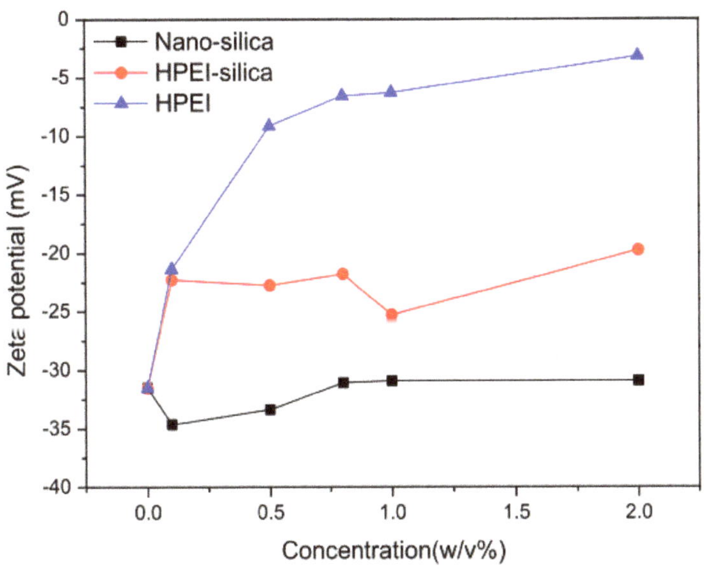

Figure 8. HPEI and nanoparticles affect sample dispersions' zeta potential [48] (Reproduced with permission from Journal of Natural Gas Science and Engineering by Elsevier, Amsterdam, The Netherlands, 2024; Order Number: 586035001894.6).

To enhance the plugging effectiveness of oil-based DF (OBDF) in shale formations, Li et al. (2020) [49] employed the continuous emulsion polymerization process to create a styrene–butadiene resin/nano-SiO_2 (SBR/SiO_2) composite. This composite was then used as a plugging agent for OBDF. Thermal stability, Fourier transform infrared spectroscopy, and SEM were used to characterize SBR/SiO_2. Inferred from the data, the SBR/SiO_2

structure seems to be as expected. The findings showed that SBR/SiO_2 may be distributed in both mineral oil and water at the same time, which is not the case for most conventional OBDF plugging agents. Because OBDF is an emulsion of water and oil, the plugging agents work best when distributed in two continuous stages. The findings show that SBR/SiO_2 may penetrate shale formation nanopores, reducing fluid invasion and enhancing wellbore stability. Figure 9 shows that the downstream pressure of 1#OBDF reached 12 MPa in 2.5 h, verifying that 1#OBDF cannot fill shale pores in the presence of plugging agents. These conventional plugging agents typically display imperfect compatibility with both mineral oil and water, leading to challenges in accomplishing uniform distribution and operative plugging in OBDF systems. On the other hand, the results pointed out that the SBR/SiO_2 composite can exhibit superior properties that permit it to be distributed in both mineral oil and water phases concurrently. This dual-phase compatibility of SBR/SiO_2 improves its efficiency as a plugging agent in OBDF by guaranteeing uniform distribution and upgraded plugging efficiency across both phases. By expanding on these conventional observations with OBDF plugging agents, the researchers intend to underscore the groundbreaking nature of the SBR/SiO_2 composite and its potential to reduce the limitations associated with traditional plugging agents in OBDF systems.

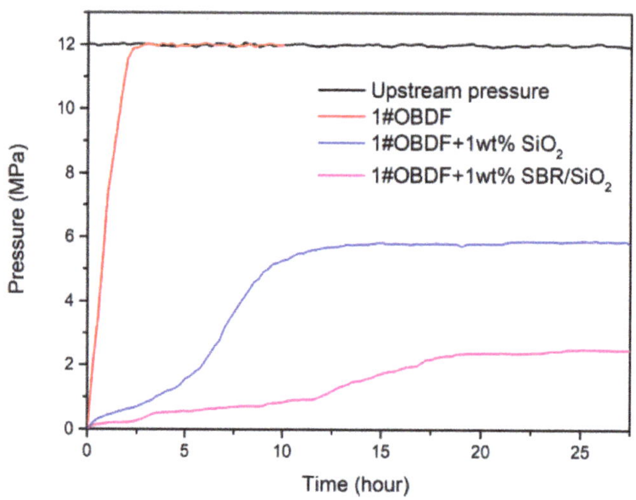

Figure 9. Time-dependent downstream pressure fluctuations of 1#OBDF with various plugging agents [49] (Reproduced with permission from Colloids and Surfaces A: Physicochemical and Engineering Aspects by Elsevier, Amsterdam, The Netherlands, 2024; Order Number: 586035022417.6).

3.2. Shale, Sustainability, and Thermal Stability

As a multi-purpose ingredient in oil-based DFs (OBMs), Jiang et al. (2020) [50] developed the super-amphiphobic nanofluid (SAN). The investigation was achieved to characterize the features of SAN and observe its impact on OBMs. The results showed that SAN may form a micro-nano structure due to its spherical shape and around 100 nm diameter. Super-amphiphobic shale surface wettability, reduced filtration volume of OBMs, decreased spontaneous imbibition, and maintained shale core compressive strength are all conceivable results of SAN's unique particle surface structure and functional groups (perfluorooctyl). It was formerly believed that OBMs equipped with SAN may be used for the drilling of unconventional gas and oil reservoir wells, thanks to their high degree of stability and the fact that they would allow less liquid to permeate the pores of shale. This improved the emulsion's stability while keeping the wellbore intact. Figure 10 displays the results, showing that SAN with super-amphiphobic and nano-diameter particle properties could successfully block liquid from entering the shale core.

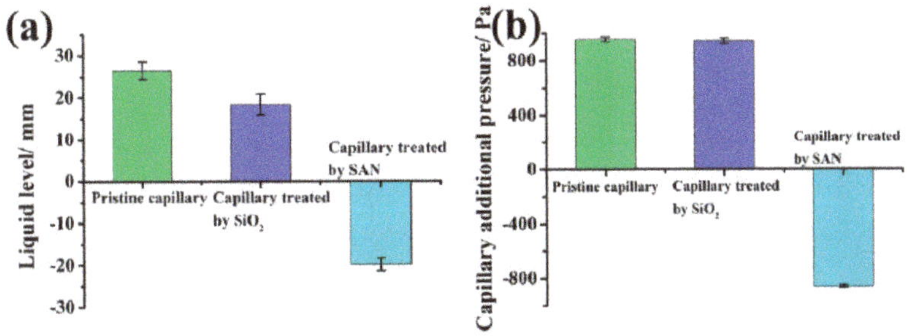

Figure 10. The impact of additives on capillary (**a**) liquid level inside the capillary and (**b**) capillary extra pressure [50] (Reproduced with permission from Colloids and Surfaces A: Physicochemical and Engineering Aspects by Elsevier, Amsterdam, The Netherlands, 2024; Order Number: 586035052693.6).

Moraveji et al. (2020) [51] developed a sustainable glycol-based DF to replace oil-based ones. First, rice husks are used to make amorphous silica nanoparticles (12, 22, and 54 nm). Water-dispersed nanoparticles are added to glycol-based mud as a liquid addition. Finally, silica NP's effects on glycol drilling fluid rheology, filtration, and shale stability are examined. Silica nanoparticles increase glycol drilling fluid rheology. This improvement depends on nanoparticle quantity and size. Silica nanoparticles reduce fluid loss and improve drilling fluid thermal stability. Silica nanoparticles may also seal Gurpi shale nanoscale pores, boosting cutting recovery and lowering glycol drilling fluid penetration. Figure 11 compares nano-based drilling fluid invasion mitigation factors. The base glycol DF's downstream pressure equals the upstream pressure after the test.

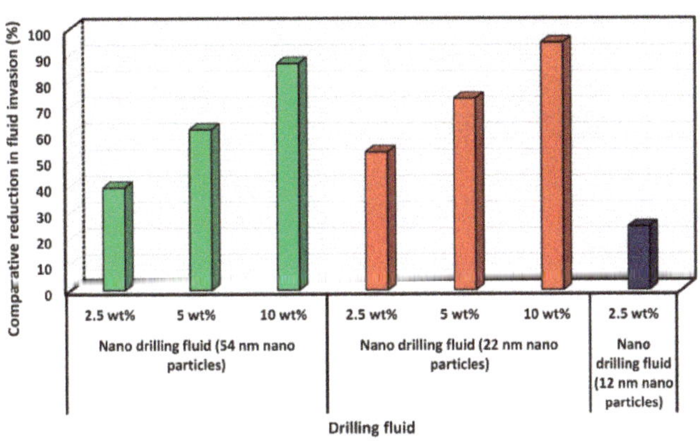

Figure 11. A comparison of nano-based drilling fluids shows that fluid invasion decreases [51] (Reproduced with permission from International Communications in Heat and Mass Transfer by Elsevier, Amsterdam, The Netherlands, 2024; Order Number: 586035077878.2).

Beg et al. (2020) [11] examined how TiO_2 nanoparticles affected drilling fluid thermal stability in two polyanionic cellulose and hydroxyethyl cellulose mud systems. In a roller oven, drilling fluids were rolled at 110 °C at 30 rpm for 16 h to mimic the environment of the wellbore. In DFP3, hot rolling decreased AV at 25 °C by ~34% without nanoparticles and by ~15% with them. The percentage drop in AV at 25 °C owing to aging was ~24% for DFH3 and ~16% for DFHN (1.0 w/v% HEC and 0.5 w/v% TiO_2). Nanoparticles made drilling fluids thermally resistant in rheology and filtration. In Figure 12, TiO_2 nanoparticles strengthened the thermal stability of API FL, which did not rise dramatically owing to hot rolling, unlike the mud samples without nanoparticles.

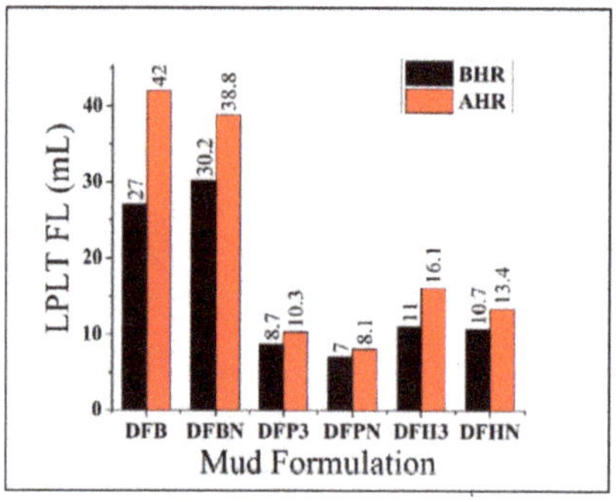

Figure 12. API filtrate loss of TiO_2 nanoparticle-based drilling mixtures before and after 110 °C and 100 psi hot rolling [11] (Reproduced with permission from Upstream Oil and Gas Technology by Elsevier, Amsterdam, The Netherlands, 2024; Order Number: 586035101908.8).

3.3. Formation Protection and Enhanced Efficiency

Hajiabadi et al. (2020) [52] modified nanosilica particles with 3-glycidoxypropyl-triethoxy silane (GPTS) and mixed GPTS and propyl silane to change their hydrophilicity. Researchers showed that surface modification of NPs may improve rheology and reduce formation damage depending on the functionalized group and NP size. Due to weak particle aggregation, surface changes were effective below 1 wt.% NPs. Plastic viscosity decreased with rising temperature due to continuous phase viscosity decrease. The yield stress exhibited greater variation with temperature due to the increased temperature sensitivity of interactions between drilling fluid components. Chemically modified NPs have caused a sharp increase in pressure, possibly due to the formation of an external impermeable barrier or internal core damage, such as a filter cake on the core face, as shown in Figure 13. The base drilling fluid has only a slight pressure increase (around 456 psi).

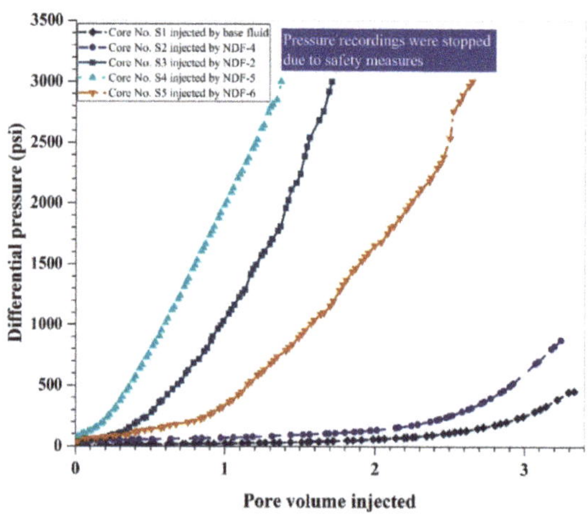

Figure 13. Coreflood experiments: differential pressures vs. pore volume injected via core samples [52] (Reproduced with permission from Journal of Petroleum Science and Engineering by Elsevier, Amsterdam, The Netherlands, 2024; Order Number: 586035124004.2).

Gudarzifar et al. (2020) [53] synthesized GONP, PAM, and GO/PAM nanocomposites with varying GONP loading (0.5, 1, 1.5, and 2 wt.% in proportional with monomer weight) as drilling-fluid additives using solution polymerization in aqueous systems. Researchers examined the effects of adding GONPs, PAM, and GO/PAM nanocomposites to water-based DF (WBDF) on the rheological properties before and after hot rolling, as well as filtration properties at low and high temperature and pressure (HPHT and LPLT), pH and mud weight, and thermal conductivity compared to WBDF. Figure 14 shows that WBDF with the GONP/PAM nanocomposite has better thermal conductivity than pure PAM.

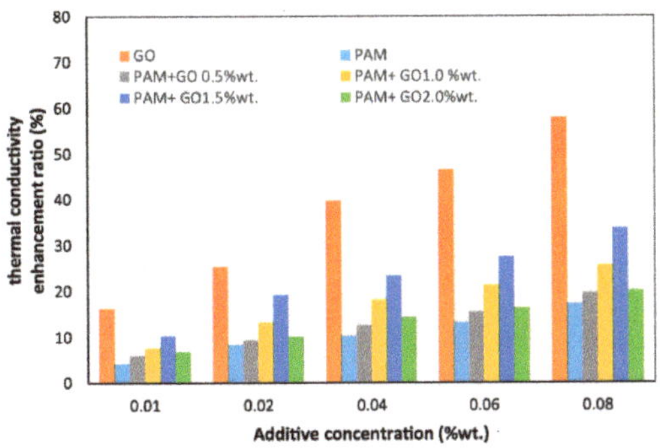

Figure 14. Additive concentration affects thermal conductivity enhancement [53] (Reproduced with permission from Powder Technology by Elsevier, Amsterdam, The Netherlands, 2024; Order Number: 586035146305.2).

At 30, 60, and 80 °C, Medhi et al. (2020) [54] studied and measured the performance improvement of NDDF with the addition of zirconium oxide (ZrO_2) NPs. The influence of ZrO_2 NPs on NDDF characteristics has been studied at various concentrations (0.5, 0.8, and 1 wt.%). The results of the viscoelastic measures and steady-state rotational tests showed that the thermal stability of the solution improved with 1 wt.% ZrO_2 NP NDDF, particularly with regard to elasticity and viscosity, and with very little filtrate loss. The inclusion of ZrO_2 NPs also improved the capacity to restore structural strength. The rheological data were used to signify and enhance the Herschel–Bulkley (HB) parameters, which were then used in the CFD study. The maximum cutting carrying capacity was found to be with 1 wt.% ZrO_2 NP NDDF, as depicted by the CFD simulation findings when contrasted with 0 wt.% (the base), 0.5 wt.%, and 0.8 wt.% ZrO_2 NP NDDF. Figure 15 elucidates that at 689 kPa (100 psi) and 1378 kPa (200 psi) differential pressure, the fluid loss volume is lowest for 1 wt.% ZrO_2 NP NDDF.

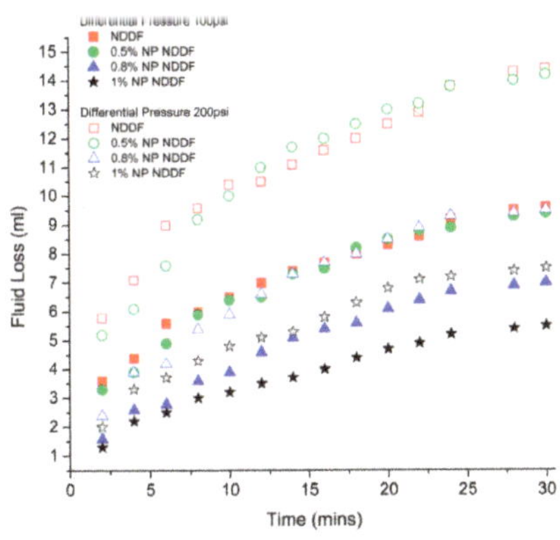

Figure 15. Various ZrO_2 NP NDDF filter loss volumes vs. time: NDDF is the ZrO_2 NP-free base. 0.5 wt.% ZrO_2 NP NDDF, etc. [54] (Reproduced with permission from Journal of Petroleum Science and Engineering by Elsevier, Amsterdam, The Netherlands, 2024; Order Number: 586036018772.6).

3.4. Formation Damage Assessment

Plugging the nano-sized holes, constraining water incursion into shale, and avoiding shale clay minerals from swelling were all attained by the application of graphene oxide (GO) by Wang et al. (2020) [55]. The graphene oxide (GO) was treated and studied using imaging microscopy, X-ray diffraction (XRD), and FTIR. To systematically weigh the protection of GO sheets on shale, a battery of tests, including uniaxial compressive strength, linear swelling, imbibition, and filtration, were accomplished. At the same time, the GO was compared to nano-silica and other extensively used shale inhibitors. While compared to the generally used inhibitors, GO presented better performance in preventing swelling of clay minerals, sealing holes of nano- and micron sizes, restraining water invasion into the interior of the shale core, and maintaining shale strength.

In addition to water-based DF, Kamali et al. (2021) [56] presented the impacts of Fe_3O_4-carboxymethyl cellulose (CMC) nanocomposite on filtration and rheological properties. The characterization of the nanocomposite was carried out using energy dispersive X-ray, SEM, FTIR, X-ray diffraction, and the co-precipitation method of synthesis. The filtration and rheological characteristics of the deionized and salty water-containing drilling fluids were then tested after adding nanocomposite at varying ratios. There was a notable reduction in filtering volume and cake thickness, as well as a significant improvement in viscosity, when 57% of a 1:3 Fe_3O_4-CMC nanocomposite ratio was added to basic DF. As shown in Figure 16, the use of Fe_3O_4-CMC NC as a rheology modifier has clear advantages over using pure CMC, as the thickness of filter cake decreases to 64% in its presence compared to a 48% increase in the presence of pure CMC (Figure 16a).

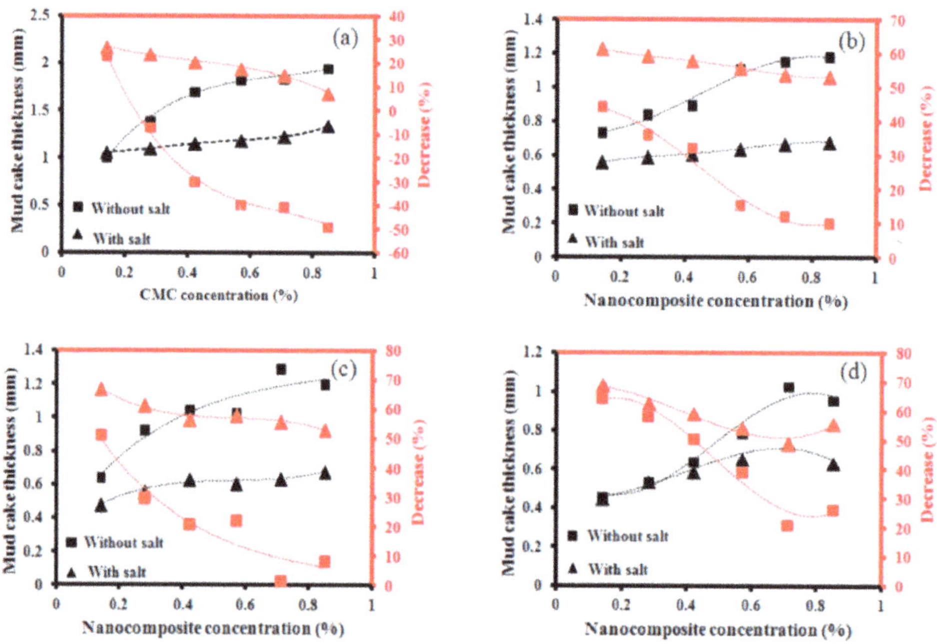

Figure 16. Mud cake thickness of (**a**) pure CMC, (**b**) 1:6, (**c**) 1:4, and (**d**) 1:3 Fe_3O_4-CMC ratio NC [56] (Reproduced with permission from Journal of Petroleum Science and Engineering by Elsevier, Amsterdam, The Netherlands, 2024; Order Number: 586036042011.8).

Because of the differences in their specifications, Hassanzadeh et al. (2021) [57] investigated how alpha and gamma alumina nanoparticles affected the stability of shale and clay. Analysis of recovered shale samples using SEM, XRD, EDAX, and AFM revealed that the mean size of the alpha alumina NPs covering the shale surface and its pores ranges from 25 to 35 nm, while the mean size of the gamma alumina NPs ranges from 20 to 30 nm. Furthermore, compared to the nano-gamma nano-sample, the shale sample retrieved by nano-alpha shows a higher percentage of remaining porosity. In addition, the EDAX findings show that the recovered shale is coated with an appropriate nano-gamma alumina. This study used five different kinds of drilling fluid in an experiment that did not include the inhibitor salt potassium chloride. It follows that the optimal sample is the water-based DF with a nano-gamma alumina content of less than 1%wt. Shale and other ion-containing formations are often more stable when alumina nanoparticles are present.

Graphene oxide (GO) shows greater performance in terms of formation damage evaluation by effectively avoiding the swelling of clay minerals through its high surface area and adsorption capacity, which allow it to cooperate with clay surfaces and constrain water uptake. Moreover, GO's two-dimensional structure qualifies it to seal nano- and micron-sized holes in shale formations more excellently than traditional inhibitors, thus reducing fluid invasion into the shale core. Also, the robust interfacial interactions between GO and shale surfaces donate to the retention of shale strength by creating a protective barrier that boosts mechanical stability. By illuminating these mechanisms, the researchers intend to offer a comprehensive description of how GO outperforms traditional inhibitors in modifying formation damage and refining shale integrity. Hajiabadi et al. (2021) [58] studied the impact of three distinct GO-based nanomaterials on the following invert emulsion drilling fluids: GO, Cu(II) salen, and Cu(II) salen@GO. A series of core flooding tests, computed tomography studies, electrical conductivity measurements, and wettability-based investigations were utilized to evaluate the corresponding formation damage on a subset of carbonate samples. While the Casson model provided the closest match to the data, it was obvious that the nano-additives used had no effect on diminishing the penetration depth. There was an obvious drop in the contact angle of over 20% in carbonate samples and 35% in sandstone samples treated with Cu(II) salen@GO additives, indicating that these additives can fix the wettability alteration problems. Adding Cu(II) salen and Cu(II) salen@GO to drilling fluid samples also improved their electrical conductivity by about 33%.

3.5. Sustainable and Targeted

When discussing the possibility of exploring the reservoirs in the Upper Assam Basins, Dutta and Das (2021) [59] focused on two angles: the NP's high cost and improving the characteristics of the DF. Iron oxide NPs were synthesized via a chemical reduction process using a UV spectrophotometer and a particle size analyzer. The filtration and rheological characteristics of the NDF were studied by adding iron oxide NPs at concentrations ranging from 5 wt.% to 40 wt.%. This study details the development of a smart DF that uses NPs exclusively in place of other commercial additives. Due to real-time feasibility and economic considerations, NDF has not been used in the reservoirs of the Upper Assam Basins so far. Along with revealing the several reasons why NDF is compatible with these reservoirs, the study also brings attention to the practical constraints that were experienced while the mud was being prepared and used in the field. According to the research, a modest quantity of iron oxide NPs significantly improved the NDF's rheology. The viscosifying properties of iron oxide NPs efficiently addressed differential sticking, poor ROP, and high ECD.

To combat the exorbitant costs associated with nanomaterial supplementation, Mirza-asadi et al. (2021) [60] developed a straightforward technique for synthesizing nanoparticles from rice husks (RHs), a byproduct of the agricultural sector. Nanoparticles of amorphous silica oxide (SiO_2) were, therefore, produced in two ways: first, by avoiding chemical materials altogether (NPS), and second, by subjecting the material to a chemical treatment (NPT). The researchers utilized X-ray fluorescence (XRF) and field emission scanning electron microscopy (FESEM) to describe the nanoparticles. The results indicated purity levels of 94.5% for NPS and 97.4% for NPT. Under two different settings, the water-based DF was used with three different concentrations of synthesized nano-silica: 3, 5, and 7% w/v.

Hot rolling (BHR) was the first test, and it was achieved at room temperature. Second, at simulated downhole temperatures of 121.11 and 148.88 °C, the after hot rolling (AHR) test was carried out. Following the outcomes, samples with a 3% concentration of NPT had better rheological characteristics than those with a 3% concentration of NPS. Drilling fluid polymers were also protected against heat breakdown by NPT nanoparticles.

Zhong et al. (2021) [61] identified the possibility of using starch nanospheres (SNSs). Methylene-bis-acrylamide (MBA) and soluble starch were crosslinked using an inverse emulsion polymerization procedure to produce SNSs. In order to determine how SNSs affected the rheological and filtration characteristics of a DF that included bentonite, the experiment followed API guidelines. Drilling fluids based on bentonite showed little effect from SNSs on their rheological characteristics. The filtration control of SNSs remained constant and effective even after thermal aging at 150 °C. When compared to modified starch filtration reducers that are already in use, they showed improved performance after aging, as well as reduced viscosity and filtration loss. Because bentonite becomes dehydrated at high temperatures, the AV values tend to go down as the temperature goes up. Figure 17 shows that the AV rose significantly from 3 mPa·s to 14 mPa·s, 22 mPa·s, and 9.5 mPa·s, respectively, when standard modified starch polymers, like HPS, CMS, and CHS, were included.

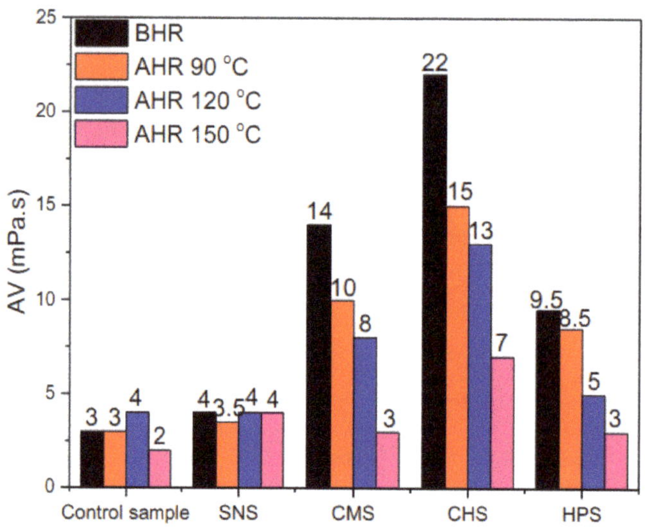

Figure 17. Various modified starch filtration reducers affect base mud rheology [61] (Reproduced with permission from Journal of Molecular Liquids by Elsevier, Amsterdam, The Netherlands, 2024; Order Number: 5860360804214).

The initial costs associated with integrating nanoparticle additions into drilling fluids vary considerably. These costs encompass procurement, formulation, and storage and handling expenses. The procurement costs can be affected by variables, such as the level of purity, the dimensions, and the production technique of the nanoparticles. Costs associated with formulation may include specialist equipment, chemicals, and procedures. To avoid contamination and guarantee safety, storage and handling expenses may require the use of specialist equipment and protocols.

The utilization of nanoparticle additions in drilling fluids is associated with a range of long-term cost reductions. These benefits include decreased drilling expenses, increased operational effectiveness, and greater adherence to environmental regulations. The use of nanoparticle additives has the potential to mitigate filtration loss, improve wellbore stability, and accelerate drilling speeds. Moreover, they may assist in minimizing drilling problems, reducing unproductive time, and curtailing waste production. The aforementioned advantages can lead to substantial financial savings and enhanced ecological performance.

To suppress shale formation in water-based DFs (WBDF), Shen et al. (2021) [62] created carboxylate cellulose nanocrystals (C-CNC). Immersion studies, a shale recovery test, and linear swelling measurements were used to investigate the inhibitive properties of C-CNC. According to the findings of the inhibition study, 1% C-CNC had an expansion of 1.41 mm, and 3% C-CNC may recover shale at a rate of 78.8%. In comparison to polyether, polyester amine, and potassium chloride, C-CNC has superior inhibitory characteristics (PEA). The research on the process showed that these nanoparticles may inhibit the hydration of clay by forming a core–shell structure with it via electrostatic contact. Furthermore, C-CNC could significantly reduce the drilling fluid's filtrate volume, but KCl and PEA negatively affected the fluid's characteristics. According to Figure 18, the stability of Na-bt was average, and its zeta potential in DI water was −38.5 mV. The zeta potential decreased to −15.4 mV, −21.7 mV, and 23.6 mV, respectively, with the addition of 5.0% KCl, 2.0% PA, and 2.0% PEA, as shown in Figure 18.

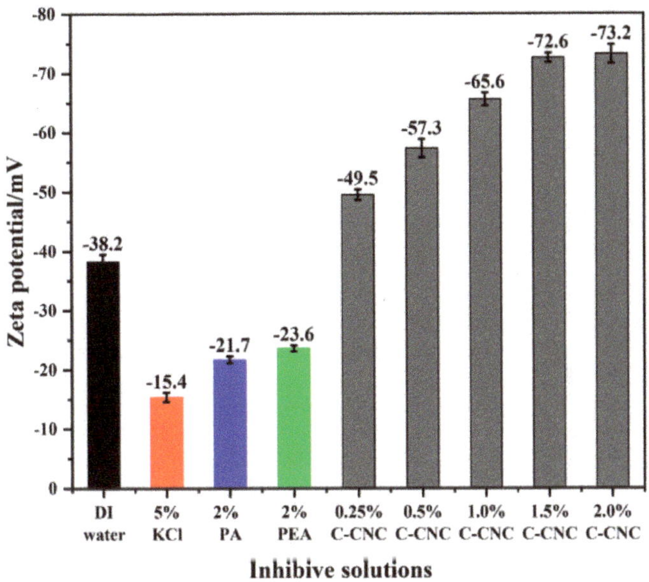

Figure 18. Na-bt dispersion zeta potential with various inhibitor doses [62] (Reproduced with permission from Colloids and Surfaces A: Physicochemical and Engineering Aspects by Elsevier, Amsterdam, The Netherlands, 2024; Order Number: 586036130436.5).

The impacts of zinc oxide NPs (ZnO NPs) on the NDDF were studied and quantified in a thorough rheological, computational fluid dynamics (CFD), and fluid-loss investigation by Medhi et al. (2021) [63]. In addition to enhancing the thixotropic and gel-forming capabilities, these NPs induce a viscoelastic solid characteristic. Adding 0.8 and 1 wt.% ZnO NP restores about 80% of the viscous structure in 180 s, while the basic NDDF only manages 44%. The working temperature of NDDF may be increased to 100 °C by adding 1 wt.% ZnO NP. In addition, adding 1 wt.% ZnO NP to NDDF decreases fluid loss by 49%. At high temperatures (80 °C), CFD simulations reveal that ZnO NP NDDFs with a concentration of 1 wt.% reduce cutting retention by 29.13%, demonstrating outstanding cutting carrying capacity. The results of the CFD pressure drop calculations are shown in Figure 19. Pressure drop magnitudes are shown to decrease with increasing temperature.

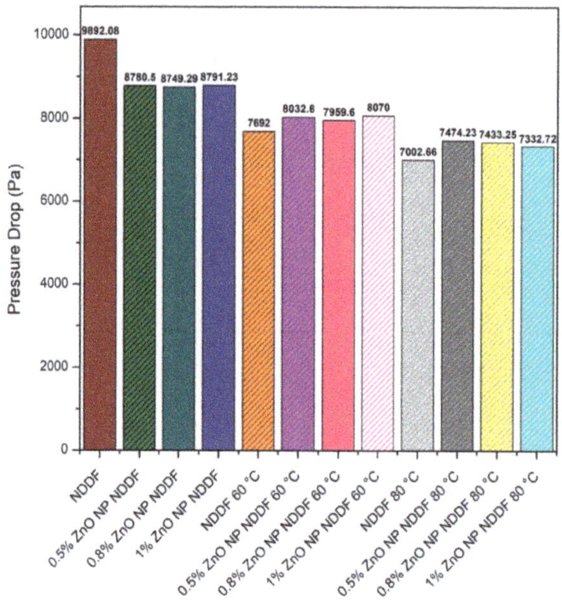

Figure 19. Pressure decrease in base and ZnO NP NDDF from CFD simulations [63] (Reproduced with permission from Journal of Natural Gas Science and Engineering by Elsevier, Amsterdam, The Netherlands, 2024; Order Number: 586037000871.7).

To lessen formation damage in horizontal wellbores, Shojaei and Ghazanfari (2022) [64] used a mix of experimental and computational methods to apply drilling fluids with nano-enhancements. To reduce formation damage caused by mud circulation, the experimental section suggests a method for finding the optimal composition of water-based samples that have been augmented with NPs of hydrophilic/hydrophobic silica and lipophilic clay. With a reported permeability of 68.4% for water-saturated porous media and 51.1% for oil-saturated porous media, samples containing 0.2 wt.% hydrophobic nano-silica demonstrated the highest functioning, according to the results. The results revealed that samples of nano-enhanced drilling fluid produced mud cakes with thinner layers and lower permeability values in much less time.

3.6. Clean-Up, Emulsions, and Water-Based Optimization

Bio-nano well clean-up fluid (BNWC) was developed by Mohammadi et al. (2022) [65] to improve the removal of WBM filter cakes. This unique formulation included a selective nanoparticle and an enzyme immobilized in potassium chloride brine. Various bulk studies show the optimized enzyme, concentration of nanoparticles, and base fluid brine; they include a precipitation test, an iodine test, and a viscosity measurement. At a differential pressure of 100 psi and a temperature of 200 °F, BNWC in potassium chloride brine demonstrated a filtering rate for HPHT that was 90% higher than that of the traditional enzyme in the same brine. The carbonate rock's wettability was proven to have changed to water-wet by measuring the contact angle, and the IFT measurements showed the increased oil mobility potential. Lastly, core flooding experiments conducted under reservoir circumstances demonstrated a damage-induced increase in the injection rate of 300% and an improvement in core permeability of 50%.

Arain et al. (2022) [66] created a hybrid nano-based inversion emulsion DF using boron nitride nanomaterials and graphene nanoplatelets to enhance the performance of oil-based DF. Along with that, we look at its viscoelastic properties and electrical stability. The shape and purity of nanoparticles are examined by characterization. There is an improvement in the rheological features and the filtration qualities of the created hybrid nano-DF. Plastic viscosity is improved by 17%, and yield point is improved by 36% using NPs at a concentration of 0.2 ppb. At 0.6 ppb, filtrate loss is reduced by a maximum of 60%.

Additionally, the viscoelastic characteristics were improved, and the nanomaterials increased the linear viscoelastic range. Moreover, variations in temperature (25–125 °C) and shear rate (0.1–1200 s^{-1}) were used to examine the flow behavior and rheological modeling. Based on the data, it seems that the hybrid nano-DF follows the Herschel–Bulkley model of non-Newtonian shear-thinning.

To find the optimal nanoparticle composition for minimal fluid loss, Dora et al. (2022) [67] studied the impacts of graphene NPs as an addition to water-based mud. An investigation of the effects of the friction factor was conducted as part of the experimental study on filtration loss from water-based mud that included graphite NPs. Nanoparticles with sizes ranging from 30 to 80 nm were introduced to fluids at concentrations ranging from 1 to 5 lb/bbl. Different compositions of nanoparticles were found to cause variations in fluid loss volume and friction coefficient. Using a basic fluid with graphite nanoparticles significantly reduced fluid loss by up to 3.5 cc, or 22%, according to the data obtained from the filter press equipment. This observation was conducted for a duration of thirty minutes.

In an experimental investigation, Mikhienkova et al. (2022) [1] examined how the characteristics of oil-based DF were affected by the addition of silicon oxide NPs at different concentrations. Drilling emulsions were tested with nanoparticle concentrations ranging from 0.25 to 2 wt.%. The typical size of the nanoparticles was 80 nm. Drilling fluids based on oil were examined for the effects of nano-sized particles on their rheological and viscosity characteristics, filtration and antifriction capabilities, and colloidal stability. It has been shown that incorporating NPs into DFS significantly improves their characteristics; a change in all parameters was seen at a very low concentration of nanoparticles.

Under high temperature and high pressure (HTHP) drilling settings, Razali et al. (2022) [68] studied the effects of carbon nanomaterial shape and graphitization on EBDF's rheology, filtering, and emulsion stability. Carbon nanotubes in cotton, graphene oxide, commercial and in-house graphene nanoplatelets, graphene nano-powder, and other carbon nanomaterials were studied. The findings demonstrated that the graphitization and shape of carbon nanomaterials influenced the EBDF behavior. Due to its huge size, low graphitic defect count, and very hydrophobic surface, graphene nano-powder was better than the other materials tested. This permitted it to augment the HTHP filtering capabilities and the stability of EBDF emulsions. Drilling fluid's emulsifier films are better protected from heat deterioration by graphene nano-powder sheets, thanks to their hydrophobic surface. Using 0.007 wt.% of a graphene concentration, the filtration experiments indicated that the filter cake thickness and filtrate reduced by 25% and 20%, respectively, due to graphene nano-powder.

An effort was undertaken by Mansoor et al. (2022) [69] to examine the impact on WBM properties of a nanofluid that was generated by dispersing copper oxide (CuO) NPs in a chia seed solution. Hence, a two-step technique is used to synthesis three samples of chia seed-based nanofluids, with the concentration of CuO NP ranging from 0.2 wt.% to 0.6 wt.%. A mixture of the produced nanofluids and water-based drilling mud is called nanofluid-enhanced water-based drilling mud or NFWBM. According to the results, the thermal stability of the WBM was significantly improved, with a viscosity reduction of approximately 61.7% at 90 °C. However, at lower temperatures, the viscosity recovered to a significant extent of approximately 14% for chia-based WBMs enhanced with 0.4 wt.% CuO nanofluid and 19% for chia-based WBMs enhanced with 0.6 wt.% CuO nanofluid. Prior to hot rolling, the API fluid loss was found to decrease from 7.2 mL to 6.8 mL, 6 mL, and 4.8 mL, respectively. In contrast, for chia-based 0.2 wt.%, 0.4 wt.%, and 0.6 wt.% CuO nanofluid enhanced water-based drilling muds (NFWBMs), the same parameter decreased from 12.4 mL to 11.4 mL, 10.2 mL, and 9.4 mL, respectively.

3.7. Taming the Heat

The findings of a comprehensive investigation into the effect of different nano-additives on the temperature dependence of the rheological behavior and viscosity of water-based DFs were reported by Minakov et al. (2023) [70]. The researchers tested nano-additives with varying concentrations, average sizes, and compositions. Drilling fluids with common ingredients were considered, such as gammaxene-based polymer solutions and water suspensions of different clay solutions. As for nano-additives, concentrations ranging from 0.25 to 3 wt.% of hydrophilic silicon and aluminum oxide nanoparticles were utilized. Nanoparticle sizes ranged between 10 to 151 nm on average. In the drilling process, fluid temperatures ranged between 25 to 80 °C. It has been indicated that drilling fluids with nanoparticles added undergo a temperature-dependent alteration in their rheological features. The consistency index and yield stress of DFs containing nanoparticles increase with increasing temperature; however, the behavior index falls.

Poly (NVP-TAAC-AMPS) (NTA) and a temperature-sensitive polymer-based nano-SiO_2 composite (SNAS) were both provided by Lai et al. (2023) [71]. One way to manipulate SNAS's lower critical solution temperature (LCST) is to change the monomer ratio. To influence the rheology of the drilling fluid, SNAS uses temperature-sensitive nanocomposites. Analytical approaches were utilized to inspect the impacts of NTA and SNAS on shale stability. These approaches included measuring shale wettability, shale cuttings recovery, fluid loss via microporous membranes, pressure transmission rate, linear swelling, strength, pore volume, and specific surface area, among others. With the shift from hydrophilicity to hydrophobicity in SNAS, the wetting angle of shale changes, and the plugging effect of SNAS is enhanced when the temperature exceeds the LCST value. When compared to using a single SNAS, the composite system's ability to create a hydrophobic zone and a tight plugging layer made it ideal for improving shale plugging performance. SNAS performs well in water-based DFs with respect to shear-thinning and thixotropy.

Research by Lysakova et al. (2023) [72] compared the impact of carbon nanotubes with different wall thicknesses on the primary performance metrics of drilling emulsions made with mineral oil. There has never been a first-of-its-kind comprehensive analysis of hydrocarbon-based solutions. The weight concentration of nanotubes with one wall ranged from 0.01% to 0.10%, whereas that of nanotubes with multiple walls ranged from 0.1% to 0.5%. We prepared and formulated stable drilling emulsions using additions of multi-walled and single-walled nanotubes. DFs based on hydrocarbons have had their rheology, filtration, clay inhibition kinetics, antifriction, and colloidal stability examined in relation to nanotube additions. Researchers have shown that drilling fluids' primary performance characteristics may be fine-tuned with the addition of even trace amounts of carbon nanotubes. Therefore, by incorporating a mere 0.025 wt.% of single-walled nanotubes into the drilling fluid, many properties are enhanced: the effective viscosity rises by about 45%, the yield stress increases by 1.7 times, the colloidal stability increases by 36%, filtration loss is reduced by 55%, and the friction factor is reduced by 20%.

3.8. Micro Marvels: Polymeric and Nanoengineered Solutions for Next-Gen Drilling Fluids

By synthesizing modified polystyrene micro-nanospheres (MPS) and altering the hydrophobicity of the shale surface, Zhang et al. (2023) [73] were able to stabilize the wellbore. We tested the MPS's inhibition and plugging capabilities by expanding linearly, recovering shale, and plugging a microporous polytetrafluoroethylene (PTFE) membrane. The findings demonstrated that MPS had excellent thermal stability, a spherical shape, and a particle size range ranging from 91 to 712 nm. When compared to KCl, polyamines, and SiO_2, the MPS showed superior inhibition and were highly compatible with drilling fluids. A 3wt.% MPS aqueous solution filtered using a PTFE microporous filter membrane lost just 42 mm of active pharmaceutical ingredient (API) in a comparison against 260 mm in a solution without MPS. After plugging, the PTFE microporous membrane's pore size was condensed. The SEM outcomes further indicated that MPS can seal the tiny holes and augment the mud cake quality. Wellbore stability might be boosted with the use of MPS.

Using a traditional emulsion polymerization practice (the one-pot method), Yang et al. (2023) [74] manufactured a polymer nanolatex (SBAA) from acrylamide, butyl acrylate, styrene, and 2-acrylamide-2-methylpropanesulfonic acid. Electron microscopy, transmission electron microscopy, and particle size distribution (PSD) investigations ascertained that SBAA is a nanoparticle with a core–shell structure and a size of around 150 nm. According to the TGA findings, the breakdown temperature of SBAA was 296 °C. This research uses transmission electron microscopy (TEM) and ultraviolet light (UV–Vis) practices to first observe the unusual self-assembly behavior of nanolatex particles. The SBAA's capacity to decrease filter loss and plug micropores in water-based drilling fluids was further measured using filtration and permeability plugging tests. The outcomes established that compared to basic bentonite fluid, SBAA reduced medium-pressure filtration loss by around 33%. Furthermore, the rate of reduction enlarged to about 41% after aging at 200 °C.

An NWBDF was developed by Lin et al. (2023) [75] with the use of nano-copper oxide (CuO) and MWCNTs as modification materials. A rotating rheometer and viscometer were utilized to investigate the rheological features and how they are impacted by the concentration and temperature of the NPs. Two NPs were studied for their effects on filtration characteristics via the use of SEM, a low-temperature filtration and a low-pressure system, and other tools. The most obvious effect on the NWBDFs is seen at a concentration of 0.05 w/v% MWCNTs, which increases the gel structure's resilience against temperature and decreases the filtering rate. In addition, a DLVO-based theoretical model is created to predict the plastic viscosity (PV) and yield point (YP) relative to temperature, taking into account the effect of the NPs. According to Figure 20, the minimal filtration was realized with a concentration of 0.1 w/v% for the CuO NPs and 0.05 w/v% for the MWCNTs.

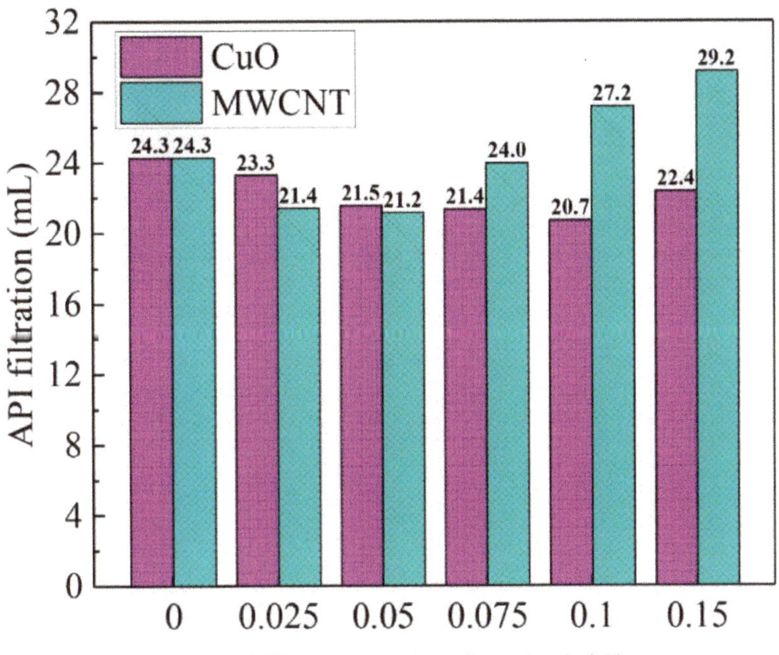

Figure 20. API filtration as a function of NPs concentration [75].

Wang et al. (2023) [76] achieved excellent suspension stability by using ultrafine barite. More crucially, a core–shell fluid loss reducer FATG that is resistant to heat and salt has been developed by altering zwitterionic polymers on the surface of nanosilica. This fluid loss reducer is then applied to WBDF, an ultra-fine clay-free water-based DF. The findings indicate that FATG-containing clay-free drilling fluid may decrease filtration loss to 8.2 mL and AV to 22 mPa·s. Even though FATG lowers viscosity, it forms a thick mud cake, which decreases filter loss. The permeability recovery assessment and linear expansion test were conducted. Bentonite has a hydration expansion inhibition rate of 72.5% and a permeability recovery rate of 77.9%; thus, it can handle the long-term fluid circulation work required for drilling.

To enhance the inhibitive, fluid loss, and rheological properties of water-based DF, Bardhan et al. (2024) [77] presented the innovative use of mesoporous nano-silica (MNS). The sol–gel method was used to manufacture MNS using cetyltrimethylammonium bromide (CTAB) to create the pores. The developed MNS has a hydrodynamic particle size of 134.47 nm and a steady zeta potential of −32.2 meV. The pore volume was 0.3415 cm^3/g, and the specific surface area was determined to be 626 m^2/g using the Brunauer–Emmett–Teller (BET) analyzer. In order to determine how thermal aging affects the drilling fluid's characteristics, samples were heated to 180 °C and then rolled under 100 psi pressure for 16 h. In addition to imparting certain inhibitive features and significantly lowering fluid loss, the experimental results show that MNS may significantly enhance the thermal properties of water-based DFs while preserving their rheological properties.

The application of wheat grain as a nano-biodegradable additive in WBDFs was examined by Ali et al. (2024) [78], with a focus on various particle sizes of wheat powder, from nano to course. XRF, FTIR, TGA, DLS, TEM, and SEM were used to characterize the nano-biopolymers (WNBPs) that were created by breaking them down into nano-sized structures using the ball mailing technique. From the API SPEC 13A standard, carboxymethyl starch (CMS), wheat powder (WP) at 75–600 μm, and nano-biopolymer (WNBPs), several DFs were created, including the reference, nano-biodegradable, modified, and biodegradable ones. Research shows that nano-biodegradable materials have a pH over 10, which protects drill pipes and other equipment from corrosion. The fluid loss was reduced from 19.5 to 14 mL when 2 wt.% WNBPs were added to the reference DF. However, the filtration rate was lowered to 11.5 mL because of the higher influence of fine WPs on the filtration characteristics.

Hydrophilic (phi-SiO_2) and hydrophobic (pho-SiO_2) nano-SiO_2 samples were used as fluid loss reducers and rheological modifiers in water-based DFs by Li et al. (2024) [79]. Findings showed that this nano-SiO_2 exhibited a range of effects on WBDFs, some of which were negative. Both phi-SiO_2 and pho-SiO_2 thickened and increased shear force at ambient temperature and 150 °C. As the concentration of these two nano-SiO_2 increased, the yield point and viscosity of four WBDFs were improved; however, the degree of improvement for pho-SiO_2 was larger than that of phi-SiO_2. At the same time, phi-SiO_2 did not reduce fluid loss of WBDFs, but it has shown high compatibility with CMC and KHm and may be used to reduce fluid loss up to 150 °C. Concentration and additive type were additional factors influencing pho-SiO_2's filtering control efficacy in WBDFs. The addition of pho-SiO_2 reduced the fluid loss volume of WBDFs by improving viscosity, increasing shear force, stabilizing the colloidal structure, and forming a hydrophobic barrier.

Table 1 demonstrates a summary of the associated studies that investigated the applications of mano-additives for improving drilling fluid properties, providing the most important findings of each study.

Table 1. A summary of research on the applications of nano-additives for improving drilling fluid properties.

Author [Ref.]	Composition/Configuration	Parameters to Be Studied	Findings/Highlighted Results
Hajiabadi et al. (2019) [34]	Enhanced inverted emulsion drilling fluid with nanometer-sized modifications.	Effect of carbon-based nanomaterials.	The Carreau model is well-supported by the drilling fluid sample behaviors, and it has been shown that the nanomaterial additions greatly enhance the crucial rheological parameters.
Boyou et al. (2019) [38]	Drilling fluids containing nano-silica and water.	Effect of nano-silica.	For several nano-enhanced water-based DFs, the nanosilica addition to the mud enhanced colloidal interactions with cuttings, which in turn improved the efficiency of cuttings transportation by 30.8% to 44.4%.
Minakov et al. (2019) [39]	Filtering drilling fluid using silicon oxide nanoparticles.	Effect of silicon oxide nanoparticles.	Drilling fluids containing nanoparticles may be filtered out by adjusting the pore diameters of ceramic filters, the size and concentration of microparticles, and the composition of the NPs themselves. Adding nanoparticles to a micro-suspension drastically decreases its filtering capacity and changes the shape and cake thickness that forms on the filter's surface.
Inturi et al. (2019) [40]	Sodium dodecyl sulfate and polyacrylamide are combined with different amounts of silicon dioxide.	Impact of combination.	When nanoparticles are added to a fluid at low velocity, the viscosity of the fluid is lowered by a factor of two. Nevertheless, because of the intricate structure, including the particle, surfactant, and polymer, the fluid exhibited shear-thickening characteristics.
Nascimento et al. (2019) [41]	Drilling fluids' apparent viscosity factors, including solid additives and particle size distribution.	Impact of particle size distribution.	In contrast to calcite suspensions, which exhibited a greater electrostatic interaction with polymers, glass and barite sphere suspensions had lower shear stress and larger zeta potential values.
Saboori et al. (2019) [42]	A nanocomposite of copper oxide and polyacrylamide is used in addition to drilling fluids, including water.	Effect of nanocomposite additive.	In terms of filtration and rheological characteristics, the nanocomposite addition performed better in deionized water-based DF; however, in salty water, it enhanced thermal conductivity more than in deionized base DF.
Perween et al. (2019) [43]	Bismuth ferrite NPs suspended in a water-based drilling fluid.	Impact of NP concentration	When the NP concentration was increased from 0% to 0.30 w/v%, AV, PV, and YP all increased by 25%. On the other hand, API filtrate loss decreased by 35% at 20 °C prior to hot rolling.

Table 1. *Cont.*

Author [Ref.]	Composition/Configuration	Parameters to Be Studied	Findings/Highlighted Results
Dejtaradon et al. (2019) [44]	Nanoparticles of zirconium oxide and copper oxide are added to a DF based on water at 25, 50, and 80 °C.	The impact of adding NPs of ZnO and CuO at varying concentrations.	Both nanoparticle-based DFs outperformed the base fluid in terms of rheological parameters; however, ZnO performed better in general than CuO.
Al-Yasiri et al. (2019) [45]	Advanced drilling fluid formulated with a unique blend of biopolymers and nanoparticles.	Effect of nanoparticles.	The modified drilling fluids outperformed the usual WBM formulations in many areas, including bit lubrication, filtrate loss, hole cleaning, and yield point.
Huang et al. (2019) [46]	Fluids for drilling that include nanoparticles of laponite.	Effect of nanoparticles.	Due to the strong interactions between the laponite and the functional groups of the ADD terpolymer, the viscosity of the drilling fluids, including laponite, was much greater than that of the fluids without laponite.
Ma et al. (2020) [47]	Nano-plugging additives in DFs.	Impact of nano-plugging additives.	To achieve a plugging efficiency of 50.96%, a small quantity of 0.3 wt.% MWCNTs-g-SPMA-2 was added.
Zhong et al. (2020) [48]	Hyperbranched polyethyleneimine grafted onto nano-silica in a water-based DF.	Effect of nanoparticles.	For nanosilica, which showed only limited physical plugging, the modified nanoparticles demonstrated both efficient chemical inhibition and physical plugging.
Li et al. (2020) [49]	The composite is made of nano-SiO_2 and styrene butadiene resin.	Effect of SBR/SiO_2.	In order to improve wellbore stability, SBR/SiO_2 may penetrate shale formation nanopores and drastically decrease fluid incursion.
Jiang et al. (2020) [50]	Drilling fluid derived from oil with super-amphiphobic nanofluid for several purposes.	Impact of SAN.	Super-amphiphobic shale surface wettability, reduced spontaneous imbibition, decreased OBM filtration volume, and maintained compressive strength of shale core are all possible outcomes of SAN.
Moraveji et al. (2020) [51]	Fluids used in drilling include amorphous silica nanoparticles.	The effect of silica nanoparticles.	Adding silica NPs to glycol DF boosted its rheological characteristics. Incorporating silica NPs into the DF not only improves its thermal stability but also reduces fluid loss. In order to improve shale-cutting recovery and decrease the penetration rate of glycol DF into Gurpi shale samples, silica NPs may efficiently block nanoscale holes of the shale.

Table 1. *Cont.*

Author [Ref.]	Composition/Configuration	Parameters to Be Studied	Findings/Highlighted Results
Beg et al. (2020) [11]	Enhancement of drilling fluids using TiO_2 nanoparticles.	Impact of TiO_2 nanoparticles.	Drilling fluids using nanoparticles have improved filtration and rheological properties, making them more resistant to heat deterioration. Mud systems' thermal stability and rheological characteristics are both improved by adding TiO_2 nanoparticles to them in addition to a traditional fluid loss reduction additive, which increases the effectiveness of the latter.
Hajiabadi et al. (2020) [52]	Fluid for drilling that has had nano-silica added to its surface.	Impact of nanoparticles.	Due to the absence of particle agglomeration, modifications of the surface were shown to be effective below the optimum concentration of 1 wt.% of NPs. Additionally, when the temperature is increased, the plastic's viscosity decreases because of the continuous phase's reduced viscosity.
Gudarzifar et al. (2020) [53]	Water-based drilling fluid enhances.	Effect of nanoparticles.	Analyses of thermal conductivity revealed that, as compared to pure PAM, WBDF with a GONP/PAM nanocomposite addition exhibited significantly higher thermal conductivity.
Medhi et al. (2020) [54]	Zirconium oxide nanoparticle-containing drilling fluid.	Effect of ZrO_2 NP.	A greater level of thermal stability was noted for 1 wt.% ZrO_2 NP NDDF, as measured by elasticity and viscosity, along with little filtrate loss.
Wang et al. (2020) [55]	Drilling fluids blended with water and graphene oxide.	Impact of graphene oxide.	In comparison to the typically used inhibitors, GO demonstrated superior performance in retaining shale strength, avoiding swelling of clay minerals, sealing holes of nano- and micron-sizes, and limiting water incursion into the shale core interior.
Kamali et al. (2021) [56]	Using a nanocomposite of CMC and Fe_3O_4 in water-based DFs.	Impact of CMC-Fe_3O_4 nanocomposite.	Along with a significant reduction in viscosity, the volume of filtration and thickness of cake were both significantly lowered when 57% of a 1:3 Fe_3O_4-CMC nanocomposite ratio was added to basic DF.
Hassanzadeh et al. (2021) [57]	Gamma alpha and DF systems based on nano-alumina.	The impacts of alumina NPs.	The optimal sample was determined to be the water-based DF that included nano-gamma alumina at a concentration of less than 1%wt. Shale and other ion-containing formations are made more stable by the addition of alumina nanoparticles.

Table 1. *Cont.*

Author [Ref.]	Composition/Configuration	Parameters to Be Studied	Findings/Highlighted Results
Hajiabadi et al. (2021) [58]	Drilling fluids changed using graphene oxide/inorganic compounds for invert emulsion models.	Effect of graphene oxide/inorganic complexes.	Adding Cu(II) salen@GO and GO to carbonate and sandstone samples, respectively, reduced the contact angle by around 20% and 35%. Adding Cu(II) salen and Cu(II) salen@GO to DF samples boosted their electrical conductivity by about 33%.
Dutta and Das (2021) [59]	Iron oxide nanoparticles introduced into smart bentonite drilling fluid.	Impact of iron oxide nanoparticles.	According to the research, a modest quantity of iron oxide NPs significantly improved the NDF's rheology. The viscosifying properties of iron oxide NPs all efficiently addressed high ECD, poor ROP, and differential sticking.
Mirzaasadi et al. (2021) [60]	Biogenic silica nanoparticles in a water-based DF.	Impact of biogenic silica NPs.	When compared to NPS, samples with a 3% concentration of NPT exhibited an improvement in all rheological characteristics. Drilling fluid polymers were also protected against heat breakdown by NPT nanoparticles.
Zhong et al. (2021) [61]	Drilling fluids, including water and starch nanospheres, have been cross-linked.	Effect of cross-linked starch nanospheres.	Drilling fluids based on bentonite showed little effect from SNSs on their rheological characteristics. When compared to modified starch filtration reducers that are already in use, they showed improved performance after age and reduced viscosity and filtration loss.
Shen et al. (2021) [62]	Hydrocarbon-based drilling lubricants, including carboxylate cellulose nanoparticles.	Effect of carboxylate cellulose nanocrystals.	C-CNC significantly reduced the drilling fluid's filtrate volume, but KCl and PEA negatively affected the fluid's characteristics.
Medhi et al. (2021) [63]	Drilling fluid containing zinc oxide nanoparticles that is safe for use.	The effects of zinc oxide nanoparticles.	A cutting retention decrease of 29.13% at high temperature values (80 °C) is shown by the good cutting carrying capacity of ZnO NP NDDFs at a concentration of 1 wt.%.
Shojaei and Ghazanfari (2022) [64]	Nano-enhanced DFs.	The impact of nanoparticles' hydrophobicity.	Lower permeability and thinner mud cake values were produced in a much shorter amount of time using nano-enhanced drilling fluid samples.
Mohammadi et al. (2022) [65]	Fluids for drilling and well cleanup that are water-based and include a new bio-nano-catalyst.	Effect of bio-nano-catalyst.	When tested under reservoir circumstances, core flooding increased the injection rate by 300% and improved the core's permeability by 50% after damage.
Arain et al. (2022) [66]	Graphene nanoplatelets with boron nitride in an inverted emulsion drilling fluid.	Nanomaterials' impact on oil-based drilling fluid's filtration and rheological characteristics.	Plastic viscosity is improved by 17%, and yield point is improved by 36% using NPs at a concentration of 0.2 ppb. At 0.6 ppb, filtrate loss is reduced by a maximum of 60%.

Table 1. *Cont.*

Author [Ref.]	Composition/Configuration	Parameters to Be Studied	Findings/Highlighted Results
Dora et al. (2022) [67]	Graphite nanoparticles suspended in a water-based mud.	Graphite nanoparticles added to water-based mud: the impact.	There was a significant reduction in fluid loss, up to 3.5 cc (i.e., 22% reduction) when a basic fluid, including graphite NPs, was utilized.
Mikhienkova et al. (2022) [1]	Hydrocarbon drilling fluid with nanoparticles.	How particles on the nanoscale affect rheological and viscosity parameters.	Adding nanoparticles to drilling fluids changes their qualities for the better; all metrics change even at a very low concentration of NPs.
Razali et al. (2022) [68]	Fluids for drilling that include carbon nanoparticles.	Effect of carbon nanomaterials.	Carbon nanomaterial shape and graphitization determined EBDF behavior. At a concentration of 0.007 wt.%, graphene nano-powder decreased the thickness of the filtrate by 20% and the filter cake by 25%.
Mansoor et al. (2022) [69]	Using chia-based copper oxide nanofluid with water-based DFs.	Effect of chia-based copper oxide nanofluid.	A notable improvement in the WBM thermal stability was noted, with a viscosity reduction of approximately 61.7% at 90 °C. For chia-based WBMs enhanced with 0.4 wt.% CuO nanofluid, the viscosity recovered to a significant extent of about 14%, and for chia-based WBMs enhanced with 0.6 wt.% CuO nanofluid, it recovered to about 19%.
Minakov et al. (2023) [70]	Hydrocarbon DFs with nano-additives.	The effect of nano-additives of different concentrations.	Depending on the temperature, the rheological characteristics of DFs that have nanoparticles added to them alter significantly. The consistency index and yield stress of DFs containing nanoparticles were seen to rise as the temperature rose.
Lai et al. (2023) [71]	Using a polymer-based nano-SiO_2 composite in water-based DFs.	The effects of SNAS on shale stability.	SNAS performs well in water-based DFs with respect to shear-thinning and thixotropy.
Lysakova et al. (2023) [72]	Drilling fluids containing hydrocarbons and carbon nanotubes, both single-walled and multi-walled.	The effect of nanotube additives.	A mere 0.025 weight % of single-walled nanotubes boost the drilling fluid's effective viscosity by about 45%, increases yield stress by 1.7 times, increases colloidal stability by 36%, reduces the friction factor by 20, and reduces filtration loss by 55%. Simultaneously, this addition reduces the fluid's loss by 55%.
Zhang et al. (2023) [73]	Drilling fluids with micro-nanospheres made of modified polystyrene and dissolved in water.	Effect of MPS.	MPS exhibited excellent thermal stability and spherical geometry with a particle size ranging from 91 to 712 nm. When compared to KCl, polyamines, and SiO_2, the MPS showed superior inhibition and were highly compatible with drilling fluids.

Table 1. *Cont.*

Author [Ref.]	Composition/Configuration	Parameters to Be Studied	Findings/Highlighted Results
Yang et al. (2023) [74]	Drilling fluids containing polymer nanolatex particles on a water basis.	Effect of polymer nanolatex particles.	When compared to basic bentonite fluid, SBAA reduced filtration loss of pressure by about 33%; after aging at 200 °C, the reduction rate increased to around 41%.
Lin et al. (2023) [75]	Water-based drilling fluids using MWCNTs.	Nanoparticle concentration and temperature impacts.	To reduce the filtration rate and increase the gel structure's resilience against temperature, MWCNTs at a concentration of 0.05 w/v% have the most noticeable effect on the NWBDFs.
Wang et al. (2023) [76]	Drilling fluid with a nanofiltration control additive and a core–shell structure that is clay-free and resistant to salt.	Effects of a nano-filtration control additive with a core–shell structure.	It is possible to decrease AV to 22 mPa·s and the filtering loss of clay-free DF, including FATG, to 8.2 mL.
Bardhan et al. (2024) [77]	Hydrocarbon drilling fluids containing mesoporous nano-silica.	Effect of mesoporous nano-silica.	By preserving rheological qualities, significantly lowering fluid loss, and imparting certain inhibitive properties, MNS may greatly boost the thermal properties of water-based DFs.
Ali et al. (2024) [78]	Wheat nano-biopolymer drilling fluids.	Effect of WNBPs.	The loss of fluid was reduced from 19.5 to 14 mL when 2 wt.% WNBPs were added to the reference DF. However, the filtration rate was lowered to 11.5 mL because of the higher influence of fine WPs on the filtration characteristics.
Li et al. (2024) [79]	Hydrophilic and hydrophobic nano-silica-based drilling fluids.	Influence of hydrophilic/hydrophobic nano-silica.	The addition of pho-SiO_2 reduced the volume of fluid loss of WBDFs by improving viscosity, increasing shear force, stabilizing the colloidal structure, and forming a hydrophobic barrier.

Table 2 introduces a summary of the relevant information on nano-additives used to improve drilling fluid performance.

Table 2. Associated information of nano-additives used in drilling fluids.

Nano-Additive	Particle Size (nm)	Concentration (wt.%)	Improved Properties	References
Nano-gamma alumina	30–50	<1	Improved shale stability	[57,80]
Nanoclay	50–100	6	Reduced fluid loss and improved rheological properties	[81,82]
Cellulose nanofibers	10–100	0.5–1	Improved filtration control and rheological features	[83]
Single-walled nanotubes (SWNTs)	0.75–1.25	0.027	Improved fluid viscosity, yield stress, friction factor, colloidal stability, and reduced fluid loss	[72]
Multi-walled carbon nanotubes (MWCNTs)	10–30	0.05	Reduced filtration rate and increased gel structure resilience	[84]
Nanosilica (SiO_2)	150–700	2	Improved rheological properties and thermal stability	[85]
Zinc oxide nanoparticles (ZnO NPs)	20–70	0.5	Improved filtration control and thermal conductivity	[9]
Iron oxide nanoparticles (Fe_3O_4 NPs)	10–40	0.5–1	Improved thermal stability and rheological properties	[86]
Titania nanoparticles (TiO_2 NPs)	<25	1–3	Improved thermal stability and lubricity	[87]

4. Critical Issues and Recommendations for Future Research

The present study assessed the impact of adding nanoparticles to improve the properties of drilling fluids. The specific benefits include reduced fluid loss, increased shear force, greater thermal stability, and a smaller filtrate volume, as well as improved viscosity. Nanoparticle production often comes with a hefty price tag. Also, since they tend to aggregate, they might block the system and produce sedimentation. As a result, certain parts would be damaged, and performance would be diminished. When subjected to extreme heat, nanofluids may not retain their stability. That is why it is crucial to fully grasp these issues in order to obtain the optimal design parameters for enhanced drilling fluids' characteristics. Listed below are some of these restrictions:

1. Reducing their environmental presence necessitates stringent controls over the production and disposal of materials used in drilling fluids, as well as the prevention of contamination.
2. Many nano-additives are still in the early phases of manufacture and marketing. Their limited supply may prevent them from being widely used in the drilling sector.
3. The effectiveness of the drilling fluid is condensed, and problems with pumping and mixing are possible because nanoparticles tend to cluster or clump together.
4. Drilling fluids are only one example of how regulatory frameworks for nanomaterials are persistently changing. Doubt and a halt to the industry's acceptance and development of nano-additives may result from this.
5. Nano-additives are problematic and costly to generate and incorporate into drilling fluids. When compared against more traditional additives, this has the potential to significantly raise the total cost of drilling operations.

6. Using nano-additives in drilling fluids has not been well examined for its potential long-term impacts on the environment and human health. Further investigations are required to make sure they can be used responsibly.

Nano-additives in drilling fluids have definite restrictions, but continuous research and development are desirable to develop them and make them more useful and sustainable. Nano-additives may yet prove to be a priceless resource for improving drilling efficiency and wellbore stability as technology advances and information gaps are filled. Accordingly, new ideas, and public education will be essential to resolve the complications above and restrictions. To fight these weaknesses and make them a more environmentally friendly and practical option, nanofluids can be added to drilling fluids to improve their qualities. To advance these systems and open the door to further research, the following bullets can be considered:

1. In addition to stabilizing wellbore formations, nano-clay can enhance the filtration control and viscosity of DFs.
2. Wellbore features and drilling problems will determine which nano-additives are most suitable for use in drilling fluids.
3. The potential health and environmental concerns of nano-additives in the long run have not been fully resolved. Before applying nano-additives in drilling fluids, it is vital to methodically evaluate these possible dangers.
4. In some nations, rules concerning the use of nano-additives in drilling fluids are still being determined. Prior to employing nano-additives, it is essential to be familiar with these limitations.
5. Cellulose nano-fibers can improve the filtration control and rheological features of drilling fluids. In addition to stabilizing the wellbore, they can reduce fluid loss.

5. Conclusions

After reviewing the data in Table 1 and analyzing each study, the following recommendations emerge. To rephrase, the review identifies the following gaps in the literature about the application of nano-additives to boost the characteristics of DFs and proposes the following solutions:

1. The fluid loss was reduced from 19.5 to 14 mL when 2 wt.% WNBPs were added to the reference DF. However, the filtration rate was lowered to 11.5 mL because of the higher influence of fine WPs on the filtration characteristics.
2. By preserving rheological qualities, significantly lowering fluid loss, and imparting certain inhibitive properties, MNS may greatly boost the thermal properties of water-based DFs.
3. To reduce the filtration rate and increase the gel structure's resilience against temperature, MWCNTs at a concentration of 0.05 w/v% have the most noticeable effect on the NWBDFs.
4. MPS exhibited excellent thermal stability and a spherical shape with a particle size ranging from 91 to 712 nm. When compared to KCl, polyamines, and SiO_2, the MPS showed superior inhibition and were highly compatible with drilling fluids.
5. A mere 0.025 wt.% of single-walled nanotubes boost the drilling fluid's effective viscosity by about 45%, increase the yield stress by 1.7 times, reduce the filtration loss by 55%, decrease the friction factor by 20%, and increase colloidal stability by 36%. Simultaneously, this addition reduces the fluid's loss by 55%.
6. Depending on the temperature, the rheological characteristics of DFs that have nanoparticles added to them alter significantly. The consistency index and yield stress of DFs containing NPS were seen to rise as the temperature rose.
7. Adding nanoparticles to drilling fluids changes their qualities for the better; all metrics change, even at a very low concentration of nanoparticles.
8. When compared to NPS, samples with a 3% concentration of NPT exhibited an improvement in all rheological characteristics. Drilling fluid polymers were also protected against heat breakdown by NPT nanoparticles.

9. Adding GO- and Cu(II) salen@GO to carbonate and sandstone samples, respectively, reduced the contact angle by around 20% and 35%.
10. Plastic viscosity is improved by 17%, and yield point is improved by 36% using nanomaterials at a concentration of 0.2 ppb. At 0.6 ppb, filtrate loss is reduced by a maximum of 60%.
11. Silica nanoparticles, carbon nanotubes, and metal oxides are among the nano-additives that have shown improvements in yield stress, plastic viscosity, and loss of filtrate.
12. The optimal sample was determined to be the water-based DF that included nano-gamma alumina at a concentration of less than 1wt.%. Shale and other ion-containing formations are made more stable by the addition of alumina nanoparticles.
13. Adding nanoparticles to a micro-suspension drastically decreases its filtering capacity and changes the thickness and shape of the cake that forms on the filter's surface.

Author Contributions: Conceptualization, M.S.A. and F.L.R.; formal analysis, M.S.A., M.T.J., F.L.R., H.T., M.K.R., M.A.A.-O., Q.R.A.-A., H.I.M. and I.E.S.; investigation, F.L.R., M.K.R., Q.R.A.-A., H.I.M. and I.E.S.; resources, M.S.A., M.T.J., F.L.R., H.T., M.K.R., M.A.A.-O., Q.R.A.-A., H.I.M. and I.E.S.; writing—original draft preparation, M.S.A., F.L.R. and M.A.A.-O.; writing—review and editing, M.A.A.-O.; visualization, H.I.M. and I.E.S.; supervision, F.L.R., H.T. and M.A.A.-O. All authors have read and agreed to the published version of the manuscript.

Funding: This research received no external funding.

Data Availability Statement: Data are contained within the article.

Conflicts of Interest: The authors declare no conflicts of interest.

Abbreviations

AFM	Atomic force microscopy
AHR	After hot rolling
API	American Petroleum Institute
AV	Apparent viscosity
BHR	Before hot rolling
BNWC	Bio-nano well clean-up fluid
CFD	Computational fluid dynamics
CMC	Carboxymethyl cellulose
CTE	Coefficient of thermal expansion
DF	Drilling fluid
ECB	Equivalent circulating density
EDAX	Energy dispersive X-ray analysis
FESEM	Field emission scanning electron microscope
FTIR	Fourier transform infrared spectrum
GO	Graphene oxide
GONP	Graphene oxide nano sheet
HB	Herschel–Bulkley
HEC	Hydroxyethylcellulose
HSE	Health, safety, and environmental
HTHP	High temperature, high pressure
LPLT	Low pressure–low temperature
MBA	Methylene-bis-acrylamide
MWCNT	Multi-walled carbon nanotube
NDDF	Non-damaging drilling fluid
NDF	Nano-based drilling fluid
NFWBM	Nanofluid enhanced water-based drilling mud
NPs	Nanoparticles
NWBDFs	Nano-water-based drilling fluids
OBDF	Oil-based drilling fluid
PAM	Polyacrylamide
PHPA	Partially hydrolyzed polyacrylamide

POCNT	PEGylated oxidized carbon nanotube
PV	Plastic viscosity
ROP	Rate of penetration
SAN	Super-amphiphobic nanofluid
SBASC	Synthetic-based acrylamide–styrene copolymer
SDS	Sodium dodecyl sulfate
SEM	Scanning electron microscope
SNSs	Starch nanospheres
TEM	Transmission electron microscope
TGA	Thermogravimetric analysis
WBM	Water-based mud
XG	Xanthan gum
XRD	X-ray diffraction
YP	Yield point

References

1. Mikhienkova, E.I.; Lysakov, S.V.; Neverov, A.L.; Zhigarev, V.A.; Minakov, A.V.; Rudyak, V.Y. Experimental Study on the Influence of Nanoparticles on Oil-Based Drilling Fluid Properties. *J. Pet. Sci. Eng.* **2022**, *208*, 109452. [CrossRef]
2. Minakov, A.V.; Zhigarev, V.A.; Mikhienkova, E.I.; Neverov, A.L.; Buryukin, F.A.; Guzei, D.V. The Effect of Nanoparticles Additives in the Drilling Fluid on Pressure Loss and Cutting Transport Efficiency in the Vertical Boreholes. *J. Pet. Sci. Eng.* **2018**, *171*, 1149–1158. [CrossRef]
3. Parizad, A.; Shahbazi, K.; Tanha, A.A. Enhancement of Polymeric Water-Based Drilling Fluid Properties Using Nanoparticles. *J. Pet. Sci. Eng.* **2018**, *170*, 813–828. [CrossRef]
4. Alsaba, M.T.; Al Dushaishi, M.F.; Abbas, A.K. Application of Nano Water-Based Drilling Fluid in Improving Hole Cleaning. *SN Appl. Sci.* **2020**, *2*, 1–7. [CrossRef]
5. Abd Kati, F. The effects of TiO_2 Nanoparticles on Flexural Strength of Self-Polymerized Resins: In Vitro Study. *J. Tech.* **2022**, *4*, 45–49. [CrossRef]
6. Dahwal, S.H.; Al-Ziyadi, H. Nanomaterials-Assisted Drilling Fluids for Application in Oil Fields—Challenges and Prospects. *Int. J. Emerg. Technol. Innov. Res.* **2023**, *10*, 567–582.
7. Abramov, A. Optimization of Well Pad Design and Drilling–Well Clustering. *Pet. Explor. Dev.* **2019**, *46*, 614–620. [CrossRef]
8. Adah, E.; Joubert, A.; Boudhan, R.; Henry, M.; Durécu, S.; Le Coq, L. Spray Scrubber for Nanoparticle Removal from Incineration Fumes from the Incineration of Waste Containing Nanomaterials: Theoretical and Experimental Investigations. *Aerosol Sci. Technol.* **2021**, *56*, 75–91. [CrossRef]
9. Ahasan, M.H.; Alahi Alvi, M.F.; Ahmed, N.; Alam, M.S. An Investigation of the Effects of Synthesized Zinc Oxide Nanoparticles on the Properties of Water-Based Drilling Fluid. *Pet. Res.* **2022**, *7*, 131–137. [CrossRef]
10. Li, Q.; Li, Y.; Cheng, Y.; Li, Q.; Wang, F.; Wei, J.; Liu, Y.; Zhang, C.; Song, B.; Yan, C.; et al. Numerical simulation of fracture reorientation during hydraulic fracturing in perforated horizontal well in shale reservoirs. *Energy Sources Part A Recovery Util. Environ. Effects* **2018**, *40*, 1807–1813. [CrossRef]
11. Beg, M.; Kumar, P.; Choudhary, P.; Sharma, S. Effect of High Temperature Ageing on TiO_2 Nanoparticles Enhanced Drilling Fluids: A Rheological and Filtration Study. *Upstream Oil Gas. Technol.* **2020**, *5*, 100019. [CrossRef]
12. Kafashi, S.; Rasaei, M.; Karimi, G. Effects of Sugarcane and Polyanionic Cellulose on Rheological Properties of Drilling Mud: An Experimental Approach. *Egypt. J. Pet.* **2017**, *26*, 371–374. [CrossRef]
13. Villada, Y.; Gallardo, F.; Erdmann, E.; Casis, N.; Olivares, L.; Estenoz, D. Functional Characterization on Colloidal Suspensions Containing Xanthan Gum (XGD) and Polyanionic Cellulose (PAC) Used in Drilling Fluids for a Shale Formation. *Appl. Clay Sci.* **2017**, *149*, 59–66. [CrossRef]
14. Chaouf, S.; El Barkany, S.; Jilal, I.; El Ouardi, Y.; Abou-salama, M.; Loutou, M.; El-Houssaine, A.; El-Ouarghi, H.; El Idrissi, A.; Amhamdi, H. Anionic Reverse Microemulsion Grafting of Acrylamide (AM) on HydroxyEthylCellulose (HEC): Synthesis, Characterization and Application as New Ecofriendly Low-Cost Flocculant. *J. Water Process Eng.* **2019**, *31*, 100807. [CrossRef]
15. Su, X.; Lian, Z.; Yuan, Y. Study on the Effect of the Oil-Water Ratio on the Rheological Properties of Hydroxyethyl Cellulose (HEC). *Geofluids* **2019**, *2019*, 7405702. [CrossRef]
16. Menezes, R.R.; Marques, L.N.; Campos, L.A.; Ferreira, H.S.; Santana, L.N.L.; Neves, G.A. Use of Statistical Design to Study the Influence of CMC on the Rheological Properties of Bentonite Dispersions for Water-Based Drilling Fluids. *Appl. Clay Sci.* **2010**, *49*, 13–20. [CrossRef]
17. Zhang, L.; Sun, H.; Han, B.; Peng, L.; Ning, F.; Jiang, G.; Chehotkin, V.F. Effect of Shearing Actions on the Rheological Properties and Mesostructures of CMC, PVP and CMC + PVP Aqueous Solutions as Simple Water-Based Drilling Fluids for Gas Hydrate Drilling. *J. Unconv. Oil Gas. Resour.* **2016**, *14*, 86–98. [CrossRef]
18. Davoodi, S.; Ramazani, S.A.A.; Soleimanian, A.; Fellah Jahromi, A. Application of a Novel Acrylamide Copolymer Containing Highly Hydrophobic Comonomer as Filtration Control and Rheology Modifier Additive in Water-Based Drilling Mud. *J. Pet. Sci. Eng.* **2019**, *180*, 747–755. [CrossRef]

19. Jain, R.; Mahto, V. Evaluation of Polyacrylamide/Clay Composite as a Potential Drilling Fluid Additive in Inhibitive Water Based Drilling Fluid System. *J. Pet. Sci. Eng.* **2015**, *133*, 612–621. [CrossRef]
20. Mech, D.; Sangwai, J.S. Effect of Molecular Weight of Polyethylene Glycol (PEG), a Hydrate Inhibitive Water-Based Drilling Fluid Additive, on the Formation and Dissociation Kinetics of Methane Hydrate. *J. Nat. Gas. Sci. Eng.* **2016**, *35*, 1441–1452. [CrossRef]
21. Tan, T.Y.; Katende, A.; Sagala, F.; Mugisa, J. Experimental Investigation of Cuttings Lifting Efficiency Using Low and High Density Polyethylene Beads in Different Hole Angles. *J. Mater. Sci. Appl.* **2017**, *3*, 71–78.
22. Zakaria, M.F.; Husein, M.; Hareland, G. Novel Nanoparticle-Based Drilling Fluid with Improved Characteristics. In Proceedings of the Society of Petroleum Engineers—SPE International Oilfield Nanotechnology Conference, Noordwijk, The Netherlands, 12–14 June 2012; pp. 232–237.
23. Abbas, A.K.; Alsaba, M.T.; Al Dushaishi, M.F. Improving Hole Cleaning in Horizontal Wells by Using Nanocomposite Water-Based Mud. *J. Pet. Sci. Eng.* **2021**, *203*, 108619. [CrossRef]
24. Akhtarmanesh, S.; Shahrabi, M.J.A.; Atashnezhad, A. Improvement of Wellbore Stability in Shale Using Nanoparticles. *J. Pet. Sci. Eng.* **2013**, *112*, 290–295. [CrossRef]
25. Vryzas, Z.; Kelessidis, V.C. Nano-Based Drilling Fluids: A Review. *Energy* **2017**, *10*, 540. [CrossRef]
26. Cheraghian, G. Nanoparticles in Drilling Fluid: A Review of the State-of-the-Art. *J. Mater. Res. Technol.* **2021**, *13*, 737–753. [CrossRef]
27. Tahr, Z.; Ali, J.A.; Mohammed, A.S. Sustainable Aspects behind Nano-Biodegradable Drilling Fluids: A Critical Review. *Geoenergy Sci. Eng.* **2023**, *222*, 211443. [CrossRef]
28. Martin, C.; Nourian, A.; Babaie, M.; Nasr, G.G. Environmental, Health and Safety Assessment of Nanoparticle Application in Drilling Mud—Review. *Geoenergy Sci. Eng.* **2023**, *226*, 211767. [CrossRef]
29. Rashid, F.L.; Mohammed, H.I.; Dulaimi, A.; Al-Obaidi, M.A.; Talebizadehsardari, P.; Ahmad, S.; Ameen, A. Analysis of heat transfer in various cavity geometries with and without nano-enhanced phase change material: A review. *Energy Rep.* **2023**, *10*, 3757–3779. [CrossRef]
30. Hussein, A.K.; Rashid, F.L.; Rasul, M.K.; Basem, A.; Younis, O.; Homod, R.Z.; Attia, M.E.H.; Al-Obaidi, M.A.; Hamida, M.B.B.; Ali, B.; et al. A review of the application of hybrid nanofluids in solar still energy systems and guidelines for future prospects. *Sol. Energy* **2024**, *272*, 112485. [CrossRef]
31. Rashid, F.L.; Hashim, A.; Dulaimi, A.; Hadi, A.; Ibrahim, H.; Al-Obaidi, M.A.; Ameen, A. Enhancement of Polyacrylic Acid/Silicon Carbide Nanocomposites' Optical Properties for Potential Application in Renewable Energy. *J. Compos. Sci.* **2024**, *8*, 123. [CrossRef]
32. Rajab, M.A.; Yahiya, M.M.; Hameed, N.A.; Al-Obaidi, M.A.; Al-Arkawazi, R.R. Wear resistance of fiber-reinforced nanoparticle hybrid mixture. *Int. J. Nanoelectron. Mater. (IJNeaM)* **2024**, *17*, 6–11.
33. Srivastava, V.; Beg, M.; Sharma, S.; Choubey, A.K. Application of Manganese Oxide Nanoparticles Synthesized via Green Route for Improved Performance of Water-Based Drilling Fluids. *Appl. Nanosci.* **2021**, *11*, 2247–2260. [CrossRef]
34. Hajiabadi, S.H.; Aghaei, H.; Kalateh-Aghamohammadi, M.; Sanati, A.; Kazemi-Beydokhti, A.; Esmaeilzadeh, F. A Comprehensive Empirical, Analytical and Tomographic Investigation on Rheology and Formation Damage Behavior of a Novel Nano-Modified Invert Emulsion Drilling Fluid. *J. Pet. Sci. Eng.* **2019**, *181*, 106257. [CrossRef]
35. Al-Shargabi, M.; Davoodi, S.; Wood, D.A.; Al-Musai, A.; Rukavishnikov, V.S.; Minaev, K.M. Nanoparticle applications as beneficial oil and gas drilling fluid additives: A review. *J. Mol. Liq.* **2022**, *352*, 118725. [CrossRef]
36. Suryam, L.V.; Pradeep, A.V.; Prasad, S.S.; Kumari, P.P. Functioning mechanism of nanomaterials and significance on machinability: A review. *Mater. Today Proc.* **2019**, *19*, 560–563. [CrossRef]
37. Martin, C.; Babaie, M.; Nourian, A.; Nasr, G.G. Designing Smart drilling fluids using modified nano silica to improve drilling operations in Geothermal wells. *Geothermics* **2023**, *107*, 102600. [CrossRef]
38. Boyou, N.V.; Ismail, I.; Wan Sulaiman, W.R.; Sharifi Haddad, A.; Husein, N.; Hui, H.T.; Nadaraja, K. Experimental Investigation of Hole Cleaning in Directional Drilling by Using Nano-Enhanced Water-Based Drilling Fluids. *J. Pet. Sci. Eng.* **2019**, *176*, 220–231. [CrossRef]
39. Minakov, A.V.; Mikhienkova, E.I.; Voronenkova, Y.O.; Neverov, A.L.; Zeer, G.M.; Zharkov, S.M. Systematic Experimental Investigation of Filtration Losses of Drilling Fluids Containing Silicon Oxide Nanoparticles. *J. Nat. Gas. Sci. Eng.* **2019**, *71*, 102984. [CrossRef]
40. Inturi, S.K.; Dash, G.R.; Kodavaty, J. Role of Silica Nano Particles in Altering Rheological Properties of Drilling Fluid in Enhanced Oil Recovery. *Mater. Today Proc.* **2019**, *17*, 354–361. [CrossRef]
41. Nascimento, D.R.; Oliveira, B.R.; Saide, V.G.P.; Magalhães, S.C.; Scheid, C.M.; Calçada, L.A. Effects of Particle-Size Distribution and Solid Additives in the Apparent Viscosity of Drilling Fluids. *J. Pet. Sci. Eng.* **2019**, *182*, 106275. [CrossRef]
42. Saboori, R.; Sabbaghi, S.; Kalantariasl, A. Improvement of Rheological, Filtration and Thermal Conductivity of Bentonite Drilling Fluid Using Copper Oxide/Polyacrylamide Nanocomposite. *Powder Technol.* **2019**, *353*, 257–266. [CrossRef]
43. Perween, S.; Thakur, N.K.; Beg, M.; Sharma, S.; Ranjan, A. Enhancing the Properties of Water Based Drilling Fluid Using Bismuth Ferrite Nanoparticles. *Colloids Surf. A Physicochem. Eng. Asp.* **2019**, *561*, 165–177. [CrossRef]
44. Dejtaradon, P.; Hamidi, H.; Chuks, M.H.; Wilkinson, D.; Rafati, R. Impact of ZnO and CuO Nanoparticles on the Rheological and Filtration Properties of Water-Based Drilling Fluid. *Colloids Surf. A Physicochem. Eng. Asp.* **2019**, *570*, 354–367. [CrossRef]
45. Al-Yasiri, M.; Awad, A.; Pervaiz, S.; Wen, D. Influence of Silica Nanoparticles on the Functionality of Water-Based Drilling Fluids. *J. Pet. Sci. Eng.* **2019**, *179*, 504–512. [CrossRef]

46. Huang, X.; Lv, K.; Sun, J.; Lu, Z.; Bai, Y.; Shen, H.; Wang, J. Enhancement of Thermal Stability of Drilling Fluid Using Laponite Nanoparticles under Extreme Temperature Conditions. *Mater. Lett.* **2019**, *248*, 146–149. [CrossRef]
47. Ma, L.; Luo, P.; He, Y.; Zhang, L.; Fan, Y.; Jiang, Z. Improving the Stability of Multi-Walled Carbon Nano-Tubes in Extremely Environments: Applications as Nano-Plugging Additives in Drilling Fluids. *J. Nat. Gas. Sci. Eng.* **2020**, *74*, 103082. [CrossRef]
48. Zhong, H.; Gao, X.; Zhang, X.; Qiu, Z.; Zhao, C.; Zhang, X.; Jin, J. Improving the Shale Stability with Nano-Silica Grafted with Hyperbranched Polyethyleneimine in Water-Based Drilling Fluid. *J. Nat. Gas. Sci. Eng.* **2020**, *83*, 103624. [CrossRef]
49. Li, W.; Jiang, G.; Ni, X.; Li, Y.; Wang, X.; Luo, X. Styrene Butadiene Resin/Nano-SiO_2 Composite as a Water-and-Oil-Dispersible Plugging Agent for Oil-Based Drilling Fluid. *Colloids Surf. A Physicochem. Eng. Asp.* **2020**, *606*, 125245. [CrossRef]
50. Jiang, G.; Ni, X.; Yang, L.; Li, W.; Li, Y.; Deng, Z. Synthesis of Superamphiphobic Nanofluid as a Multi-Functional Additive in Oil-Based Drilling Fluid, Especially the Stabilization Performance on the Water/Oil Interface. *Colloids Surf. A Physicochem. Eng. Asp.* **2020**, *588*, 124385. [CrossRef]
51. Moraveji, M.K.; Ghaffarkhah, A.; Agin, F.; Talebkeikhah, M.; Jahanshahi, A.; Kalantar, A.; Amirhosseini, S.F.; Karimifard, M.; Mortazavipour, S.I.; Sehat, A.A.; et al. Application of Amorphous Silica Nanoparticles in Improving the Rheological Properties, Filtration and Shale Stability of Glycol-Based Drilling Fluids. *Int. Commun. Heat Mass Transf.* **2020**, *115*, 104625. [CrossRef]
52. Hajiabadi, S.H.; Bedrikovetsky, P.; Mahani, H.; Khoshsima, A.; Aghaei, H.; Kalateh-Aghamohammadi, M.; Habibi, S. Effects of Surface Modified Nanosilica on Drilling Fluid and Formation Damage. *J. Pet. Sci. Eng.* **2020**, *194*, 107559. [CrossRef]
53. Gudarzifar, H.; Sabbaghi, S.; Rezvani, A.; Saboori, R. Experimental Investigation of Rheological & Filtration Properties and Thermal Conductivity of Water-Based Drilling Fluid Enhanced. *Powder Technol.* **2020**, *368*, 323–341.
54. Medhi, S.; Chowdhury, S.; Kumar, A.; Gupta, D.K.; Aswal, Z.; Sangwai, J.S. Zirconium Oxide Nanoparticle as an Effective Additive for Non-Damaging Drilling Fluid: A Study through Rheology and Computational Fluid Dynamics Investigation. *J. Pet. Sci. Eng.* **2020**, *187*, 106826. [CrossRef]
55. Wang, K.; Jiang, G.; Li, X.; Luckham, P.F. Study of Graphene Oxide to Stabilize Shale in Water-Based Drilling Fluids. *Colloids Surf. A Physicochem. Eng. Asp.* **2020**, *606*, 125457. [CrossRef]
56. Kamali, F.; Saboori, R.; Sabbaghi, S. Fe3O4-CMC Nanocomposite Performance Evaluation as Rheology Modifier and Fluid Loss Control Characteristic Additives in Water-Based Drilling Fluid. *J. Pet. Sci. Eng.* **2021**, *205*, 108912. [CrossRef]
57. Hassanzadeh, A.; Alihosseini, A.; Monajjemi, M.; Nazari-Sarem, M. Nano-Alumina Based (Alpha and Gamma) Drilling Fluid System to Stabilize High Reactive Shales. *Petroleum* **2021**, *7*, 142–151. [CrossRef]
58. Hajiabadi, S.H.; Aghaei, H.; Ghabdian, M.; Kalateh-Aghamohammadi, M.; Esmaeilnezhad, E.; Choi, H.J. On the Attributes of Invert-Emulsion Drilling Fluids Modified with Graphene Oxide/Inorganic Complexes. *J. Ind. Eng. Chem.* **2021**, *93*, 290–301. [CrossRef]
59. Dutta, D.; Das, B.M. Development of Smart Bentonite Drilling Fluid Introducing Iron Oxide Nanoparticles Compatible to the Reservoirs of Upper Assam. *Upstream Oil Gas. Technol.* **2021**, *7*, 100058. [CrossRef]
60. Mirzaasadi, M.; Zarei, V.; Elveny, M.; Alizadeh, S.M.; Alizadeh, V.; Khan, A. Improving the Rheological Properties and Thermal Stability of Water-Based Drilling Fluid Using Biogenic Silica Nanoparticles. *Energy Rep.* **2021**, *7*, 6162–6171. [CrossRef]
61. Zhong, H.; Kong, X.; Chen, S.; Grady, B.P.; Qiu, Z. Preparation, Characterization and Filtration Control Properties of Crosslinked Starch Nanospheres in Water-Based Drilling Fluids. *J. Mol. Liq.* **2021**, *325*, 115221. [CrossRef]
62. Shen, X.; Jiang, G.; Li, X.; He, Y.; Yang, L.; Cui, K.; Li, W. Application of Carboxylated Cellulose Nanocrystals as Eco-Friendly Shale Inhibitors in Water-Based Drilling Fluids. *Colloids Surf. A Physicochem. Eng. Asp.* **2021**, *627*, 127182. [CrossRef]
63. Medhi, S.; Gupta, D.K.; Sangwai, J.S. Impact of Zinc Oxide Nanoparticles on the Rheological and Fluid-Loss Properties, and the Hydraulic Performance of Non-Damaging Drilling Fluid. *J. Nat. Gas. Sci. Eng.* **2021**, *88*, 103834. [CrossRef]
64. Shojaei, N.; Ghazanfari, M.H. Reduction of Formation Damage in Horizontal Wellbores by Application of Nano-Enhanced Drilling Fluids: Experimental and Modeling Study. *J. Pet. Sci. Eng.* **2022**, *210*, 110075. [CrossRef]
65. Mohammadi, M.K.; Riahi, S.; Boek, E.S. Developing Novel Bio-Nano Catalyst Well Clean up Fluid to Remove Formation Damage Induced by Polymeric Water-Based Drilling Fluids. *J. Pet. Sci. Eng.* **2022**, *210*, 109809. [CrossRef]
66. Arain, A.H.; Ridha, S.; Mohyaldinn, M.E.; Suppiah, R.R. Improving the Performance of Invert Emulsion Drilling Fluid Using Boron Nitride and Graphene Nanoplatelets for Drilling of Unconventional High-Temperature Shale Formations. *J. Mol. Liq.* **2022**, *363*, 119806. [CrossRef]
67. Dora, T.K.; Vamsi Krishna, K.; Ismail Iqbal, M.; Ranjan, A. An Experimental Analysis on Nanoparticles Role in Drilling Fluids. *Mater. Today Proc.* **2022**, *68*, 717–721. [CrossRef]
68. Razali, S.Z.; Yunus, R.; Kania, D.; Rashid, S.A.; Ngee, L.H.; Abdulkareem-Alsultan, G.; Jan, B.M. Effects of Morphology and Graphitization of Carbon Nanomaterials on the Rheology, Emulsion Stability, and Filtration Control Ability of Drilling Fluids. *J. Mater. Res. Technol.* **2022**, *21*, 2891–2905. [CrossRef]
69. Mansoor, H.H.A.; Devarapu, S.R.; Samuel, R.; Sangwai, J.S.; Ponmani, S. Investigation of Chia Based Copper Oxide Nanofluid for Water Based Drilling Fluid: An Experimental Approach. *J. Nat. Gas Sci. Eng.* **2022**, *107*, 104775. [CrossRef]
70. Minakov, A.V.; Pryazhnikov, M.I.; Mikhienkova, E.I.; Voronenkova, Y.O. Systematic Experimental Study of the Temperature Dependence of Viscosity and Rheological Behavior of Water-Based Drilling Fluids with Nano-Additives. *Petroleum* **2023**, *9*, 534–544. [CrossRef]

71. Lai, N.; Fan, W.; Zhang, X.; Liu, L.; Zhou, X.; Chen, S. Temperature-Sensitive Polymer Based Nano-SiO_2 Composite Multi-Component Synergistic Improvement of Shale Stability in Water-Based Drilling Fluids. *Geoenergy Sci. Eng.* **2023**, *224*, 211498. [CrossRef]
72. Lysakova, E.; Skorobogatova, A.; Neverov, A.; Pryazhnikov, M.; Zhigarev, V.; Voronin, A.; Shabanova, K.; Minakov, A. Comparative Analysis of the Effect of Single-Walled and Multi-Walled Carbon Nanotube Additives on the Properties of Hydrocarbon-Based Drilling Fluids. *Colloids Surf. A Physicochem. Eng. Asp.* **2023**, *678*, 132434. [CrossRef]
73. Zhang, C.; Sun, J.; Huang, X.; Zhang, Y.; Zong, J.; Lv, K.; Dai, J. Use of Modified Polystyrene Micro-Nano Spheres to Improve the Inhibition and Plugging Performance of Water-Based Drilling Fluids. *Colloids Surf. A Physicochem. Eng. Asp.* **2023**, *668*, 131409. [CrossRef]
74. Yang, J.; Jiang, G.; Wang, G.; Yang, L.; He, Y.; Dong, T.; Yuan, X. Performance Evaluation of Polymer Nanolatex Particles as Fluid Loss Control Additive in Water-Based Drilling Fluids. *Geoenergy Sci. Eng.* **2023**, *223*, 211462. [CrossRef]
75. Lin, Y.; Tian, Q.; Lin, P.; Tan, X.; Qin, H.; Chen, J. Effect of Nanoparticles on Rheological Properties of Water-Based Drilling Fluid. *Nanomaterials* **2023**, *13*, 2092. [CrossRef] [PubMed]
76. Wang, G.; Li, W.; Qiu, S.; Liu, J.; Ou, Z.; Li, X.; Ji, F.; Zhang, L.; Liu, S.; Yang, L.; et al. Application of a Core-Shell Structure Nano Filtration Control Additive in Salt-Resistant Clay-Free Water-Based Drilling Fluid. *Polymers* **2023**, *15*, 4331. [CrossRef]
77. Bardhan, A.; Vats, S.; Prajapati, D.K.; Halari, D.; Sharma, S.; Saxena, A. Utilization of Mesoporous Nano-Silica as High-Temperature Water-Based Drilling Fluids Additive: Insights into the Fluid Loss Reduction and Shale Stabilization Potential. *Geoenergy Sci. Eng.* **2024**, *232*, 212436. [CrossRef]
78. Ali, J.A.; Abbas, D.Y.; Abdalqadir, M.; Nevecna, T.; Jaf, P.T.; Abdullah, A.D.; Rancová, A. Evaluation the Effect of Wheat Nano-Biopolymers on the Rheological and Filtration Properties of the Drilling Fluid: Towards Sustainable Drilling Process. *Colloids Surf. A Physicochem. Eng. Asp.* **2024**, *683*, 133001. [CrossRef]
79. Li, X.; Wang, K.; Lu, Y.; Shen, X.; Zhang, H.; Peng, J.; Jiang, S.; Duan, M. Compatibility and Efficiency of Hydrophilic/Hydrophobic Nano Silica as Rheological Modifiers and Fluid Loss Reducers in Water-Based Drilling Fluids. *Geoenergy Sci. Eng.* **2024**, *234*, 212628. [CrossRef]
80. Alihosseini, A.; Hassan Zadeh, A.; Monajjemi, M.; Nazary Sarem, M. Effectiveness of Alumina Nanoparticles in Improving the Rheological Properties of Water-Based Drilling Mud. *Iran. J. Oil Gas Sci. Technol.* **2021**, *10*, 12–27.
81. Mohammadi, M.; Kouhi, M.; Sarrafi, A.; Schaffie, M. Studying rheological behavior of nanoclay as oil well drilling fluid. *Res. Chem. Intermed.* **2015**, *41*, 2823–2831. [CrossRef]
82. Vipulanandan, C.; Mohammed, A.; Samuel, R.G. Fluid loss control in smart bentonite drilling mud modified with nanoclay and quantified with Vipulanandan fluid loss model. In Proceedings of the Offshore Technology Conference, Houston, TX, USA, 30 April–3 May 2018.
83. Liu, C.; Li, M.; Mei, C.; Chen, W.; Han, J.; Yue, Y.; Ren, S.; French, A.D.; Aita, G.M.; Eggleston, G.; et al. Cellulose nanofibers from rapidly microwave-delignified energy cane bagasse and their application in drilling fluids as rheology and filtration modifiers. *Ind. Crops Prod.* **2020**, *150*, 112378. [CrossRef]
84. Ismail, A.R.; Aftab, A.; Ibupoto, Z.H.; Zolkifile, N. The novel approach for the enhancement of rheological properties of water-based drilling fluids by using multi-walled carbon nanotube, nanosilica and glass beads. *J. Pet. Sci. Eng.* **2016**, *139*, 264–275. [CrossRef]
85. da Luz, R.C.; Paixão, M.V.; Balaban, R.D. Nanosilica-chitosan hybrid materials: Preparation, characterization and application in aqueous drilling fluids. *J. Mol. Liq.* **2019**, *279*, 279–288. [CrossRef]
86. Alvi, M.A.; Belayneh, M.; Bandyopadhyay, S.; Minde, M.W. Effect of iron oxide nanoparticles on the properties of water based drilling fluids. *Energies* **2020**, *13*, 6718. [CrossRef]
87. Ahmad, H.M.; Iqbal, T.; Kamal, M.S.; Al-Harthi, M.A. Influence of hydrophobically modified polymer and titania nanoparticles on shale hydration and swelling properties. *Energy Fuels* **2020**, *34*, 16456–16468. [CrossRef]

processes

MDPI

Article

A Dynamic Simulation of a Piezoelectric Energy-Harvesting System Integrated with a Closed-Loop Voltage Source Converter for Sustainable Power Generation

Ahmed K. Ali [1,*], Ali Abdulwahhab Abdulrazzaq [2] and Ali H. Mohsin [3]

1 Institute of Technology, Middle Technical University, Baghdad 10074, Iraq
2 Technical Instructors Training Institute, Middle Technical University, Baghdad 10074, Iraq; dr.ali.abdulwahhab@mtu.edu.iq
3 Air Conditioning and Refrigeration Techniques Engineering Department, College of Engineering and Technologies, Al-Mustaqbal University, Babylon 51001, Iraq; ali.hasanain.mohsin@uomus.edu.iq
* Correspondence: engineer28ahmed@gmail.com or ahmed_khalid@mtu.edu.iq

Abstract: Numerous recent studies address the concept of energy harvesting from natural wind excitation vibration to piezoelectric surfaces, aerodynamic losses, and electromagnetic dampers. All these techniques require a connection to an energy-management circuit. However, the simulation model for energy conversion and management dedicated to this task has not yet been described. This paper presents a model-based simulation for an energy conversion system using piezoelectric energy-harvester system (PEHS) technology. A controlled pulse width modulation (PWM) rectifier, a closed-loop buck-boost converter, and a piezoelectric transducer comprise a dynamic mathematical model of a PEHS. The control blocks of the closed-loop buck-boost converter use the perturbation and observation (P&O) algorithm based on maximum power point tracking (MPPT), which adapts the operational voltage of the piezoelectric source to deliver the maximum power to load. A simulation program is employed to perform mathematical analysis on various wind vibration scenarios, piezoelectric sources without PWM converters, and piezoelectric vibration sources connected to a closed-loop P&O converter. The crucial results of this paper demonstrated that the proposed dynamic PEHS model effectively fed low-power electronic loads by directly adjusting the output voltage level to the set voltage, even under different vibration severity levels. As a result, the proposed PEHS dynamic model serves as a guideline for researchers in the development of self-powered sensors, which contributes to understanding sustainable energy alternatives.

Keywords: controlled PWM rectifier; P&O MPPT; closed-loop converter; piezoelectric energy harvesting; sustainable renewable energy

1. Introduction

Daily activities and production processes frequently demonstrate inefficiencies in energy use, resulting in the inadequate harnessing of significant amounts of energy and its release into the atmosphere as various forms of waste heat. This event results in substantial waste and environmental degradation. Since the initial observation of piezoelectric electricity (PE), extensive research efforts have been dedicated to exploring its diverse applications [1–5]. On the other hand, wind-induced vibrations offer a valuable source of potential for electrical power generation [6]. The vibrational force is used as input to generate AC electricity that changes sinusoidally, which is further rectified to be practical for power applications [7]. The outline of the state-of-the-art in power-management circuits for PEHS was reported in [8]. It covered various power-management techniques for different PEHS circuits. In the same context, the current work contributes to integrating these techniques with a new control strategy using dynamic simulation to improve PEHS performance.

Ericka et al. [9] presented an approach for energy harvesting with a circular piezoelectric membrane to optimize energy production derived from mechanical vibrations. Their system has a significantly improved conception of piezoelectric materials and geometric arrangements, laying the groundwork for more efficient energy-harvesting devices.

Wang et al. [10] focused on the enhancement of energy-harvesting systems using a piezoelectric circular diaphragm array. This array structure improved energy production and efficiency, making it possible to collect vibration energy on a larger scale and making piezoelectric harvesters more useful. Barton et al. [11] studied energy harvesting with a nonlinear oscillator, tackling the constraints of linear systems in low-frequency and wideband vibrations. Using nonlinear dynamics, they improved the energy-harvesting efficiency in fluctuating conditions, advancing more versatile and functional energy systems. Mitcheson et al. [12] explored the adjustment of resonance frequency and damping in electromagnetic energy harvesters using power electronic circuits.

Wu et al. [13] introduced a piezoelectric energy harvester that adjusts its resonance frequency in response to wind speed fluctuations. This technique enhances the offshore wind-energy-gathering efficiency by dynamically adapting to fluctuating cases, marking a significant advancement in optimizing energy conversion in changeable conditions. Meng et al. [14] conducted a study that provided significant insights into the effectiveness of various geometric designs via a comparative analysis of two geometric spheres and cylinder shapes for piezoelectric energy-harvester devices.

Wang et al. [15] conducted an experimental and analytical investigation to study the design of a nonlinear sickle-shaped cantilever beam piezoelectric energy harvester for wind energy applications. Their study enhanced the design of piezoelectric energy harvesters by demonstrating the efficacy of nonlinear geometric shapes in optimizing energy conversion. Kan et al. [16] developed a new axially retractable bracket-shaped piezoelectric vibrator that uses magnetic force for wind energy harvesting. They discussed a practical application of magnetic stimulation to enhance the vibrational responsiveness of piezoelectric materials. The suggested design presents an innovative method for enhancing the energy-harvesting efficiency by integrating piezoelectric material technologies and voltage control circuits.

Zhao et al. [17] described the combination of two approaches involving a nanogenerator on the sole of a shoe to use energy from walking. They presented a new device architecture that combines flexible thermoelectric and piezoelectric materials to increase the overall power output while reducing the effect on shoe functionality. Additionally, Wang et al. [18] presented an innovative electrical power generation device that uses high-frequency dynamic piezoelectric deformation through shear caused by friction. Despite these advances, research initiatives concentrating on power generation via piezoelectric and thermal means remain relatively limited. Various mechanisms have been used to increase the efficiency of smart tiles, including wearable biomechanical energy harvesting, triboelectric pairs, and mechanical impact mechanisms. Krishnasamy et al. [19] proposed the conversion of the input vibration force, which is dependent on the vibration frequency, as a common method for harvesting piezoelectric energy.

Yingyong et al. [20] presented a system that uses piezoelectric cantilevers (PZCs) connected to the moving surface of a tile to generate alternative electricity. The principle of the presented system comes from the human footstep, because people walking by can produce a mechanical impact, which generates kinetic energy. To maintain frequency movement, a fixed mass is placed at the end of the PZCs. Panthongsy et al. [21] suggested an approach using the principle of a magnetic tension mechanism to increase vibration in PZCs. In this configuration, a permeant magnet attached to one end and an iron piece fastened to the moving surface of the tile act as activators for the piezoelectric consoles. When individuals step on the tile, the moving surface descends, reducing the distance between the iron piece and the permeant magnet, thus causing attraction. On the basis of the passage, the tile surface returns to its original position, separating the magnet from the iron piece and inducing vibration in the piezoelectric console. In the same context, many mechanical methods have been used to harness the energy of smart tiles, although with considerable friction. Duarte et al. [22] presented the rack-and-pinion mechanical approach, in which the gear connected to the moving surface of the tile interacts with a pinion gear connected to a generator. With tile stimulation, gear rotation drives the gear pinion, thereby activating the generator to produce energy. In this instance, the applied force directly stimulates the piezoelectric cells, facilitating energy production. The study presented by Puscasu et al. [23] adopted a similar approach by placing PE cells beneath the tile surface, using pedestrian-induced force to drive energy production. Their findings emphasized the capacity of piezoelectric flooring as a sustainable solution for urban lighting, especially in regions with significant pedestrian activity. Building on this, Kim et al. [24] investigated the utilization of the triboelectric effect within a tile structure and explored the dielectric effects to improve energy harvesting. The electromagnetic effect uses coils around the perimeter of the tile and a central disc that holds magnets. The disc rotates due to the applied force, which creates flux on the coils [25,26].

In a deeper delve into the field, Balato et al. [27] explored techniques in resonant piezoelectric vibration energy harvesters (RPVEHs). These devices generate electrical power from vibrating sources, but their optimal operating point changes over time as input vibrations vary. This study focuses on the most common RPVEH architecture, which employs a diode bridge rectifier and a DC/DC converter. Additionally, it provides experimentally validated guidelines for parameter selection in two popular maximum power point tracking (MPPT) techniques (perturb and observe and open-circuit voltage). The simulation results are analyzed to assess the efficacy of the proposed modeling method. A previous study [28] explored the modeling of a piezoelectric bender to convert ambient vibrations into electrical energy. It compared an analytical approach based on beam theory with two existing models. Lefeuvre et al. [29] delved into a design buck-boost converter for sensor-less power optimization in PEHS. Their proposed buck-boost converter circuit improved the energy extraction efficiency of piezoelectric materials without the need for an external feedback sensor. This sensor-less power-management approach proved effective in improving PEHS performance, particularly in situations where vibration sources were inconsistent. Although they addressed a crucial issue in converter design, their discussion remains limited as they solely concentrated on energy extraction without guaranteeing stability in the produced power. Building on this, Mitcheson et al. [30] introduced a different approach for controlling the resonant frequency and damping of energy harvesters using power electronics circuits. While their focus was on frequency tuning, the current research incorporates this aspect while placing a strong emphasis on enhancing piezoelectric efficiency, developing electromechanical models, and optimizing the rectification-conversion process.

In PEHSs, converter circuits play a critical role in the management of usable power. Cai and Zhu [31] delved into improving the performance of electromagnetic harvesters by developing the converter circuit with microcontrollers. Their integration approach is similar to the current study, but the process of converting the vibration energy source to usable power is different. The wind force is still an unstable source of vibrational energy. Cai et al. [32] highlighted the importance of bistable processes and nonlinear energy sinks for better energy collection from wind-induced vibrations, particularly in offshore areas where conditions are harsh and changeable. Although Cai et al. concentrated on optimizing the structure using bistable energy sinks, the current study adopts a systems approach by designing a simulation model that converts vibrational force into a reliable electrical output, appropriate for many applications. The combination of PE with advanced power electronics represents a distinct but equally groundbreaking advancement in the realm of energy harvesting. To the best of our knowledge, the current study makes novel contributions by presenting a new dynamic simulation model using wind-induced vibrations with piezoelectric materials, an area that has not been thoroughly explored in previous literature [17–28].

Previous studies have proposed many useful models for PE energy harvesting, but few of them have concentrated on investigating the dynamic simulation of these models.

However, this paper delves into the theoretical modeling of PE vibration energy harvesting and the associated DC–DC boost converter. The aim of this paper is to present a new approach to the use of wind-induced vibrations for energy harvesting by developing a detailed dynamic simulation model. The simulation analysis focuses on achieving a constant DC output of the energy harvested. The following is a summary of the key contributions:

- An approach to harnessing the energy vibrations caused by wind, which can be harnessed as a source of electrical power, is explored. More precisely, we present a comprehensive and detailed dynamic mathematical model for a PEHS that combines a piezoelectric transducer, a regulated PWM rectifier, and a closed-loop buck-boost converter. This model accurately depicts the intricate correlation between wind-induced mechanical vibrations and the subsequent conversion of electrical energy, offering a more precise portrayal of real-life conditions.
- This paper emphasizes the efficiency of PEHS in response to vibrations, which makes them suitable for energy-harvesting performance across a spectrum of low and high vibration frequencies.
- An electromechanical dynamic simulation model is developed to represent the piezoelectric materials harvester, which translates the vibrational force into the electrical output. The process entails converting the vibration of the piezoelectric sheet sinusoidally, varying AC output into rectified DC output suitable for power applications. This approach broadens the current understanding of the dynamic simulation and mathematical representation of PEHS across various excitation frequencies, both high and low. It consolidates the simulations of vibration originating from diverse wind energy sources, such as those generated by cars passing on a highway.
- The piezoelectric model is also interconnected with an electrical circuit that has a PWM single-phase bridge rectifier and a DC–DC boost converter that is controlled by a P&O MPPT to obtain greater stability in power output.

2. Piezoelectric Harvester Model

The electromechanical model diagram for a piezoelectric-based energy harvester consists of many components and primarily incorporates wind vibrations. Figure 1 depicts the functional schematic of a wind vibration energy harvester that utilizes piezoelectric technology.

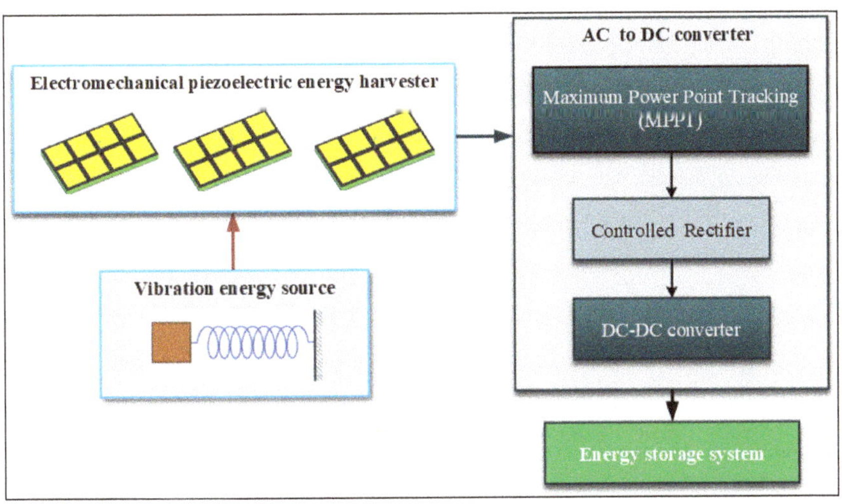

Figure 1. Diagram of the piezoelectric energy harvester for wind-generated vibrations.

1. Vibration source: Wind can cause vibrations in a structure. This structure can be a cantilever beam or any other resonant structure specifically built to oscillate due to wind-induced vibrations [33].
2. Piezoelectric materials: The direct piezoelectric effect converts mechanical strain into a train of electrical charges [34].
3. Electrical output: The piezoelectric material generates an electrical charge, which it immediately collects, stores, or uses as electrical power.
4. Energy storage management: This circuit divides a group of capacitors or batteries into rectifiers, voltage regulators, a DC-to-DC converter, and possibly an energy storage device [7].

A piezoelectric transducer converts mechanical vibrations from a wind force into an oscillating train of electrical charges. The sinewave peak-to-peak level relationship describes the changeable values of these charges, which depend on the severity of the wind vibration [35]. The controlled PWM rectifier, in conjunction with a closed-loop buck-boost converter, converts this train of sinewave cycles into DC voltages.

The piezoelectric vibration energy harvester uses a complex second-order mass spring damper model to fully describe its dynamic behavior [28,36]. Within this model, the harvester frame is composed of a mass, denoted as M, suspended by a spring characterized by its stiffness constant, μ_k, and a damper with a coefficient μ_D. This damper embodies various sources of mechanical damping, such as wind-induced vibrations and air friction. The application of an external force to the system induces movement in the mass, which in turn causes deformation in the connected piezoelectric element.

The principle of the piezoelectric effect governs this idea by generating an electric voltage when mechanical stress is applied to the piezoelectric material.

Crucially, we assumed that the external vibrational force acting on the system follows a sinusoidal pattern [37]. Thus, variations in wind speed and direction that exert aerodynamic forces on the turbine structure can be assumed in a sinusoidal manner, and we can express this force mathematically, as follows:

$$F_{wind}(t) = F_0 \sin(\omega t + \theta) \tag{1}$$

where F_0 represents the amplitude of the force, ω is the angular frequency of the sinusoidal vibrational force, t denotes time, and θ is the angle between the direction of the wind and the direction of motion of the harvester frame. This equation concisely captures the sinusoidal nature of the external vibrational force, which in turn induces harmonic motion in the harvester frame, facilitating efficient energy harvesting [37]. The equation of motion in a second-order mass–spring damper system describes the relative motion of the mass with respect to the frame, represented as $x(t)$ [38]. When the influence of wind as an external force is considered, the equation becomes $F_{wind}(t)$, resulting in:

$$F_{wind}(t) = M\frac{d^2x(t)}{dt^2} + \mu_D\frac{dx(t)}{dx} + \mu_k x(t) \tag{2}$$

where $x(t)$ represents the displacement of the mass from its equilibrium position. The acceleration of the mass is $\frac{d^2x(t)}{dt^2}$. The velocity of the mass is $\frac{dx(t)}{dx}$.

In order to calculate the formal transfer function of the system, first it needs to express the equation of motion in terms of the Laplace transform and then solve for the output, $X(s)$, in terms of the input force, $F_{wind}(s)$. When all initial conditions are zero, the transfer function, $H(s)$, is defined as follows:

$$H(S) = \frac{1}{Ms^2 + \mu_D s + \mu_k} \tag{3}$$

In the case of nonzero initial conditions, the transfer function can be expressed as:

$$H(S) = \frac{Ms^2}{Ms^2 + \mu_D s + \mu_k} \quad (4)$$

The mass, M, and stiffness, μ_k, of the system define the natural frequency:

$$\omega_n = \sqrt{\frac{\mu_k}{M}} \quad (5)$$

where ω_n is the natural frequency, which represents the frequency at which a system oscillates in the absence of damping $(\mu_D = 0)$. The mass and stiffness of the system dictate its intrinsic attributes. The damping ratio, ζ, can be expressed in terms of the damping coefficient, μ_D, as:

$$\zeta = \frac{\mu_D}{2\sqrt{M\mu_k}} \quad (6)$$

The damping ratio, ζ, reveals the rate at which oscillations diminish after the system is subjected to a disturbance. It compares actual damping, μ_D, with the critical damping, $2\sqrt{M\mu_k}$, which represents the lowest level of damping required to avoid oscillations.

3. Modeling Interfaced Electrical Circuits

This section explores the methodology of modeling interconnected electrical circuits, involving transfer functions, impedance analysis, and a state–space model, to predict the PEHS precisely.

3.1. PWM Full-Wave Rectifier

A power electronic circuit, known as a pulse width modulation (PWM) single-phase bridge rectifier, converts the AC voltage to the control DC voltage in the output voltage. The rectifier employs the PWM technique to regulate the output voltage. When the duty cycle of the PWM signal supply to the bridge switches is varied, the effective voltage of the load can be adjusted. The rectifier circuit typically uses MOSFETs or IGBTs as switches. To regulate the output voltage, the PWM signal controls these switches.

Figure 2 shows the principal circuit configuration of a single-phase voltage source PWM converter. The switches T_1 through T_4 are fully controllable. On the AC side, the inductance is denoted as L_m, where R_m represents the corresponding resistance. The DC side includes a filter capacitor labeled C_d. Furthermore, the terms U_s and i_s refer to the voltage and current of the AC supply, whereas U_{ab} and U_{Dc} indicate the voltage on the AC side and the output voltage on the DC side, respectively [39].

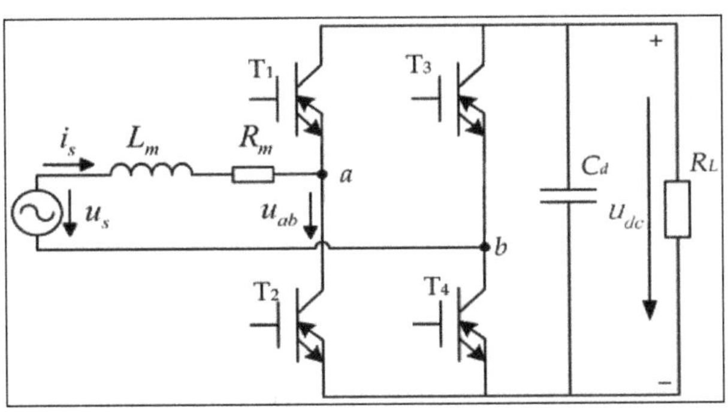

Figure 2. Essential circuit layout of a single-phase voltage source PWM converter featuring an RC-IGBT.

The rectifier employs PWM to regulate the operation status of four switches. Generally, the average output voltage over the bridge in a phase-controlled rectifier is often defined as:

$$U_{\text{Phase}-\text{DC}} = \frac{U_{peak}}{\pi}(1 + \cos(\alpha)) \quad (7)$$

where $U_{\text{Phase}-\text{DC}}$ is the phase-controlled rectifier output DC volage, U_{peak} is the peak of input AC voltage, and α represents the firing angle. The firing angle is used to control turn-on devices, such as thyristors (SCRs), in a fully controlled rectifier to determine when in the AC waveform the two complementary switches are turned on [40]. Thus, the output DC voltage in the phase-controlled rectifier is controlled by the firing angle, α, which achieved a maximum output voltage when $\alpha = 0°$ and a minimum voltage when $\alpha = 180°$.

Power electronics applications, such as DC–DC converters and PWM rectifiers, commonly use PWM-controlled circuits. The following formula provides the output voltage of a PWM-controlled converter [41]:

$$\begin{gathered} T_{\text{off}} U_{\text{DC}} = T_s U_{\text{in}} \\ D = \frac{T_{\text{On}}}{T_s} \end{gathered} \quad (8)$$

where U_{in} is the input voltage, and D represents the duty cycle of the PWM signal. A PWM converter controls the output voltage by varying the duty cycle of the PWM waveform. The duty cycle, D, is defined as the ratio of T_{On}, which is the duration for which the PWM signal is on (high), to the sample time, T_s, which is the total period of the PWM signal (sum of the on and off times). In this study, we proposed that the voltage-second equation depends on the T_{On}, so it can be written as:

$$1/T_{\text{on}} U_{\text{DC}} = T_s U_{\text{in}} \quad (9)$$

The duty cycle typically ranges from 0 to 1, where 0 represents the PWM signal being fully off and where 1 represents the PWM signal being fully on. Adjusting the duty cycle between 0 and 1 allows fine control of the output voltage within this range [40]. In real-time scenarios, it is advisable to avoid a duty cycle of D = 1 since it may potentially result in an infinite DC voltage. However, the voltage across the DC load can be adjusted to meet the desired specifications by selecting an appropriate duty ratio [41]. In order to find the average DC voltage of the PWM rectifier for the main part of the vibration voltage, the fully controlled rectified voltage (Equation (7)) must be sub-stationed at a maximum output voltage, where $\alpha = 0°$ with respect to the PWM switching into Equation (9). Therefore, the average DC output voltage of a single-phase PWM bridge rectifier can be expressed as:

$$U_{\text{DC}} = \frac{2U_{max_{a-b}}}{\pi} D \quad (10)$$

where $U_{max_{a-b}}$ represents the maximum amplitude of the vibration voltage. In the PWM rectifier, each PWM waveform has a D that determines the on/off duration of the corresponding switch, T_1–T_4. The duty cycles, D{T_1–T_4}, are typically adjusted to ensure that the output voltage is regulated [42].

3.2. DC–DC Converter

The DC–DC converter converts the DC input voltage source into either a step-up or a step-down of the DC output voltage. In piezoelectric vibrational energy-harvester applications, a buck-boost converter is a widely used type of DC-to-DC converter to regulate the output voltage. The operation mode increases or decreases depending on the duty cycle of the PWM signal. The typical buck-boost converter consists of an inductor, a power electronic switch, a diode, and a capacitor, as illustrated in Figure 3. Typically, a capacitor is placed across the output to reduce the voltage ripple and stabilize the output voltage. The control circuit, which is an important part of the buck-boost converter, includes a PWM generator that is responsible for controlling the switches by generating a changeable duty cycle to control the switches. In a closed-loop system, the output voltage remains at a certain level on the basis of feedback from the output voltage [40].

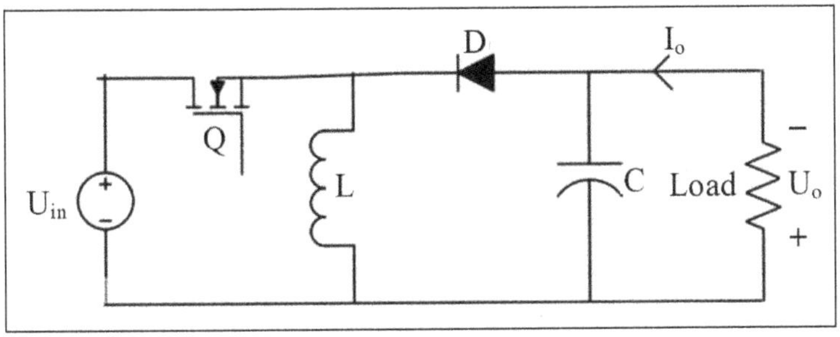

Figure 3. A traditional buck-boost converter circuit.

The expression for the output voltage of a buck-boost converter is defined as:

$$U_{B-DC} = \frac{D}{1-D} U_{DC} \tag{11}$$

where U_{B-DC} is the output voltage.

On the basis of the output voltage equation, the circuit controls the output voltage by regulating the duty cycle. Even in cases where the input voltage varies, the converter can regulate the output voltage by modifying the duty cycle.

3.3. *Maximum Power Point Tracking (MPPT) Algorithm*

The MPPT technique, which is based on perturbation and observation (P&O) algorithms, obtains the optimal output voltage and enhances the power transfer efficiency from the DC–DC converter to the load. The P&O MPPT technique typically determines whether the system operates at its maximum power point [27]. The P&O MPPT technique uses the rectified output voltage and output current signals to ascertain the operation point that can transfer the maximum power by adjusting the conversion output voltage. In this paper, the P&O MPPT aims to ensure that the energy-harvesting system operates at maximum power. The PWM generator block uses the signal from the P&O MPPT block to produce the reference duty cycle, allowing adaptable control for the DC–DC converter output voltages. The P&O MPPT algorithm consistently monitors the rectifier output voltage and current, even in the presence of varying wind vibration frequencies. Figure 4 illustrates the sequential operation of the MPPT technique, which is based on the P&O algorithm [43]. On the basis of the flowchart for the P&O MPPT algorithm, the duty cycle appears to play a crucial role in determining the energy transfer from the piezoelectric device to the load. By simultaneously adjusting the duty cycle via the P&O algorithm, the system ensures that it consistently operates at maximum energy conversion efficiency.

The implemented P&O MPPT block diagrams, which combine closed-loop control of a DC-to-DC converter, are shown in Figure 5. The implementation solely utilizes Simulink blocks—it does not incorporate scripting code. The flowchart of the P&O algorithm provides recommendations for marking each block with its role in the implanted P&O MPPT. The MPPT based on the P&O algorithm starts by accepting voltage and current readings from the rectifier outputs U_{DC} and I_{DC}, respectively.

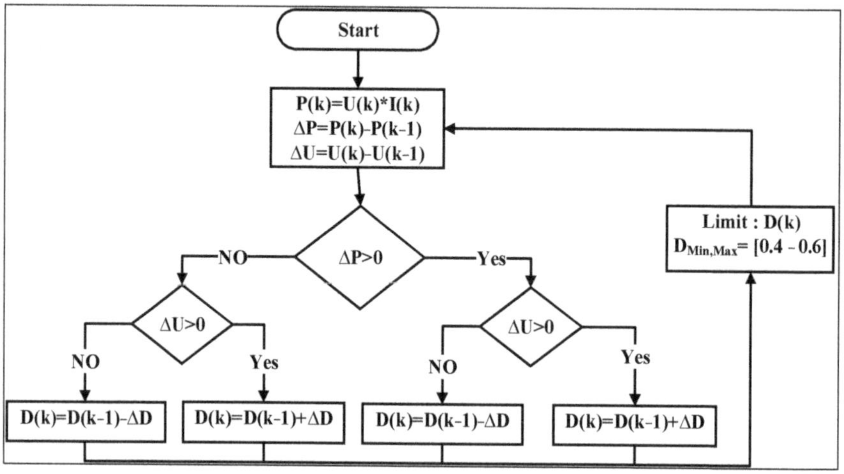

Figure 4. The flowchart of the perturb and observe (P&O) MPPT algorithm system.

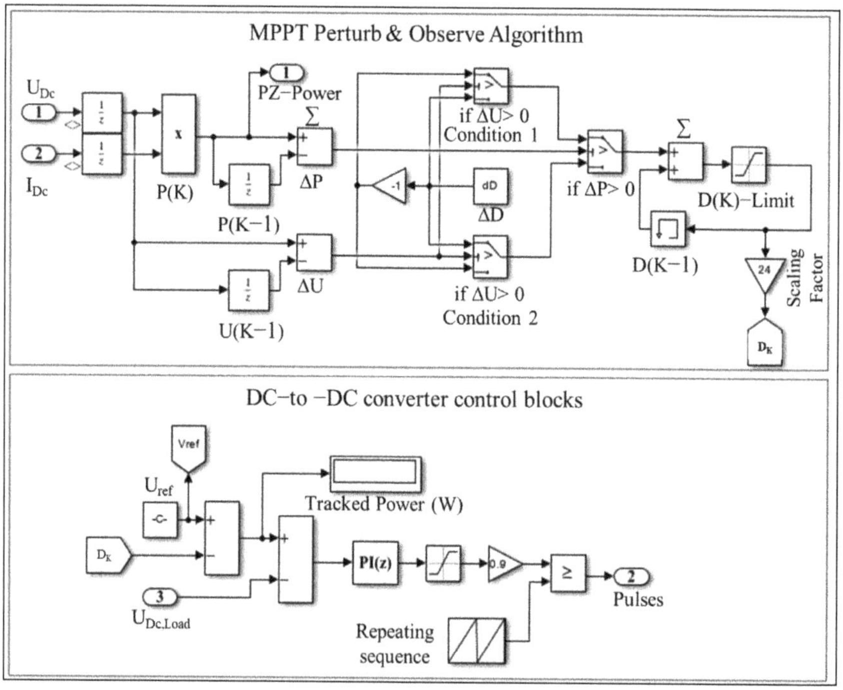

Figure 5. Implementation of the perturb and observe (P&O) MPPT algorithm with a closed-loop controller for a DC-to-DC converter.

The unit delay function block implements the previous sample $(k-1)$. The P&O algorithm uses the condition switch block to manage its three if–else conditions. Additionally, the block ΔD_B enables users to specify the perturbation step size of the duty cycle. An adder with a memory block manages the functions for the duty cycle's increase and decrease, ensuring consistency with the previous duty cycle, $D_B(k-1)$. A limit block, $D_B(k-1)$, is used to restrict the duty cycle within the range of 0.4–0.6, as reported in [43].

For D(k), multiplication by a constant gain of 24 represents the desired V_{ref} for the MPPT controller. This scaling factor is critical for ensuring that the system operates at the optimal voltage level to extract the maximum power from the vibration source, as well as aligns with the desired operating point.

In a closed-loop controller implementation using a buck-boost converter, the D_k generated from the P&O MPPT algorithm is multiplied by a constant gain to match the level of the control signal, U_D. The control signal is subtracted from the V_{ref} set reference voltage to satisfy the requirements of the load or storage system. The PI controller receives the loop error by subtracting the measured converter output voltage from the reference voltage, which adjusts the duty cycle for the PWM generator. On the basis of the feedback from the voltages, the controller continuously modifies the operating point of the DC-to-DC converter to maintain system operation at or near the reference voltage [40].

4. Simulation of Wind Vibrations Integrated with Piezoelectric Materials

In this section, the simulations explain how the proposed integration of PEHS with wind-induced mechanical vibrations effectively generated electrical energy. The goal was to improve the PEHS design. Table 1 lists the parameters used in the simulation analysis. The simulation focused on the first five cycles of the fundamental frequency for the vibrational wind force.

Table 1. Simulation parameters for the PEHS with vibrational wind force.

Definition	Value
Amplitudes of wind force (N)	200
Wind vibration speed (Hz)	50
Translational damper N(m/s)	100
Translational spring (N/m)	400
Mass for mechanical part (kg)	3.6
Rectifier switching frequency (Hz)	5000
Filter capacitor (μf)	4500
DC–DC converter switching frequency (Hz)	20,000
DC–DC converter inductance (μH)	45
DC–DC converter capacitor (μf)	100
R load (Ω)	4000

4.1. AC-to-DC Converter Based on Wind Energy Harvesting

A translational motion sensor was inserted to extract and convert signals between two mechanical translational nodes in accordance with velocity and position. The piezo stack is a piezoelectric sensor model. The piezo stack uses the translation damper and the translation spring to model its mechanical components. Figure 6 illustrates the electromechanical model of the wind energy harvester. In the piezoelectric model, the mass–spring damper system received a sinusoidal force input, whereas a sensor detected the electrical output of the piezoelectric element. This force caused displacement in the spring and damper, which in turn generated vibrations in the piezoelectric element. A sinusoidally varying AC voltage source was used to represent the electrical voltage that the piezoelectric effect generated.

The proposed framework simplified the process of exploring the effects of changing the mechanical vibrating source input parameters, such as displacement and frequency. The physical parameters, such as velocity, mass, force applied, and damping factor, can describe the wind-induced vibration. Several properties of the piezoelectric material underwent changes when subjected to a vibration source.

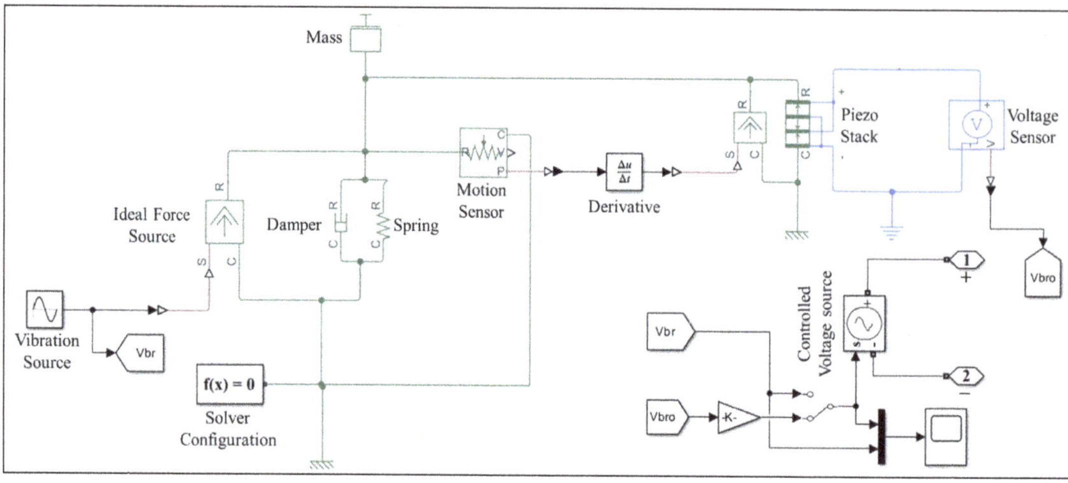

Figure 6. Piezoelectric wind-energy-harvesting simulation blocks.

These properties included the effective mass size, number of layers, charge constant, dielectric constant, elastic compliance, damping, and mechanical quality factor [34]. The piezoelectric material produced an electrical charge when exposed to mechanical stress [44]. The amount of electrical response depends on the displacement of pulsing of mechanical stress applied. When stress was applied to a piezoelectric material, the positive and negative charge centers within the crystal structure moved. This movement created an electric polarization and an electric potential across the material [8]. Thus, Figure 7 shows a comprehensive explanation of the mechanical and electrical AC-to-DC converters in Simulink.

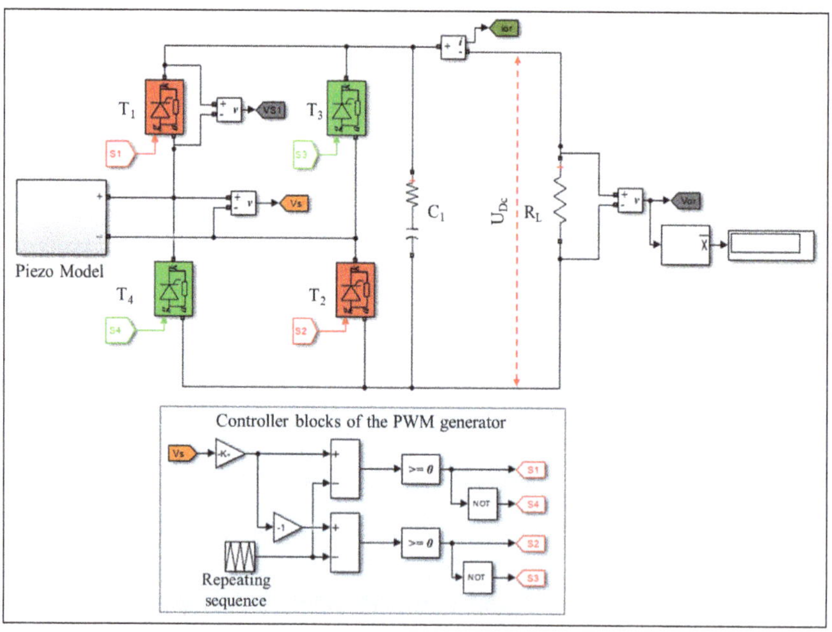

Figure 7. Simulation blocks of the AC–DC converter.

The rectifier switches used a PWM full-wave rectifier to reduce total harmonic distortions in the output waveform [40]. The rectifier passed the DC output voltage through a capacitive filter to minimize pulsation and reduce distortion in the DC voltage.

Table 2 shows the simulation results for the maximum and minimum voltages of an AC–DC converter with two excitation frequency ranges. The maximum voltage and minimum voltage were recorded at 0.09 and 1 s, respectively.

Table 2. Minimum and maximum voltages of an AC–DC converter with various vibration frequencies.

Amplitude of Wind Force (m/s^2)	Excitation Frequency Range	Vibration Frequency (Hz)	Δt = 1.16 s	
			V_{DCMax}	V_{DCMin}
200	Low	0.5	71	7.6
		1	71	7.6
		1.5	71	7.6
		2	71	7.6
	High	15	75	45
		25	80	58
		40	80	66
		50	80	70
		150	82	80

The data listed in Table 2 show the impact performance across a spectrum of low and high vibration frequencies. A low excitation frequency and high vibration frequency were used to demonstrate the variability in the studied simulation model in terms of better aligning various excitation frequency sources.

In the context of PEHS technology, using a sinusoidal waveform with a frequency greater than 5 Hz was more relevant to the ideal [35]. In such cases, PEHS often operates at higher frequencies to maximize the energy conversion efficiency. In addition, the system might be exposed to higher-frequency components, such as those resulting from rotational resonances or other environmental factors.

Vehicle-generated wind can produce excitation frequencies exceeding 15 Hz due to differences in vehicle size and speed, unlike natural wind, which typically has a dominant vibration frequency below 5 Hz [45]. This provides a comprehension performance analysis for the PEHS model. It confirms its suitability for simulating more realistic vibration scenarios, such as vibration sources of wind from passing cars on the highway, which can generate excitation frequencies. The dominant frequency can vary depending on factors such as wind speed, terrain, and the structure's height [35]

Therefore, in order to better align with the typical wind excitation scenarios, additional simulations at frequencies below 5 Hz were conducted, which revealed how the system performed under more realistic conditions, where the vibration wind excitation frequency was much lower. At low ranges of excitation frequencies of 0.5 and 1 Hz, Figures 8 and 9 show the output voltage, current, and switch pulse of the AC–DC converter based on the input sinusoidal frequencies.

Figures 10–12 show the output voltage, current, and switch pulse of the AC–DC converter, respectively, on the basis of input sinusoidal frequencies of 15, 50, and 150 Hz. The closed-loop buck-boost converter then received this DC voltage and adjusted the DC voltage level to meet the load requirements.

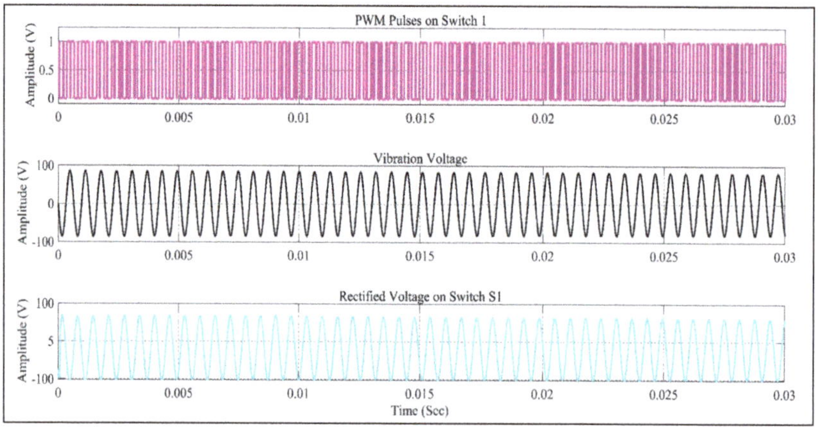

(**a**) PWM signal and voltage characteristics at Switch 1.

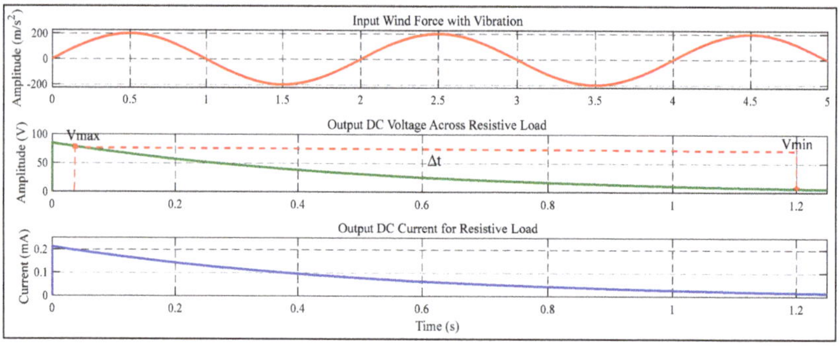

(**b**) The impact of vibration-induced wind force and cell output voltage current across resistive loads.

Figure 8. Rectifier voltage, current, and vibration force at a vibration frequency of 0.5 Hz.

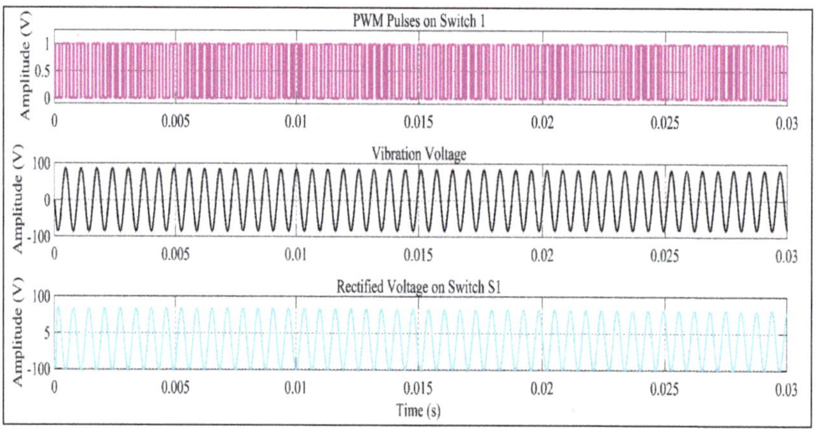

(**a**) PWM signal and voltage characteristics at Switch 1.

Figure 9. *Cont.*

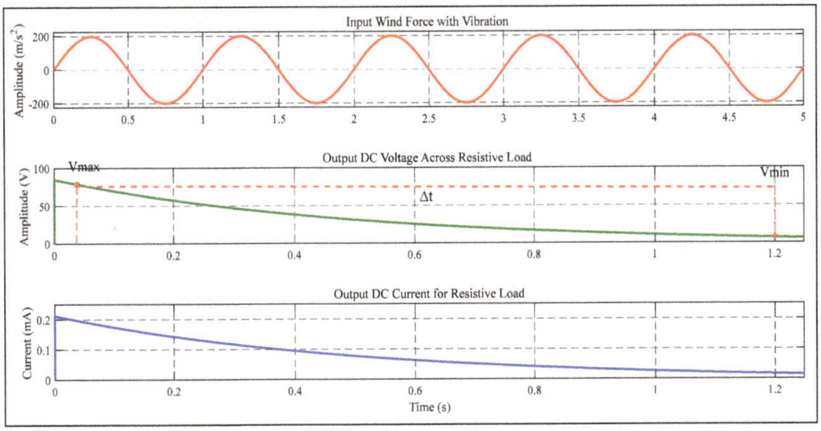

(**b**) The impact of vibration-induced wind force and cell output voltage current across resistive loads.

Figure 9. Rectifier voltage, current, and vibration force at a vibration frequency of 1 Hz.

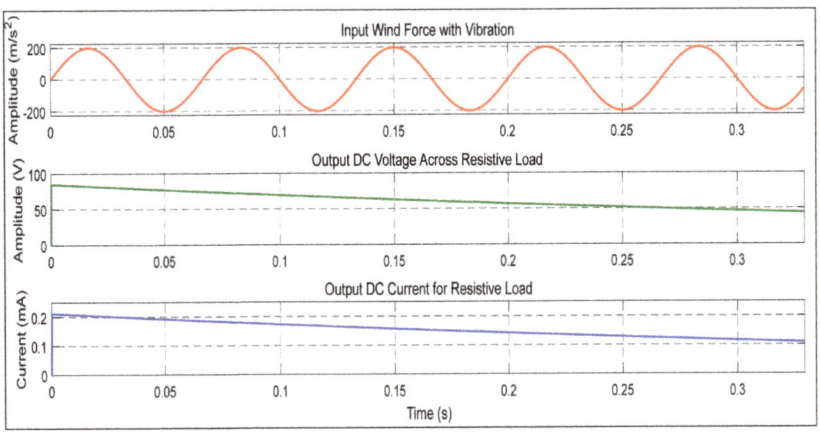

(**a**) PWM signal and voltage characteristics at Switch 1.

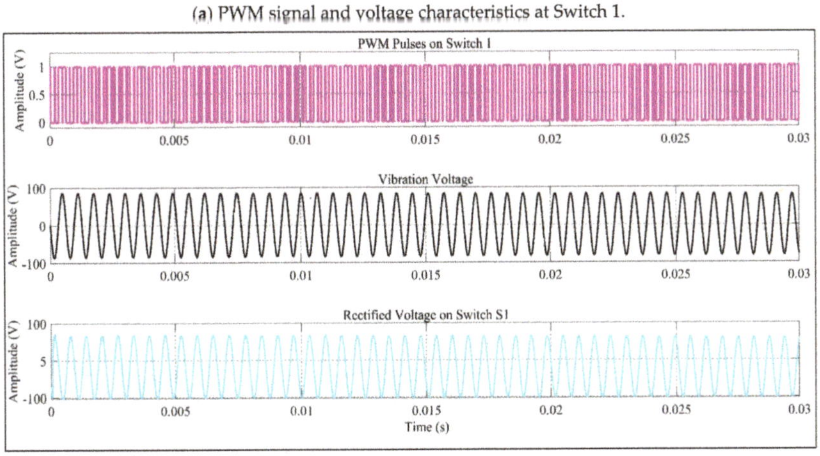

(**b**) The impact of vibration-induced wind force and cell output voltage current across resistive loads.

Figure 10. Rectifier voltage, current, and vibration force at a vibration frequency of 15 Hz.

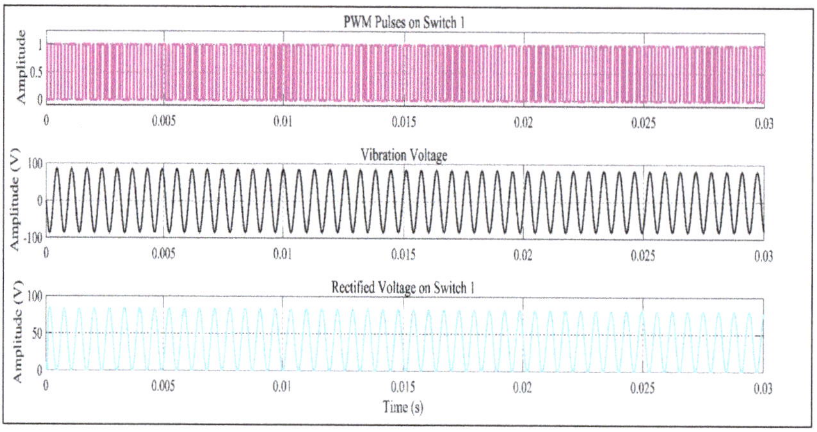

(**a**) PWM signal and voltage characteristics at Switch 1.

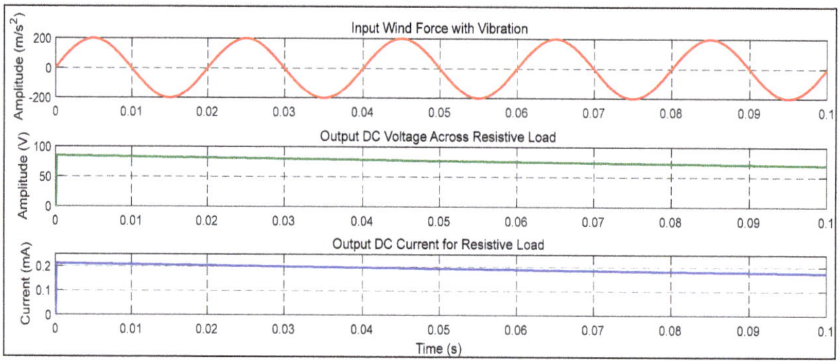

(**b**) The impact of vibration-induced wind force and cell output voltage current across resistive loads.

Figure 11. Rectifier voltage, current, and vibration force at a vibration frequency of 50 Hz.

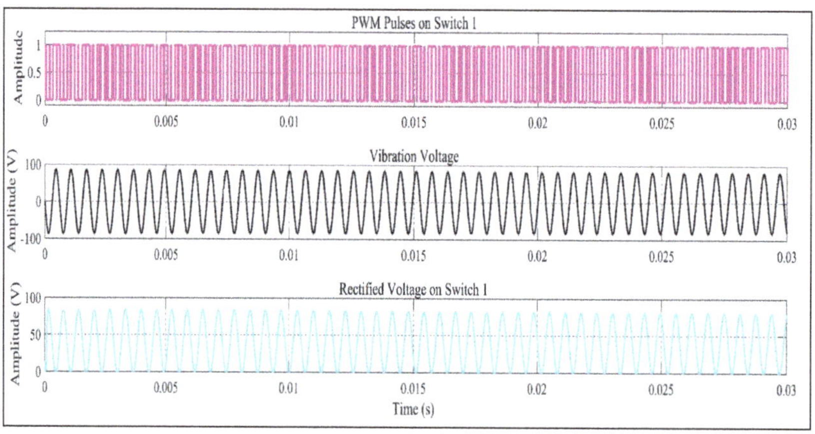

(**a**) PWM signal and voltage characteristics at Switch 1.

Figure 12. *Cont.*

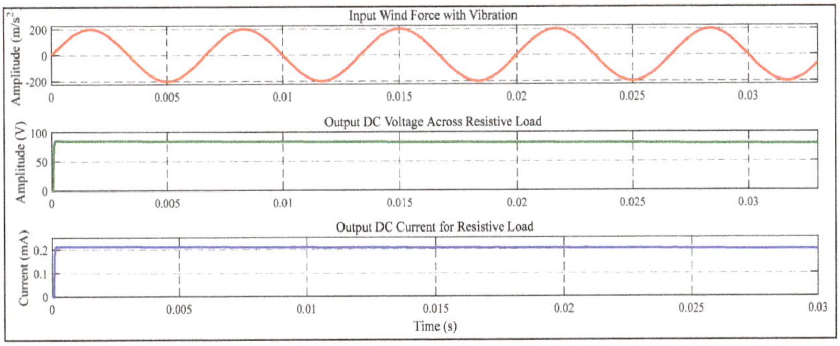

(**b**) The impact of vibration-induced wind force and cell output voltage current across resistive loads.

Figure 12. Rectifier voltage, current, and vibration force at a vibration frequency of 150 Hz.

Figure 13 shows the simulation blocks of the closed-loop buck-boost converter. The evolution of closed-loop controllers for DC-to-DC converters played a significant role in the overall performance of the proposed energy conservation system.

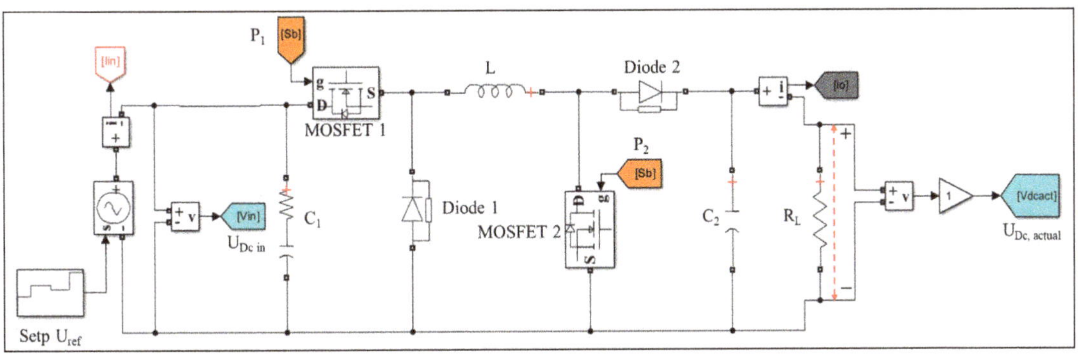

Figure 13. Closed-loop buck-boost converter.

Figure 14 illustrates the performance of the DC-to-DC converter controller, which continuously modified the output voltage according to the reference voltage, set at 24 V.

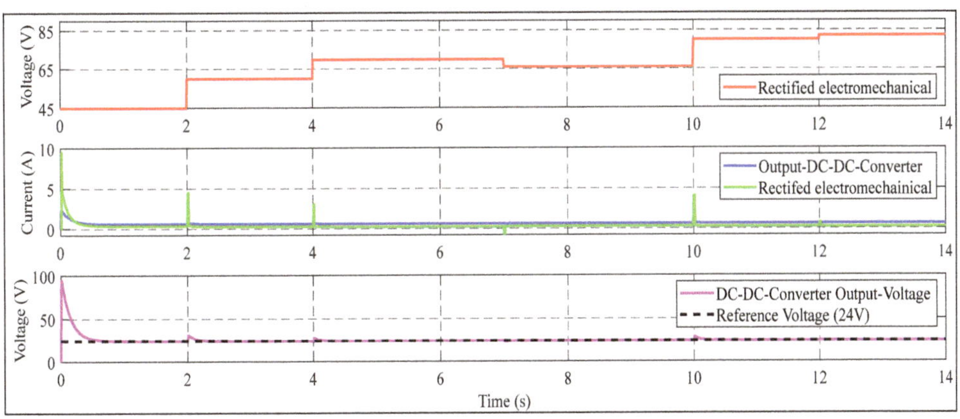

Figure 14. The performance of the DC-to-DC converter controller with a reference voltage set at 24 V.

4.2. Simulation of the Proposed Integrated Piezoelectric Energy-Harvesting Cell

To simulate an integrated PEHS, we combine an AC–DC converter, a buck-boost converter based on the P&O MPPT algorithm, and a piezoelectric energy-harvesting source. The closed-loop DC–DC converter changed its duty cycle on the basis of the input voltage and current from a full-wave rectifier, whereas the controller continuously tracked the set reference voltage. This made it flexible for changing environmental conditions, but there could be oscillations around the maximum power point.

Figure 15 illustrates the wind vibration energy-harvester system, which fed the current and voltage outputs from the PWM rectifier to the P&O MPPT buck-boost converter, specifically focusing on the conversion of the mechanical impact of wind into electrical energy.

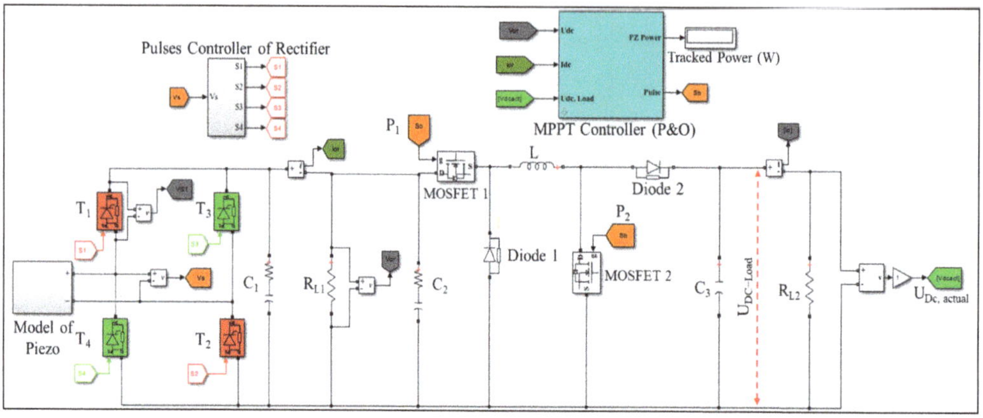

Figure 15. Wind vibration energy harvester with closed-loop controllers in buck-boost converters (P&O MPPT).

The closed-loop P&O MPPT DC–DC converter is one of the most frequently used converters [27]. The DC–DC converter continuously regulated the operating point to ensure that it operated at its maximum level, which depended on the rectified voltage and current of the PEHS sources. In addition, the closed loop was responsible for setting the required output voltage. Figure 16 shows the DC output voltage of the PEHS over time under a constant wind impact force and a 50 Hz vibration frequency. This can provide good insight into the performance of wind-energy-harvester systems that track the set reference voltage.

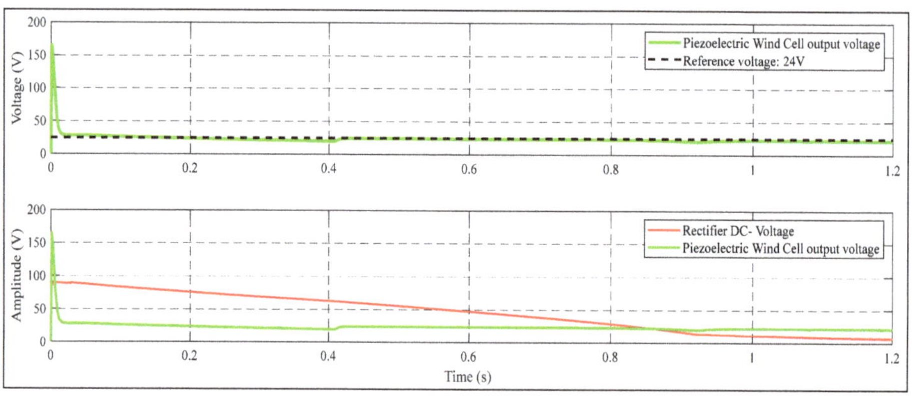

Figure 16. DC output voltage of PEHS with a closed-loop P&O MPPT buck-boost converter.

5. Discussion

The proposed design of the PEHS approach, which incorporates a closed-loop buck-boost converter based on the P&O MPPT algorithm, represents a substantial improvement in the area of vibrational energy-harvesting technology. The proposed design of the PEHS operates at a wide range of vibration frequencies, which, together with its effective power conversion and consistent output, make it a very favorable option for sustainable energy production.

Figures 10–12 show the rectifier voltage, current, and vibration force for input frequencies of 15 Hz, 50 Hz, and 150 Hz, respectively. The results indicated the ability of the proposed PEHS to efficiently transform generated voltages under different vibration frequencies. As shown in Figures 8 and 9, the AC-to-DC converter section stability increased the rectified voltage, which meant that it could start transferring power even at lower vibration frequencies. These results align with previous studies on energy-harvesting devices, which validates the rectifier's appropriateness for situations that include low-frequency vibrations. However, more research needs to prioritize investigating the rectifier's performance at low vibration frequencies, varying wind force amplitudes, and long-term vibration. Additionally, it should explore the integration of the rectifier with energy storage devices to enhance its suitability in dynamic and real-world settings.

The output demonstrated greater and more consistent voltages at a frequency of 50 Hz, suggesting that this frequency range is best for achieving peak performance. The system exhibited stable operation at a frequency of 50 Hz, which is typically used in several practical applications.

As shown in Figure 11, the AC-to-DC converter section effectively managed higher frequencies, resulting in a notable amplification in the output voltage. However, the heightened wind vibration force may induce supplementary mechanical strains, although the AC-to-DC converter seemed to handle it effectively. As shown in Figure 14, the proposed buck-boost converter control approach exhibited notable performance in terms of the output voltage matching the reference voltage set. Despite fluctuations in the input voltage level, the controller continuously maintained an output voltage that reflected the robust and reliable performance of the suggested control method. Furthermore, the ability of the closed-loop controller to promptly adjust the output voltage signified an effective feedback system, which is crucial for sustaining stability in dynamic circumstances. The closed-loop system ensured a consistent and steady DC output, which is crucial for ensuring a dependable power supply to the load. Despite fluctuations in the input voltage level, the controller continuously maintained an output voltage that reflected the strong and dependable performance of the proposed control method.

Furthermore, as shown in Figure 16, the ability to control and quickly modify the output voltage signified an effective feedback system, which is crucial for sustaining stability in dynamic circumstances. The closed-loop system guaranteed a steady DC output, which is essential for achieving optimal power delivery to the load.

Moreover, Figure 17 and Table 3 present the recommended electric circuit diagram and real-time circuit components for the proposed PEHS design, respectively. Figure 17 shows the diagram of the suggested circuit design, which provides a visual representation of the organization and interconnections of the necessary components for the system. Table 4 provides a comprehensive list of the individual components used in the design, including piezoelectric transducers, PWM-controlled rectifiers, DC–DC converters, and lithium batteries. The circuit diagram in Figure 17 and the detailed component breakdown in Table 3 provide a comprehensive reference for constructing the proposed PEHS.

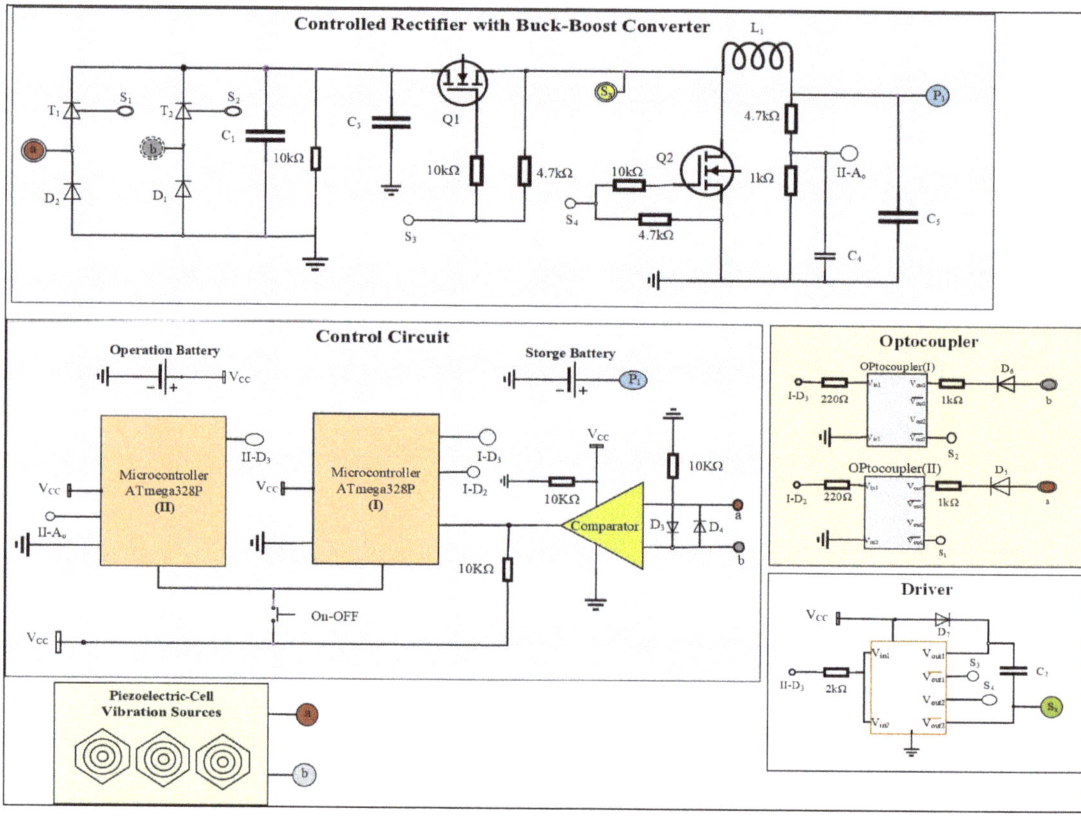

Figure 17. Comprehensive diagram of the recommended design for the proposed PEHS.

Table 3. The recommended electric components in the proposed design of the PEHS.

Section	Components	Location	Manufacturer and Specifications	Number	Approximal Cost (in USD)
Source of energy	Piezo transducer	-	-	12	36 (3 each)
	Microcontroller	I	ATmega328P	1	2.5
	Optocoupler	I, II	PC817	2	1.4 (0.7 each)
	Thyristor (SCR)	T_1, T_2	TYN1225	2	5 (2.5 each)
Rectifier	Comparator	-	LM339	1	0.6
	Diode	D_1, D_2, D_3, D_4, D_5, D_6	IN4007	6	1.5 (0.25 each)
	Resistor	-	Ω to Ωk	9	1.8 (0.2 each)
	DC capacitors	C_1	470 µf	1	0.5 (0.5 each)
	Microcontroller	II	ATmega328P	1	2.5
DC–DC Auto Voltage Converter + MPPT	Capacitors	-	C_3, C_5 = 3300 µf C_4 = 10 pf C_2 = 100 µf	2	1
	Resistor	-	-	7	1.4 (0.2 each)
	MOSFET	Q_1, Q_2	IRFZ44Z1	2	1.5 (0.75 each)
	MOSFET-driver	driver	IR2103	1	0.5
	diode	D_7	IN4007	1	0.25
	Inductor	L1	220 mH	1	1.25
Energy storage	Lithium battery	Storge battery	3.3 V, 350 mA	1	7.5
Operation battery	Lithium battery	Operation battery	5 V, 200 mA	1	7.5

Table 4. Comparison of the main parts of PEHS established on a closed-loop P&O MPPT DC–DC converter against energy harvesters' existing state-of-the-art systems.

Reference No.	Rectifier Structure	DC–DC Converter	SimulationDesign	Experimental Design
Ajitsaria et al., 2007, [28]	Uncontrolled bridge diodes	Not used	Yes	No
Boisseau et al., 2011, [34]	None	None	Yes	No
Balato et al., 2018, [27]	Uncontrolled bridge diodes	P&O MPPT closed-loop boost converter	Yes	Yes
Proposed system	PWM rectifier bridge IGBT	P&O MPPT closed-loop buck-boost converter	Yes	No

The estimated costs for components can vary depending on the provider and geographical location. At our location in Iraq, the total estimated cost of all components would be approximately USD 73.

5.1. Limitations of the Proposed PEHS

The most striking limitation is the source of vibration, which has a significant effect on the proposed PEHS. The effectiveness of the PEHS is intrinsically contingent upon the presence and regularity of ambient wind-induced vibrations or vibration sources. In environments where these vibrations occur sporadically or are feeble, the system can have difficulty producing an adequate amount of electricity, thereby restricting its reliability in such environments.

Piezoelectric materials are susceptible to environmental conditions, such as temperature and humidity, which can have an impact on the energy conversion efficiency and durability as time passes. The PEHS primarily targets low-power applications, leading to restricted power output. This constraint may limit its use in situations that require higher energy outputs, restricting it to certain categories of low-power electronic equipment. Moreover, another limitation is energy storage, as well as the cycle of charging and discharging batteries.

5.2. A Comparison with Previous State-of-the-Art and Broader Contexts

The findings indicated that the proposed design of the PEHS, which incorporates a closed-loop perturbation and an observed MPPT buck-boost converter, was remarkably efficient in converting the mechanical vibration force at various frequencies into a consistent supply of electrical energy. This feature is crucial for applications in domains with fluctuating and uncertain sources of vibrational energy, such as equipment, structures, and transportation systems.

Previous studies have often concentrated on either low-frequency or high-frequency vibration energy-harvesting devices, which yield different levels of achievement. The present paper demonstrated a significant enhancement in its ability to handle a broad variety of vibration frequencies while maintaining a consistent output voltage.

This increase effectively bridged the gap between various operating situations. In addition, the use of advanced MPPT techniques improved the overall efficiency, which is a significant improvement over simpler and less effective systems. The broad and significant implications of this study can be summarized as follows:

- Improved energy solutions. The improved energy solution system efficiently converted vibrational energy into electrical energy. This design can offer dependable power sources for remote sensors, IoT devices, and other low-power applications, reducing reliance on batteries or traditional power sources.
- Sustainable energy sources. The use of ambient vibrations for energy harvesting is compatible with sustainable energy objectives, offering an environmentally friendly substitute for traditional energy sources while diminishing eco-friendly alternatives.

- Scalability and flexibility. The suggested design of the PEHS has the potential to be expanded to handle larger applications, enabling the provision of power solutions for more extensive infrastructure. This technology's ability to accommodate a broad variety of frequencies makes it extremely flexible for many industrial and commercial applications.

Table 4 compares the PEHS with a closed-loop P&O MPPT DC converter with available state-of-the-art systems. The comparison table contains the following information: types of rectifiers, DC-to-DC converters, and design applicability.

The proposed energy-harvesting system stands out because of its use of a PWM-IGBT bridge rectifier with a P&O MPPT closed-loop buck-boost converter. This structure can provide efficiency in power conversion management compared with systems employing uncontrolled bridge diodes with DC–DC converters.

In contrast to the system explained in [27], the proposed PEHS did not undergo full experimental testing, despite partial experimental testing via simulation and validation, which provided a thorough evaluation of its performance mathematically. This limitation underscores the importance of further experimental validation to assess its viability in real time as a prototype design. However, the proposed system shows promise because of its advanced components and comprehensive assessment through simulation compared with existing state-of-the-art systems.

Results in Table 5 show the improved efficacy of the proposed dynamic simulation model of the PEHS compared to existing PEHS models in the literature. The suggested PEHS operates within a larger, lower frequency range of 15–50 Hz, producing an elevated AC output voltage of 75 V and a substantially increased power output of 500 mW. These measurements far surpass those of the other systems examined, including those of Ericka et al. [9] and Barton et al. [11], which operate at elevated frequencies and produce diminished voltages and power.

Table 5. Comparison of the main characteristics of the proposed PEHS with existing systems in the literature.

Reference No.	Vibration Frequency (Hz)	AC Output Voltage of Piezoelectric Transducer	Maximum Power Output (mW)	R Load (kΩ)
Ericka et al., 2005, [9]	2580	18.36	56	1.8
Barton et al., 2009, [11]	80	1.2	2	333
Wang et al., 2022, [10]	150	17	27	11
Proposed system simulation	15~50	75	500	4

Figure 18 illustrates the effectiveness of the suggested PEHS diagram in comparison to other PEHS models, emphasizing the maximum actual power and open-circuit voltage. In Figure 18a, the proposed circuit diagram demonstrates a significant increase in power production, surpassing current methods, especially at lower vibration frequencies, making it more appropriate for real-world energy-harvesting applications.

The proposed circuit diagram assumes a charging current of 175 mA, which is safer and less stressful for the lithium battery. In addition, the PEHS diagram produces 500 mW of real power delivery to a lithium battery.

Figure 18b shows a significant increase in the open-circuit voltage of the suggested PEHS, indicating improved efficiency in the conversion of mechanical energy to electrical energy. These enhancements are based on improvements in the suggested circuit design, making the system more efficient for powering external loads and energy storage, ensuring a dependable and sustainable power source.

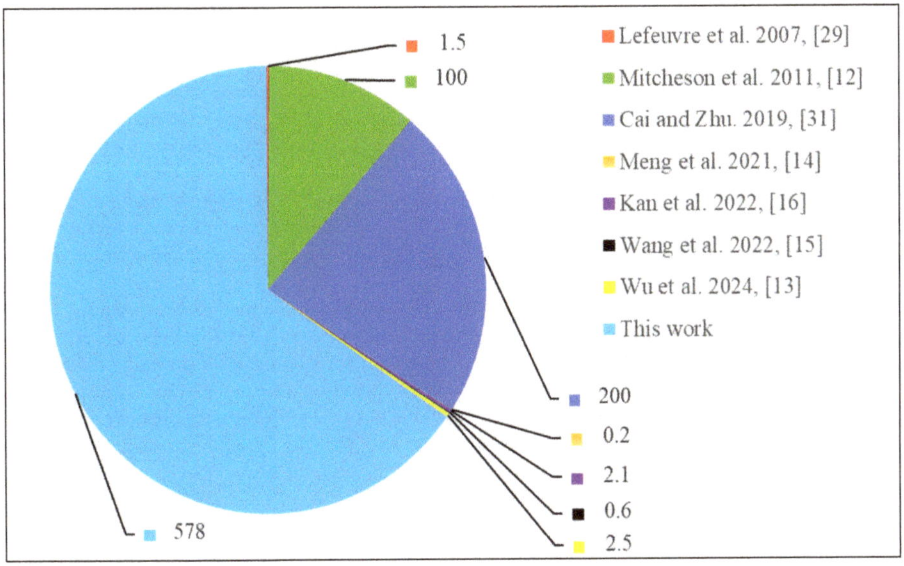

(**a**) Maximum real power produced (mW).

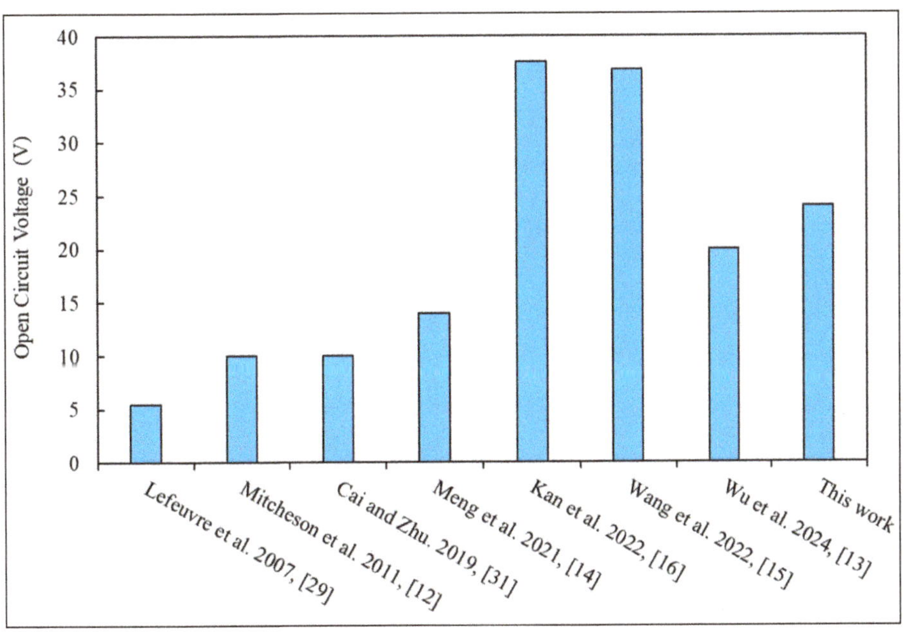

(**b**) Open-circuit voltage of piezoelectric transducer.

Figure 18. Performance comparison of the suggested PEHS circuit diagram with systems available in the literature from the point of view of real power and open-circuit voltage.

6. Conclusions

This article introduced a dynamic simulation model for a piezoelectric vibration energy-harvesting system based on wind vibration energy as a source of mechanical energy. The proposed model has integrated electrical circuits, such as an AC-DC converter, a closed-loop buck-boost converter, and an MPPT algorithm. A detailed description of the dynamic simulation performance was provided, in addition to a brief overview of each part of the system. Specifically, the MPPT-equipped piezoelectric harvester generated a significantly higher DC electrical output voltage when it encountered a resistive load of 4 kΩ. This makes it a good choice for a wide range of low-voltage applications.

Notably, the closed-loop AC-to-DC converter based on the perturb and observe MPPT harvester demonstrated more stable output voltages, and the closed-loop DC-to-DC converter provided better tracking of the reference set values. This demonstrates that the vibration energy harvester with a closed-loop AC-to-DC converter, as well as the perturbation and observation MPPT, produced better performance than the systems available in the literature.

To improve the results of this study, some recommendations for future extensions of this work are provided. The same simulation should be extended with a practical wind spectrum excitation dataset, which provides deeper insights into real-world performance. Using wind data and conducting field experiments to verify the accuracy of the simulation results and enhance the system's design under real wind conditions would be beneficial. Additional studies might investigate the optimization of electrical components using actual wind excitation situations, resulting in enhanced efficiency and dependability.

Author Contributions: A.K.A.: conceptualization, writing (original draft), methodology, investigation, writing (review and editing), software, and validation. A.A.A.: review and editing. A.H.M.: review and validation. All authors have read and agreed to the published version of the manuscript.

Funding: This research received no external funding.

Data Availability Statement: As the corresponding authors of this paper, we declare that we will upload all the simulation files and supporting data after the paper's publication.

Acknowledgments: The authors express their profound appreciation to the Middle Technical University, Institute of Technology, Baghdad, Iraq, for their significant aid and support over the course of this work. Their administrative and technical assistance has been critical to the successful completion of this study.

Conflicts of Interest: The authors declare that they have no conflicts of interest.

References

1. Adom, P.K. The Transition between Energy Efficient and Energy Inefficient States in Cameroon. *Energy Econ.* **2016**, *54*, 248–262. [CrossRef]
2. Adom, P.K.; Adams, S. Energy Savings in Nigeria. Is There a Way of Escape from Energy Inefficiency? *Renew. Sustain. Energy Rev.* **2018**, *81*, 2421–2430. [CrossRef]
3. Amjadi, G.; Lundgren, T. Is Industrial Energy Inefficiency Transient or Persistent? Evidence from Swedish Manufacturing. *Appl. Energy* **2022**, *309*, 118324. [CrossRef]
4. Butt, O.M.; Ahmad, M.S.; Che, H.S.; Rahim, N.A. Usage of On-Demand Oxyhydrogen Gas as Clean/Renewable Fuel for Combustion Applications: A Review. *Int. J. Green Energy* **2021**, *18*, 1405–1429. [CrossRef]
5. Koszalka, G.; Hunicz, J. Comparative Study of Energy Losses Related to the Ring Pack Operation in Homogeneous Charge Compression Ignition and Spark Ignition Combustion. *Energy* **2021**, *235*, 121388. [CrossRef]
6. Pao, L.Y.; Johnson, K.E. Control of Wind Turbines Recent Developments. *Control Syst. Mag.* **2011**, *31*, 44–62. [CrossRef]
7. Akkaya Oy, S.; Özdemir, A.E. Piezoelectric-Based Low-Power Wind Generator Design and Testing. *Arab. J. Sci. Eng.* **2018**, *43*, 2759–2767. [CrossRef]
8. Dell'Anna, F.G.; Dong, T.; Li, P.; Wen, Y.; Yang, Z.; Casu, M.R.; Azadmehr, M.; Berg, Y. State-of-the-Art Power Management Circuits for Piezoelectric Energy Harvesters. *IEEE Circuits Syst. Mag.* **2018**, *18*, 27–48. [CrossRef]
9. Ericka, M.; Vasic, D.; Costa, F.; Poulain, G. Predictive Energy Harvesting from Mechanical Vibration Using a Circular Piezoelectric Membrane. In Proceedings of the IEEE Ultrasonics Symposium, Rotterdam, The Netherlands, 18–21 September 2005; Volume 2, pp. 946–949.
10. Wang, W.; Yang, T.; Chen, X.; Yao, X. Vibration Energy Harvesting Using a Piezoelectric Circular Diaphragm Array. *IEEE Trans. Ultrason. Ferroelectr. Freq. Control* **2012**, *59*, 2022–2026. [CrossRef]
11. Energy, R.; Barton, D.A.W.; Burrow, S.G.; Clare, L.R. Energy harvesting from vibrations with a nonlinear oscillator. In Proceedings of the ASME 2009 International Design Engineering Technical Conferences & Computers and Information in Engineering Conference, San Diego, CA, USA, 30 August–2 September 2009.

12. Dicken, J.; Mitcheson, P.D.; Stoianov, I.; Yeatman, E.M. Power-Extraction Circuits for Piezoelectric Energy Harvesters in Miniature and Low-Power Applications. *IEEE Trans. Power Electron.* **2012**, *27*, 4514–4529. [CrossRef]
13. Wu, W.; Pan, Z.; Zhou, J.; Wang, Y.; Ma, J.; Li, J.; Hu, Y.; Wen, J.; Wang, X. Wind-Speed-Adaptive Resonant Piezoelectric Energy Harvester for Offshore Wind Energy Collection. *Sensors* **2024**, *24*, 1371. [CrossRef] [PubMed]
14. Meng, X.K.; Lu, Z.; Dai, H.L.; Abdelkefi, A. Comparative Experimental Investigation and Effectiveness of Sphere- and Cylinder-Based Piezoelectric Energy Harvesters. *Smart Mater. Struct.* **2021**, *30*, 08LT01. [CrossRef]
15. Yu, G.; He, L.; Wang, S.; Tian, X.; Zheng, X.; Cheng, G. Experiments and Analysis of a Nonlinear Sickle-Shaped Cantilever Beam Piezoelectric Energy Harvester for Wind Energy Harvesting. *Phys. Status Solidi Appl. Mater. Sci.* **2022**, *219*, 2200363. [CrossRef]
16. Kan, J.; Wang, J.; Wu, Y.; Chen, S.; Wang, S.; Jiang, Y.; Zhang, Z. Energy Harvesting from Wind by an Axially Retractable Bracket-Shaped Piezoelectric Vibrator Excited by Magnetic Force. *Energy* **2022**, *240*, 122495. [CrossRef]
17. Zhao, Z.; Dai, Y.; Dou, S.X.; Liang, J. Flexible Nanogenerators for Wearable Electronic Applications Based on Piezoelectric Materials. *Mater. Today Energy* **2021**, *20*, 100690. [CrossRef]
18. Wang, P.; Xiao, Y.; Wu, N. Electrical Power Generation Using Dynamic Piezoelectric Shear Deformation under Friction. *Acta Mech. Solida Sin.* **2021**, *34*, 977–988. [CrossRef]
19. Krishnasamy, M.; Shinde, J.R.; Mohammad, H.P.; Deepesh, U.; Lenka, T.R. Design and Simulation of Smart Flooring Tiles Using Two-Phased Triangular Bimorph Piezoelectric Energy Harvester. In Proceedings of the 2020 IEEE-HYDCON International Conference on Engineering in the 4th Industrial Revolution, HYDCON 2020, Hyderabad, India, 11–12 September 2020; pp. 19–22. [CrossRef]
20. Yingyong, P.; Thainiramit, P.; Jayasvasti, S.; Thanach-Issarasak, N.; Isarakorn, D. Evaluation of Harvesting Energy from Pedestrians Using Piezoelectric Floor Tile Energy Harvester. *Sens. Actuators A Phys.* **2021**, *331*, 113035. [CrossRef]
21. Panthongsy, P.; Isarakorn, D.; Hamamoto, K.; Janphuang, P. Performance and Behavior Analysis of Piezoelectric Energy Harvesting Floor Tiles. In Proceedings of the 2019 5th International Conference on Engineering, Applied Sciences and Technology, ICEAST 2019, Luang Prabang, Laos, 2–5 July 2019; pp. 1–4. [CrossRef]
22. Duarte, F.; Ferreira, A.; Fael, P. Road Pavement Energy–Harvesting Device to Convert Vehicles' Mechanical Energy into Electrical Energy. *J. Energy Eng.* **2018**, *144*, 04018003. [CrossRef]
23. Puscasu, O.; Counsell, N.; Herfatmanesh, M.R.; Peace, R.; Patsavellas, J.; Day, R. Powering Lights with Piezoelectric Energy-Harvesting Floors. *Energy Technol.* **2018**, *6*, 906–916. [CrossRef]
24. Kim, K.B.; Cho, J.Y.; Hamid, J.; Ahn, J.H.; Do Hong, S.; Woo, S.B.; Sung, T.H. Optimized Composite Piezoelectric Energy Harvesting Floor Tile for Smart Home Energy Management. *Energy Convers. Manag.* **2018**, *171*, 31–37. [CrossRef]
25. Zhang, Y.; Luo, A.; Wang, Y.; Dai, X.; Lu, Y.; Wang, F. Rotational Electromagnetic Energy Harvester for Human Motion Application at Low Frequency. *Appl. Phys. Lett.* **2020**, *116*, 053902. [CrossRef]
26. Islam, E.; Abdullah, A.M.; Chowdhury, A.R.; Tasnim, F.; Olivares, C.; Lozano, K.; Uddin, M.J. Hybrid Triboelectric-Electromagnetic Energy Tile for Biomechanical Energy Scavenging. *Nano Energy* **2020**, *77*, 105250. [CrossRef]
27. Balato, M.; Costanzo, L.; Lo Schiavo, A.; Vitelli, M. Optimization of Both Perturb & Observe and Open Circuit Voltage MPPT Techniques for Resonant Piezoelectric Vibration Harvesters Feeding Bridge Rectifiers. *Sens. Actuators A Phys.* **2018**, *278*, 85–97. [CrossRef]
28. Ajitsaria, J.; Choe, S.Y.; Shen, D.; Kim, D.J. Modeling and Analysis of a Bimorph Piezoelectric Cantilever Beam for Voltage Generation. *Smart Mater. Struct.* **2007**, *16*, 447–454. [CrossRef]
29. Lefeuvre, E.; Audigier, D.; Richard, C.; Guyomar, D. Buck-Boost Converter for Sensorless Power Optimization of Piezoelectric Energy Harvester. *IEEE Trans. Power Electron.* **2007**, *22*, 2018–2025. [CrossRef]
30. Mitcheson, P.D.; Toh, T.T.; Wong, K.H.; Burrow, S.G.; Holmes, A.S. Tuning the Resonant Frequency and Damping of an Electromagnetic Energy Harvester Using Power Electronics. *IEEE Trans. Circuits Syst. II Express Briefs* **2011**, *58*, 792–796. [CrossRef]
31. Cai, Q.L.; Zhu, S. Enhancing the Performance of Electromagnetic Damper Cum Energy Harvester Using Microcontroller: Concept and Experiment Validation. *Mech. Syst. Signal Process.* **2019**, *134*, 106339. [CrossRef]
32. Cai, Q.; Hua, Y.; Zhu, S.; Zhang, X.; Zuo, H. Bistable Energy-Harvesting Track Nonlinear Energy Sink in Offshore Wind Turbines. *Mech. Syst. Signal Process.* **2024**, *215*, 111407. [CrossRef]
33. Williams, C.B.; Yates, R.B. Analysis of a Micro-Electric Generator for Microsystems. *Sens. Actuators A Phys.* **1996**, *52*, 8–11. [CrossRef]
34. Boisseau, S.; Despesse, G.; Ricart, T.; Defay, E.; Sylvestre, A. Cantilever-Based Electret Energy Harvesters. *Smart Mater. Struct.* **2011**, *20*, 105013. [CrossRef]
35. Tian, W.; Ling, Z.; Yu, W.; Shi, J. A Review of MEMS Scale Piezoelectric Energy Harvester. *Appl. Sci.* **2018**, *8*, 645. [CrossRef]
36. Durgadevi, S.; Anbazhagan, R.; Harini, R.; Vimonisha, A. A Review of Performance Optimization of Footstep Energy Harvester. In Proceedings of the International Conference on Communication, Computing and Internet of Things, IC3IoT 2022, Chennai, India, 10–11 March 2022; pp. 1–5. [CrossRef]
37. Wickenheiser, A.M.; Reissman, T.; Wu, W.J.; Garcia, E. Modeling the Effects of Electromechanical Coupling on Energy Storage through Piezoelectric Energy Harvesting. *IEEE/ASME Trans. Mechatron.* **2010**, *15*, 400–411. [CrossRef]
38. Wei, C.; Jing, X. A Comprehensive Review on Vibration Energy Harvesting: Modelling and Realization. *Renew. Sustain. Energy Rev.* **2017**, *74*, 1–18. [CrossRef]

39. Khojet, S.; Khil, E.; Youssef, A.B.; Slama Belkhodja, I. Single Phase PWM Rectifier Sensor Fault Tolerant Control Based on Residual Generation. *J. Electr. Syst.* **2018**, *14*, 190–204.
40. Rashid, M.H. *Power Electronics Handbook*; Butterworth-Heinemann: Oxford, UK, 2017.
41. Mitra, A.; Chowdhuri, S. Analysis of Single Phase PWM Rectifier for Different Applications. *J. Inst. Eng. India Ser. B* **2017**, *98*, 161–169. [CrossRef]
42. Huang, X.; Chang, D.; Ling, C.; Zheng, T.Q. Research on Single-Phase PWM Converter with Reverse Conducting IGBT Based on Loss Threshold Desaturation Control. *Energies* **2017**, *10*, 1845. [CrossRef]
43. Tan, R.H.G.; Er, C.K.; Solanki, S.G. Modeling of Photovoltaic MPPT Lead Acid Battery Charge Controller for Standalone System Applications. In Proceedings of the E3S Web of Conferences, Lisboa, Portugal, 19–21 October 2020; Volume 182, pp. 1–7. [CrossRef]
44. Tadmor, E.B.; Kósa, G. Electromechanical Coupling Correction for Piezoelectric Layered Beams. *J. Microelectromechanical Syst.* **2003**, *12*, 899–906. [CrossRef]
45. Morbiato, T.; Borri, C.; Vitaliani, R. Wind Energy Harvesting from Transport Systems: A Resource Estimation Assessment. *Appl. Energy* **2014**, *133*, 152–168. [CrossRef]

processes

MDPI

Article

Upgrading Sustainable Pipeline Monitoring with Piezoelectric Energy Harvesting

Zainab Kamal Mahdi [1,*], Riyadh A. Abbas [1], Manaf K. Hussain Al-Taleb [1], Adnan Hussein Ali [2] and Esam M. Mohamed [3]

1 Electrical Engineering Department, College of Engineering, University of Wasit, Wasit 00964, Iraq; riyadhabbas@uowasit.edu.iq (R.A.A.); manaf@uowasit.edu.iq (M.K.H.A.-T.)
2 Institute of Technology, Middle Technical University, Baghdad 10074, Iraq; aaddnnaann63@gmail.com
3 Mechanical Power Technical Engineering Department, College of Engineering and Technology, Al-Mustaqbal University, Hilla 51001, Iraq; esam.muhe@uomus.edu.iq
* Correspondence: zainabkmahdi@uowasit.edu.iq

Abstract: This study presents the design and implementation of a piezoelectric power harvesting device to capture vibrational energy from pipelines to self-powered IoT devices. The device utilizes key components along with the PPA-1001 piezoelectric sensor, the STM32F103C8T6 microcontroller, and LTC-3588 energy harvesting power supply. Experimental results verified the system's performance in harvesting power within a specific frequency range of 10 Hz to 50 Hz, with the foremost overall performance at 30 Hz. The device generated the highest voltage of 3.3 V, delivering a power output of 2.18 mW, which is sufficient to power low-power electronic devices. The device maintained solid performance across a temperature range of 40 °C to 50 °C, underscoring its robustness in various environmental situations. The findings highlight the capacity of this form of generation to offer a sustainable power source for remote pipeline tracking, contributing to stronger protection and operational efficiency.

Keywords: piezoelectric energy harvesting; energy harvesting; self-powered IoT

1. Introduction

The transportation of fluids such as water, gas, and oil over vast distances relies heavily on pipeline networks. Ensuring the safety and integrity of these pipelines is crucial to prevent leaks, corrosion, and other potential hazards that could lead to environmental damage and economic losses [1–3]. However, powering monitoring systems for these pipelines, especially in remote locations with limited access to the electrical grid, presents a significant challenge [4]. Energy harvesting has emerged as a promising solution to power pipeline monitoring devices without relying on wired connections or batteries, by utilizing locally available energy sources [4–6].

Flow-induced vibration (FIV) is a common phenomenon in pipelines due to turbulent fluid flow. These vibrations can be harnessed to generate electricity using piezoelectric materials [5,7]. Piezoelectric materials have the unique ability to convert mechanical stress into electrical energy, making them ideal for capturing vibrational energy from pipelines [7]. When subjected to mechanical stress, piezoelectric materials produce an electrical charge, which can be harvested and used to power electronic devices.

Previous studies have explored various approaches to harvesting energy from pipelines. In addition to advancements in pipeline-based energy harvesting techniques, there are exciting developments in the field of piezoelectric materials themselves. Despite these advancements in pipeline-based energy harvesting, a significant gap remains in integrating these technologies into practical, self-sustaining monitoring systems. While individual components have been studied, there is a lack of comprehensive solutions that combine energy harvesting, real-time data logging, and adaptive power management in a single, efficient system for pipeline monitoring. This research aimed to address this gap by presenting an integrated approach that overcomes the limitations of previous studies.

The novel contribution lies in the following.

(a) The integration of the PPA-1001 piezoelectric sensor with the LTC-3588 energy harvesting power supply, optimizing energy capture across a wide range of pipeline vibrations;
(b) The use of the STM32F103C8T6 microcontroller for intelligent power management, enhancing the system's efficiency;
(c) Development of a custom data logging application for real-time visualization and analysis of both the harvested energy and the pipeline's conditions;
(d) Demonstrating robust performance across a specific frequency range (10–50 Hz) and temperature range (40–50 °C), making it more adaptable to the real-world conditions of pipelines.

By addressing these aspects comprehensively, the research aimed to bridge the gap between theoretical energy harvesting concepts and practical, deployable solutions for pipeline monitoring. Recent studies have shown great potential for new materials, particularly glycine, in sensing and energy harvesting applications. For instance, Wang et al. demonstrated the possibility of using biodegradable piezoelectric nanofiber membranes made of PVA-GLY in air filtration masks, highlighting the versatility of piezoelectric material applications [8]. In a more pipeline-relevant context, Okosun et al. proved the effectiveness of glycine-based sensors in detecting and distinguishing the severity of water pipe leaks in real time, outperforming traditional PVDF sensors in sensitivity [9].

For a deeper understanding of glycine's piezoelectric properties, Guerin et al. provided a fundamental study on the piezoelectric coefficients of different glycine crystal polymorphs, emphasizing the importance of the β-polymorph for sensing applications [10]. Building on this knowledge, Zhang et al. developed a novel fabrication method for large-area, high-quality β-glycine nanocrystalline films, leading to significantly improved piezoelectric performance [11].

These advancements in piezoelectric materials open new avenues for enhancing the efficiency and sensitivity of energy harvesting and pipeline monitoring systems. While the current study focused on using the PPA-1001 piezoelectric sensor, these recent research findings point to the potential for future improvements in the system's performance through the integration of advanced piezoelectric materials such as glycine. For instance, one study investigated the use of serpentine-shaped piezoelectric cantilever beams attached to pipes for energy collection. The research demonstrated that the power harvester may produce an open-circuit voltage of 10.28–15.45 V with 1 g of vibration acceleration [3]. Another observer tested a circular plate-type piezoelectric power harvester, which may generate 1.22 V and 13.8 μW at a fluid pressure amplitude of 1724 Pa [12].

While these studies have made significant contributions to the field, this work introduces a comprehensive, integrated system that combines energy harvesting, real-time data logging, and adaptive power management. This holistic approach represents a significant advancement in sustainable pipeline monitoring technology. Unlike previous research that focused on single aspects of pipeline monitoring, this system uniquely combines high efficiency in a specific frequency range (10–50 Hz) with robust performance across varying temperatures (40 °C and 50 °C), making it more versatile for real-world applications.

This innovative design incorporates the PPA-1001 piezoelectric sensor in conjunction with the LTC-3588 energy harvesting power supply, which optimizes energy capture from a wide range of pipeline vibrations. The integration of the STM32F103C8T6 microcontroller enables intelligent power management and data processing, allowing for more efficient use of harvested energy. Furthermore, developing a custom data logging application that provides real-time visualization and analysis of the harvested energy and pipeline conditions, enhanced the system's usefulness for monitoring and maintenance purposes.

This research presents a substantial step towards sustainable pipeline monitoring by demonstrating the feasibility of harvesting energy from pipelines' vibrations to power remote IoT devices. By utilizing piezoelectric energy harvesting sensors, this design aimed to reduce reliance on traditional energy sources and diminish environmental impact. The system's ability to operate effectively in various environmental conditions and its potential to provide continuous power to monitoring devices make it a promising solution for enhancing pipelines' safety and operational efficiency.

2. Methodology

2.1. Structure of the System

The components of the system were as follows:

1. Microcontroller (stm32f103c8t6) [13],
2. Piezoelectric energy sensor (PPA-1001) [14],
3. Vibration sensor (Adxl345) [15],
4. Temperature sensor (LM35) [16],
5. Piezoelectric energy harvesting power supply (LTC-3588) [17].

2.2. Hardware Design

Figure 1 illustrates the complete layout for a vibration-based electricity harvesting machine, with the PPA-1001 piezoelectric sensor serving as the number one power supply. The PPA-1001 sensor is strategically located to capture mechanical vibrations from a pipe and convert those vibrations into electric signals. These signals are routed to the LTC3588 power harvester, a specialized integrated circuit designed to optimize the low-stage power produced by the piezoelectric sensor. The LTC3588 correctly regulates and conditions the entering signals, converting them into a stable 3.3 V output, which is essential for the steady operation of the downstream electronic components. To further ensure voltage balance, a 10 μF capacitor was attached, which helped smooth out any fluctuations and reducing voltage fluctuations. This strong voltage is then fed into a battery charger module, designed to keep an output voltage in the range of 3.7 to 4.2 V. This range is specifically crucial because it ensures that the machine remains within secure running limits even while providing enough power to the connected components. The charged battery serves as a reliable power supply, imparting energy to the microcontroller unit (MCU), which, in this example, is the STM32F103C8T6. The STM32F103C8T6 is an effective and versatile MCU that acts as the central processing unit, coping with the flow of information and controlling the system's operations.

The STM32F103C8T6 is interfaced with the ADXL345 accelerometer through an I2C interface, which makes use of the SDA (serial data) and SCL (serial clock) lines for communication. The ADXL345 accelerometer is tasked with measuring the acceleration and vibrations of the pipe, offering real-time information that is critical for tracking the mechanical behavior of the system. This information is processed through the MCU to determine the device's overall performance and make any vital adjustments. Additionally, the system consists of the LM35 temperature sensor, which is connected to one of the analog inputs of the STM32F103C8T6. The LM35 provides accurate temperature readings of the pipe, which can be crucial for monitoring the thermal conditions that can affect the system's performance. The MCU utilizes the temperature information alongside the statistics on vibration, ensuring comprehensive tracking and management of the system's operational parameters. The battery, which is constantly charged by way of the harvested power, guarantees that the system stays operational even in the absence of direct vibrations. This layout guarantees the energy's sustainability, imparting non-stop electricity to the system and permitting it to function efficaciously in far-flung or tough environments wherein upkeep and energy supply are constrained. This layout effectively harvests and stabilizes energy from mechanical vibrations and also integrates specific monitoring and management mechanisms, making it a robust answer for long-term, independent operation in field operations. The design and assembly of the system is depicted in Figure 2.

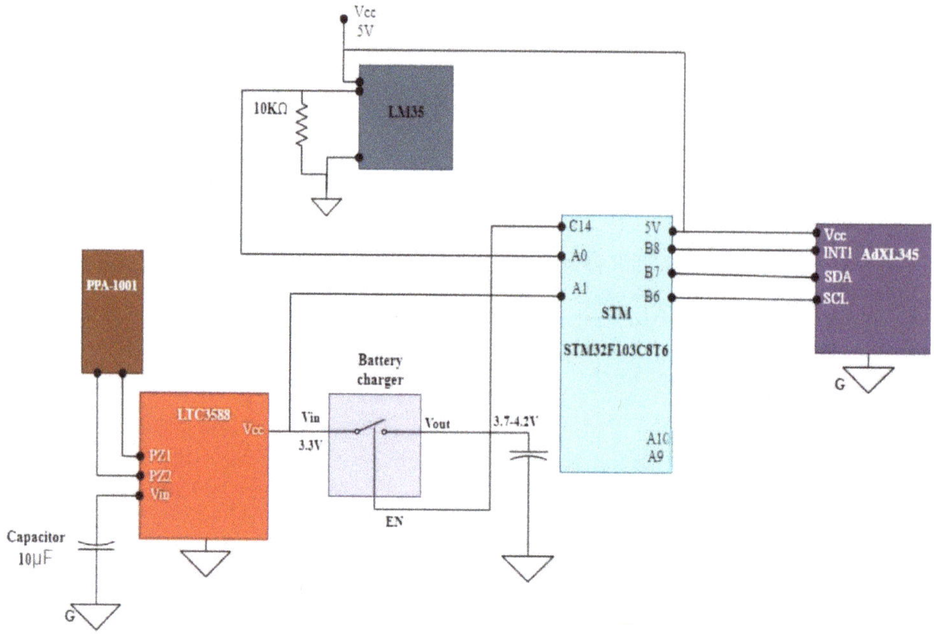

Figure 1. The mainboard diagram of the designed energy harvesting system.

Figure 2. The assembled structure of the energy harvesting system.

2.3. Working Mechanism

The flowchart in Figure 3 introduces an in-depth step-by-step system for dealing with energy harvesting and battery charging in a piezoelectric device. It starts with the machine initializing and immediately reading the acceleration information from the ADXL345 sensor, which measures the vibrations in the *x*-, *y*-, and *z*-axis of the pipeline. These measurements are critical, as they immediately have an impact on the amount of mechanical power that can be harvested. Following this, the machine reads the voltage output from the LTC3588, an efficient energy harvester that converts the vibrations into electricity and regulates the output voltage.

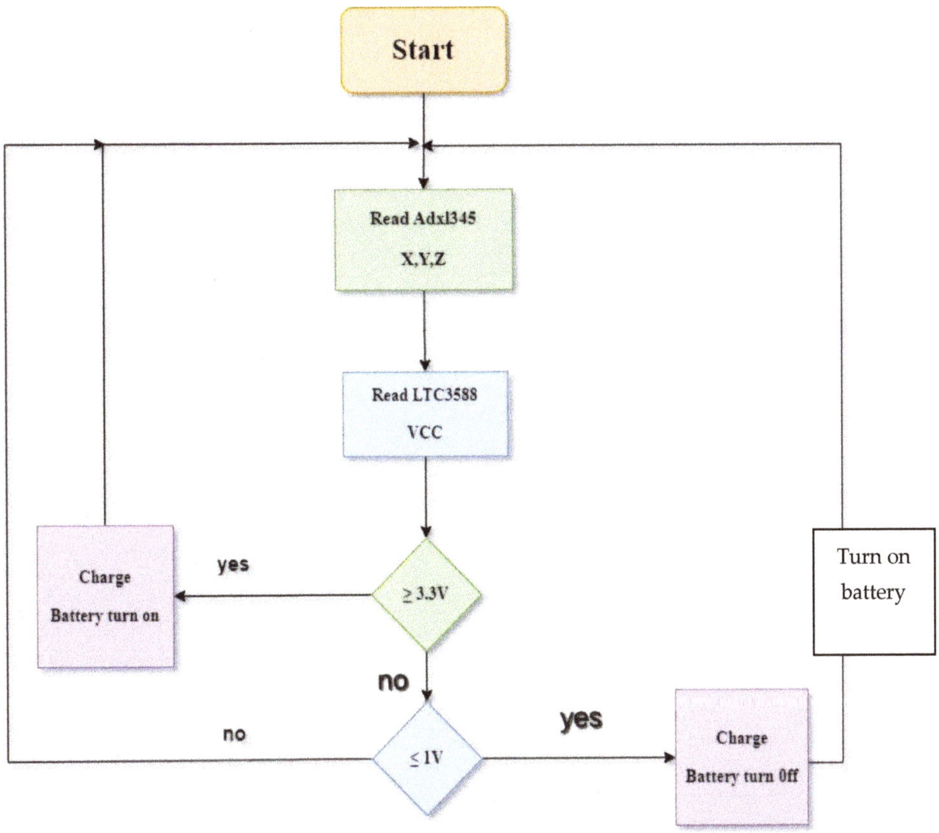

Figure 3. A flowchart illustrating the working mechanism.

The system then compares the harvested voltage against two crucial thresholds: 3.3 V and 1 V. If the voltage reaches or exceeds 3.3 V, this suggests that sufficient electricity has been accumulated, prompting the system to turn on the battery charging system. This threshold guarantees that the electricity being stored is adequate for powering the connected IoT devices and forestalls premature charging that could cause inefficient electricity storage.

If the voltage is much less than 3.3 V but still above 1 V, the system recognizes that the degree of energy is not the most efficient for charging, so it maintains tracking without starting up the charging system. This guarantees that the battery is most effectively charged when sufficient power is to be had, to avoid partial or inefficient charging cycles. However, if the voltage drops to 1 V or beneath, the device routinely turns off the battery charging system. This motion prevents the depletion of the saved power and protects the battery from being charged with inadequate power, which could lead to inefficiencies or damaged capacity. The loop then resets, usually assessing the conditions and responding dynamically to the ranges of electricity detected by the system. Smart management ensures that the energy harvesting system operates optimally, holding power and maintaining the longevity and performance of the battery and the overall system.

3. Results

3.1. Results of Data Logging and Analysis via Graphical Interface

The energy harvesting application is a graphical user interface (GUI), shown in Figure 4, that was built using MATLAB and was designed to collect, record, and analyze data from the Adxl345, LM35, and PPA-1001 sensors. The software's interface is broken down into four distinct sections: ADXL345 settings, control panel, data settings, and real-time data. The ADXL345 settings allow users to configure the accelerometer's operational parameters such as the measurement range, data output frequency, and resolution.

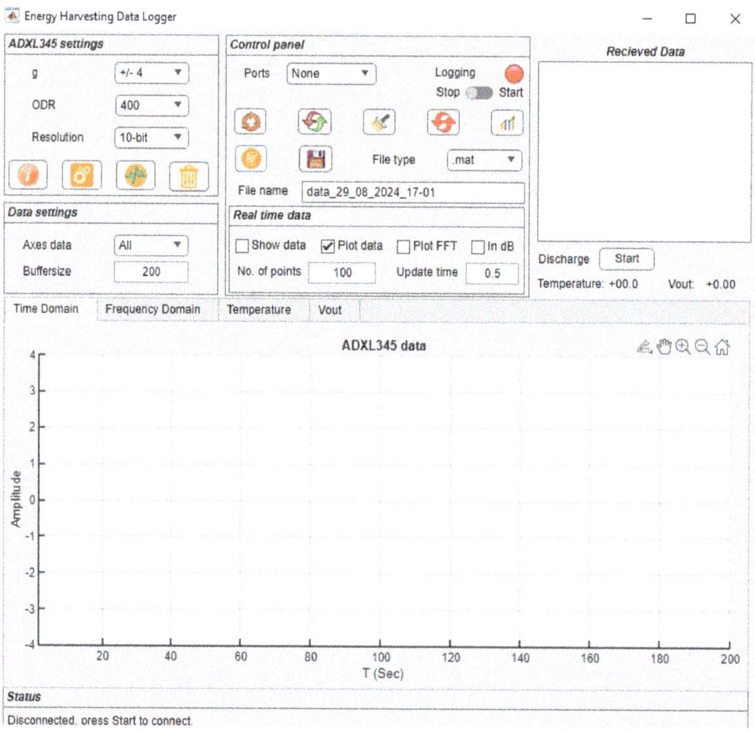

Figure 4. Energy harvesting data logger.

Figure 4 depicts the interface of the energy harvesting data logger application. The interface includes several sections as described below.

1. ADXL345 Settings

- These set the acceleration range (e.g., ±2 g to ±16 g), the output data rate (100 Hz to 3200 Hz), and the resolution (e.g., 10-bit or 13-bit). Initialization, configuration, live plot, and saving settings are available.

2. Control Panel

- This manages the sensors' connections via COM ports, start/stop data logging, saving data in different formats, and configuring the file names.

3. Data Settings

- This selects the axes (x, y, z) to be displayed, and sets the buffer size for processing the data.

4. Real-Time Data

- This enables a live data display, plots data in real time, performs FFT analysis, and controls the update rate and number of data points for visualization.

The experimental results received from the designed piezoelectric power harvesting device were collected and analyzed. The results were divided into many categories, each highlighting unique elements of the device's overall performance, along with time-domain analysis, frequency-domain analysis, temperature, and the voltage harvested by the piezoelectric sensor.

3.2. Time-Domain Analysis

In Figure 5, the graph shows vibration data from an ADXL345 accelerometer, displaying amplitudes over 300 s across three axes: x (blue), y (orange), and z (yellow). The vertical axis ranged from -4 to 4, representing the acceleration in units of g-force (g). The z-axis experienced the strongest vibrations, with amplitudes ranging from 0 to 3 g, indicating significant movement in this direction. The y-axis showed moderate vibrations, fluctuating between -1.5 g and 1.5 g, while the x-axis exhibited minimal vibrations, staying between -1 g and 1 g. This suggests that the system's primary vibration occurred in the z direction, with less pronounced movement in the y direction and the least in the x direction.

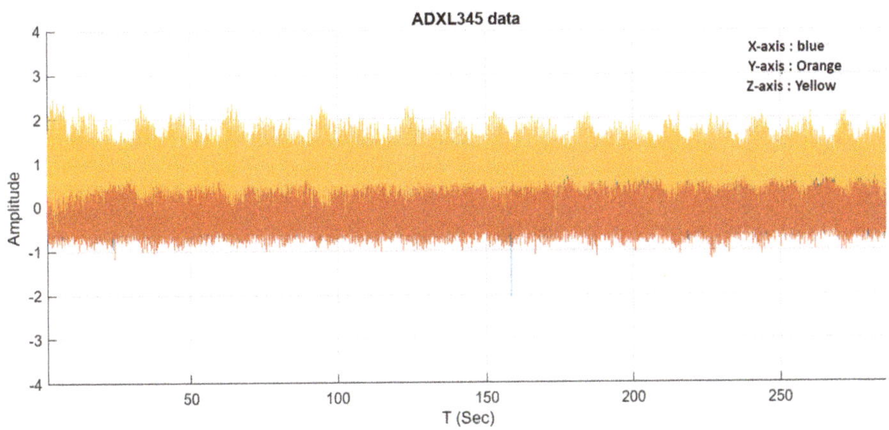

Figure 5. Time-domain analysis.

3.3. Frequency-Domain Analysis

Frequency-domain analysis was conducted using the fast Fourier transform (FFT). The system was most efficient at harvesting power from the pipeline's vibrations within a particular frequency varying between 10 Hz and 50 Hz. The results showed that 30 Hz was one of the key frequencies in which the system displayed first-rate overall performance. At these frequencies, the system was most efficient at changing kinetic energy into electric power, which meant that those frequencies matched the natural frequency of the pipeline's vibrations and were perfect for achieving maximum power harvesting performance. These results confirmed the significance of designing the device to function inside this particular frequency range to ensure first-rate overall performance in realistic contexts.

To better understand the vibration analysis, refer to Figure 6, which illustrates the frequency-domain analysis for each axis. Each axis represents a direction, and by analyzing them separately, researchers can understand how vibration impacts movement along each direction.

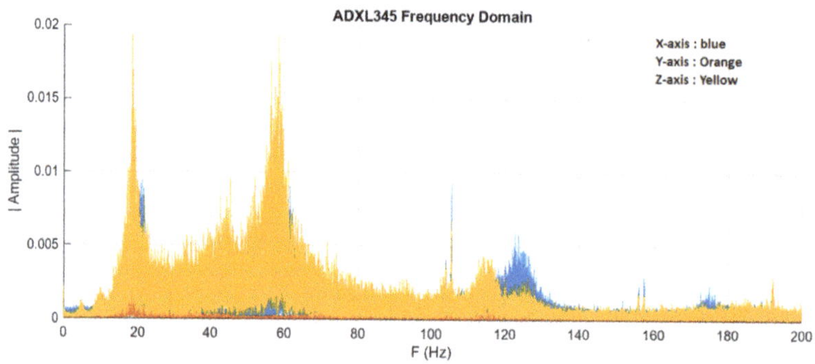

Figure 6. Frequency-domain analysis.

The z-axis experienced the strongest vibrations. This suggests that whatever is being measured vibrated most significantly in the direction of the z-axis.

The y-axis showed less vibration than the z-axis.

The x-axis showed the least amount of vibration.

This detailed analysis of vibration across different axes provided insights into the directional nature of vibrations and could be valuable in various applications such as understanding the device's orientation or analyzing structural movements.

3.4. Temperature Analysis

The temperature analysis highlighted the piezoelectric energy harvesting system's thermal stability across various temperatures. The LM35 temperature sensor recorded values indicating that the pipeline's temperature fluctuated between 40 °C and 50 °C (with a mean of 47 °C) during the testing period, as shown in Figure 7. Despite those variations, the harvested voltage remained extraordinarily strong, suggesting that temperature adjustments inside this range did not appreciably affect the tool's overall performance. This stability is vital for ensuring the reliability of the energy harvesting machine in various environmental situations, in which temperature fluctuations are commonplace. The capability to preserve normal electricity output despite temperature variations underscores the robustness and rational applicability of the system in real-world pipeline tracking situations.

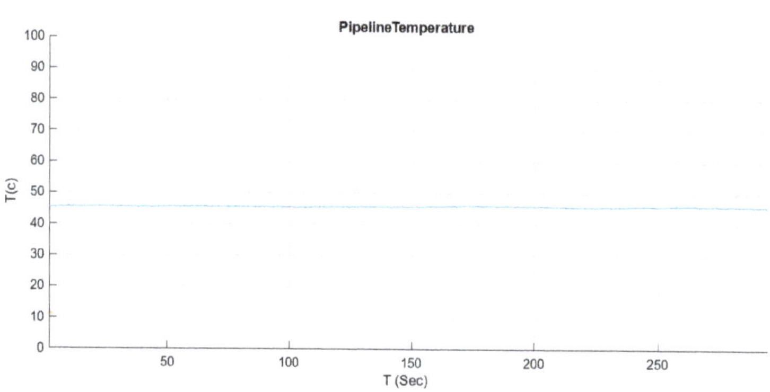

Figure 7. Temperature analysis.

3.5. Voltage and Output Power Acquired from the Sensor

In the system analyzed in Figure 8, the output voltage of the power harvesting process suggested a clear cyclic sample characterized by way of gradual charging and discharging phases. During the procedure, the voltage first rose steadily and reached the highest level of 3.3 V, which represented the maximum energy storage potential of the machine. When this threshold was reached, the system entered the discharge phase, namely the use of the saved energy, resulting in a slow reduction in the voltage from 3.3 V to 0. This cycle of charging and discharging was repeated over and over, highlighting the performance of the device in changing mechanical vibrations into usable electrical energy. The repeatability of the technique guaranteed that electricity was continuously supplied even in environments with fluctuating or intermittent vibration, permitting the related digital components to keep operating.

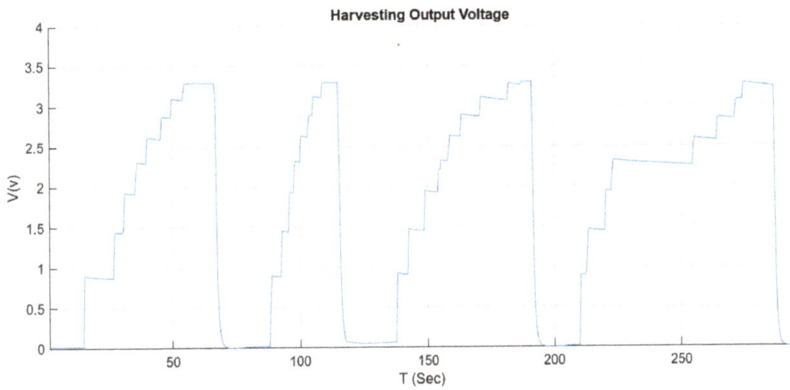

Figure 8. Harvested output voltage.

In this energy harvesting device, the voltage range was from 0 V to 3.3 V whilst driving a 5 kΩ load. The current related to this load was 0.66 mA. When the device reached the peak voltage of 3.3 V, it gave an energy output of about 2.18 mW to the load. This electricity was received from the mechanical vibrations detected by means of the piezoelectric sensor and transformed into electric power.

The importance of the 2.18 mW output power lies in its capability to powering low-energy electronics consisting of microcontrollers and sensors. The steady current of 0.66 mA ensured that the power supply remained dependable and consistent even when the voltage fluctuated during the electricity generation process. This let the system effectively provide a strong energy supply, which is crucial for the continuous operation of linked devices, especially in environments in which traditional power assets are unavailable or impractical. Therefore, the harvested electricity does not simply power the direct load, but can also be stored for later use, thereby improving the overall efficiency and sustainability of the system.

4. Discussion

- The experimental results of the piezoelectric energy harvesting system exhibited its great effectiveness in collecting pipelines' vibration energy. The time-domain evaluation suggested that the voltage generated with the aid of the piezoelectric sensor would stay stable through the years, making sure non-stop and dependable powering of linked IoT devices. This stability is critical to maintaining the operation of self-powered devices, particularly in environments in which pipeline vibration is a consistent element. The frequency-domain analysis also showed that the device was optimized to utilize electricity within a specific frequency range that coincided with the natural vibrational frequency of the pipeline, confirming the efficiency of the system's design in real-world applications. In addition, the evaluation of temperature showed that the system maintained its performance under extreme temperature conditions, which is important for use in different environments. The generator's potential to operate effectively in an extensive variety of thermal situations contributes to its robustness and makes it appropriate for monitoring pipelines in numerous climates. In addition, the voltage output discovered in the experiments confirmed a correlation between the amplitude of the vibration and the energy generated. This correlation highlights the system's ability to adequately power IoT devices, especially in conditions wherein pipelines are subject to strong vibrations.

- The results verified the effectiveness of the piezoelectric electricity harvesting system in harnessing pipelines' vibration energy and offering a sustainable electricity supply for IoT devices. The device's adaptability to extreme frequency and temperature situations suggested its capacity for widespread application within the field of pipeline monitoring and may help enhance the safety and performance of industrial procedures.
- This study's findings on the optimal frequency range for energy harvesting (10–50 Hz) aligned with the work of Rahman et al. [12], who reported efficient energy harvesting from pipeline vibrations across a similar frequency range. However, the system demonstrates improved performance at 30 Hz, which could be attributed to the unique combination of the PPA-1001 sensor and the LTC-3588 power supply.
- The temperature stability of the system (40–50 °C) compared favorably with the findings of Yazawa et al. [4], who reported efficient thermoelectric energy harvesting from pipelines at higher temperatures. The piezoelectric approach offers the advantage of operating effectively at lower, more common pipeline temperatures.
- The power output of 2.18 mW achieved by the system at 3.3 V represents a significant improvement over the 13.8 μW reported by Bakhtiar and Khan [7] for their pulsating fluid flow energy harvester. This increased power output enhanced the potential for powering a wider range of IoT devices in pipeline monitoring applications.
- The system's ability to maintain stable performance across temperature variations addressed a key challenge identified by Shen et al. [1] in their work on wireless sensor systems for monitoring oil pipelines. This stability is crucial for ensuring reliable long-term operation in varying environmental conditions.

The deployment of the energy harvesting system involves several key steps:

1. Site assessment: prior to installation, a comprehensive evaluation of the pipeline's vibration traits was carried out using accelerometers to find out the most appropriate placement locations for maximum power harvesting.
2. Installation: the PPA-1001 piezoelectric sensor was securely fixed to the pipeline's floor with a specialized adhesive that guaranteed the transmission of real vibrations whilst withstanding environmental situations. The sensor was orientated to align with the dominant axis of vibration diagnosed at an earlier stage in the online assessment.
3. Power management setup: The LTC-3588 power harvesting electricity supply and battery charging module were housed in a weatherproof enclosure close to the sensor. This setup included surge protection and voltage regulation components to ensure secure transport of electricity to the IoT devices.
4. Integration of the microcontroller and sensors: The STM32F103C8T6 microcontroller, ADXL345 accelerometer, and LM35 temperature sensor were integrated completely in a compact, low-power circuit board. This board was also contained within a weatherproof enclosure.
5. Data logging system: The custom information logging device was connected to a nearby low-power PC or transmitted to a cloud-based system for remote monitoring. This system was configured to provide real-time visualization of the records and indicatorsm primarily based on predefined thresholds.
6. Calibration and testing: post-setup, a series of calibration tests was conducted to ensure correct data collection and the most efficient energy harvesting. This consisted

of verifying the device's overall performance across the anticipated range of the pipelines' working conditions.

7. Maintenance protocol: A routine maintenance agenda was set up, together with periodic cleaning of the sensors' surface, checking of the electric connections, and recalibration of the sensors to ensure certain long-term reliability.

5. Conclusions

In conclusion, this study demonstrated the efficacy of a novel piezoelectric energy harvesting system for powering IoT devices in pipeline monitoring applications. The key findings include the following.

1. Optimal energy harvesting: the system efficiently harvested energy in the 10–50 Hz frequency range, with peak performance at 30 Hz, generating a maximum voltage of 3.3 V and a power output of 2.18 mW.
2. Temperature resilience: stable performance was maintained across a temperature range of 40–50 °C, showcasing the system's robustness in varying environmental conditions.
3. Real-time monitoring capability: The integration of the ADXL345 accelerometer and the LM35 temperature sensor, coupled with the custom data logging application, enabled comprehensive real-time pipeline monitoring.
4. Energy storage solutions: The incorporation of a battery charging module allowed for energy storage, ensuring continuous operation during periods of low vibration.
5. Practical deployment: The system demonstrated potential for real-world applications, offering a sustainable power solution for remote pipeline monitoring that would reduce reliance on traditional power sources.

This study opens several key avenues for future research as follows.

1. Field studies: extending field trials to assess the system's durability and performance in different pipeline environments;
2. Optimization of energy harvesting: investigating advanced materials and designs to enhance the efficiency of energy conversion;
3. Scalability and IoT integration: exploring scalability for larger networks and standardizing protocols for seamless IoT integration.

Author Contributions: Software, R.A.A.; Writing—original draft, Z.K.M.; Supervision, M.H.K., A.H.A. and E.M.M. All authors have read and agreed to the published version of the manuscript.

Funding: This research received no external funding.

Data Availability Statement: Dataset available on request from the authors.

Conflicts of Interest: The authors declare no conflict of interest.

References

1. Shen, Y.; Liu, X.; Liu, C.; Guo, H.; Yang, X.; Du, W. A Low Power Consumption Wireless Sensor System with Wireless Power Harvesting for Oil Pipeline Monitoring. In Proceedings of the 2018 International Conference on Microwave and Millimeter Wave Technology (ICMMT), Chengdu, China, 7–11 May 2018; pp. 1–3. [CrossRef]
2. Abbas, D.T.; Hammood, D.A.; Azemi, S.N. Minimizing Energy Consumption Based on Clustering & Data Aggregation Technique in WSN (MECCLADA). *J. Tech.* **2023**, *5*, 10–19.
3. Fairuz, W.N.M.; Nawi, I.M.; Ahmad, M.R.; Kannan, R. Experiment and performance analysis of serpentine-shaped cantilever beam for pipeline vibration-based piezoelectric energy harvester prototype development. *Clean Energy* **2024**, *8*, 111–134. [CrossRef]
4. Yazawa, K.; Feng, Y.; Lu, N. Conformal heat energy harvester on Steam4 pipelines for powering IoT sensors. *Convers. Manag.* **2021**, *244*, 114487. [CrossRef]
5. Usman, B.; Adamu, S.S.; Jimoh, T. Harvesting Energy from Flow Induced Vibration in Fixed-Fixed and Fixed-Pinned Suported Pipleline: Numerical Approach. In Proceedings of the 2019 IEEE PES/IAS PowerAfrica, Abuja, Nigeria, 20–23 August 2019; pp. 458–463.
6. Shukla, H.; Piratla, K.R.; Desai, H.; Sorber, J. Powering pipeline monitoring sensors using locally available energy. In *Pipelines*; American Society of Civil Engineers: Reston, VA, USA, 2019; pp. 311–320.

7. Bakhtiar, S.; Khan, F.U. Energy harvesting from pulsating fluid flow for pipeline monitoring systems. In Proceedings of the 2019 International Symposium on Recent Advances in Electrical Engineering (RAEE), Islamabad, Pakistan, 28–29 August 2019; Volume 4, pp. 1–6. [CrossRef]
8. Wang, Z.; Zhang, Z.; Peng, Z.; Yang, X.; Li, X.; Shan, Y.; Liu, B.; Xu, X.; Gao, Y.; Yang, Z. Self-charging and long-term face masks leveraging low-cost, biodegradable and sustainable piezoelectric nanofiber membrane. *Nano Mater. Sci.* 2024, *in press*. [CrossRef]
9. Okosun, F.; Guerin, S.; Celikin, M.; Pakrashi, V. Flexible amino acid-based energy harvesting for structural health monitoring of water pipes. *Cell Rep. Phys. Sci.* **2021**, *2*, 100434. [CrossRef]
10. Guerin, S.; Stapleton, A.; Chovan, D.; Mouras, R.; Gleeson, M.; McKeown, C.; Noor, M.R.; Silien, C.; Rhen, F.M.; Kholkin, A.L.; et al. Control of piezoelectricity in amino acids by supramolecular packing. *Nat. Mater.* **2018**, *17*, 180–186. [CrossRef] [PubMed]
11. Zhang, Z.; Li, X.; Peng, Z.; Yan, X.; Liu, S.; Hong, Y.; Shan, Y.; Xu, X.; Jin, L.; Liu, B.; et al. Active self-assembly of piezoelectric biomolecular films via synergistic nanoconfinement and in-situ poling. *Nat. Commun.* **2023**, *14*, 4094. [CrossRef] [PubMed]
12. Rahman, W.U. Modeling and Simulation of Flow-Based Circular Plate Type Piezoelectric Energy Harvester for Pipeline's Monitoring. In Proceedings of the 2019 22nd International Multitopic Conference (INMIC), Islamabad, Pakistan, 29–30 November 2019; pp. 1–6.
13. STMicroelectronics. *STM32F103x8 STM32F103xB Medium-Density Performance Line Arm®-Based 32-Bit MCU with 64 or 128 KB Flash, USB, CAN, 7 Timers, 2 ADCs, 9 com. Interfaces (Rev. 19)*; STMicroelectronics: Geneva, Switzerland, 2023.
14. Mide Technology Corporation. *PPA Products: Datasheet & User Manual (Rev. 002)*; Mide Technology Corporation: Woburn, MA, USA, 2016.
15. Analog Devices. *ADXL345: 3-Axis, ±2 g/±4 g/±8 g/±16 g Digital Accelerometer (Rev. G)*; Analog Devices Inc.: Wilmington, MA, USA, 2022.
16. Texas Instruments. *LM35 Precision Centigrade Temperature Sensors*; National Semiconductor Corporation: Santa Clara, CA, USA, 2000.
17. Linear Technology Corporation. *LTC3588-1 Nanopower Energy Harvesting Power Supply (Rev. C)*; Linear Technology Corporation: Milpitas, CA, USA, 2010.

processes

MDPI

Article

Development of Fiber-Reinforced Polymer Composites for Additive Manufacturing and Multi-Material Structures in Sustainable Applications

Rawabe Fatima Faidallah [1], Muammel M. Hanon [2,*], Nihal D. Salman [2], Yousef Ibrahim [3], Md. Noman Babu [3], Tayser Sumer Gaaz [4], Zoltán Szakál [3] and István Oldal [3]

1 Mechanical Engineering Doctoral School, Szent István Campus, MATE University, Páter Károly u. 1, 2100 Gödöllő, Hungary; faidallah.rawabe.fatima.2@phd.uni-mate.hu

2 Baquba Technical Institute, Middle Technical University (MTU), Muasker Al-Rashid Street, Baghdad 10074, Iraq; nihal.d.salman@mtu.edu.iq

3 Institute of Technology, Szent István Campus, MATE University, Páter Károly u. 1, 2100 Gödöllő, Hungary; josef98993@gmail.com (Y.I.); nomanbabu.mithu@gmail.com (M.N.B.); szakal.zoltan@uni-mate.hu (Z.S.); oldal.istvan@uni-mate.hu (I.O.)

4 Prosthetics and Orthotics Engineering Department, College of Engineering and Technologies, Al-Mustaqbal University, Babylon 51001, Iraq; tayser.sumer.gaaz@uomus.edu.iq

* Correspondence: muammel.m.hanon@mtu.edu.iq; Tel.: +964-7750540374

Abstract: This study investigates the mechanical properties of carbon and natural fiber-reinforced Polylactic Acid (PLA) and Polyethylene Terephthalate Glycol (PETG) composites produced via Additive Manufacturing (AM), focusing on Material Extrusion (MEX). The performance of filaments made from pre-consumer recycled PLA (rPLA) and PETG, with varying weight percentages of hemp and jute short fibers, was evaluated through tensile testing. Comparisons were made between the original filaments (PLA, carbon fiber-reinforced PLA [CF–PLA], and PETG) and their recycled versions. Multi-material compositions—neat PLA and PETG, single-graded (PLA + CF–PLA, PETG + CF–PETG), and multi-gradient (PLA + CF–PLA + PLA, PETG + CF–PETG + PETG)—were analyzed for mechanical properties. Optical microscope images of multi-material specimens were captured before and after fracture to assess failure mechanisms. The results indicate that the original CF–PETG filaments achieved a tensile strength of 50.14 MPa, which is higher than rPLA, PLA, and CF–PLA by 2%, 70%, and 6.7%, respectively. The re-manufactured PLA filaments reinforced with 7 wt% hemp fibers exhibited a tensile strength of 38.8 MPa, representing a 29% increase compared to the original PLA filaments and a 26% improvement over recycled PLA. Additionally, incorporating 7% jute fiber into PETG resulted in a tensile strength of 62.38 MPa, reflecting a 12% improvement over the original PETG filaments and a 15% increase compared to the recycled PETG filaments. Among specimens produced by AM, CF–PLA and rPLA demonstrated the highest tensile and compressive strengths. However, multi-material composites showed reduced mechanical performance compared to neat PLA and PETG, highlighting the need for improved interlayer adhesion. This study emphasizes the importance of optimizing material combinations and fiber reinforcement to enhance the mechanical properties of composites produced through AM.

Keywords: natural fiber; recycled filaments; tensile testing; additive manufacturing; multi-material composite

1. Introduction

Additive manufacturing (AM) has revolutionized the production of polymer-based components, enabling the creation of complex structures for a wide range of applications, from small-scale items to large industrial structures. Various commercial techniques, including polymer jetting, material extrusion, and vat polymerization, are employed to fabricate these components [1,2]. Material Extrusion (MEX), a widely adopted technique, processes polymeric filaments based on digital models [3,4]. In MEX, raw polymer filament is fed through a heated nozzle, melted, and deposited layer by layer to build 3D objects. The simplicity, precision, and cost-effectiveness of MEX make it accessible for a broad range of applications [5]. Although primarily used with thermoplastics, MEX can also process metals, ceramics, and composites [6,7]. However, commonly used filament materials, such as Polylactic Acid (PLA), Acrylonitrile Butadiene Styrene (ABS), and Polycarbonate (PC), often exhibit mechanical limitations compared to fiber-reinforced composites, such as those incorporating carbon or glass fibers. These composites significantly enhance the mechanical properties of printed parts [8–10].

Inter-layer adhesion remains a critical challenge in improving the mechanical properties of additively manufactured parts. Various studies have explored enhancement strategies, such as surface treatments, bonding agents, optimized printing conditions, and speed adjustments, to address this issue [11]. Mechanical testing of AM specimens, including tensile testing with dog-bone-shaped samples, is often conducted according to ISO 527 and ASTM D695 standards [12]. However, variations in AM processes and design parameters can affect the accuracy of these tests, leading researchers to explore alternative methods of evaluating filament tensile properties [13–17]. For instance, a study designed a hook-shaped fixture using AM and PLA material [18]. This approach allowed for robust filament installation, but issues with slippage were noted. Rahimizadeh et al. [19] used thimbles to secure filaments, which effectively prevented slippage during testing. Ferreira et al. [15] employed a clamping mechanism that wrapped the filament around a bollard, although challenges arose with less ductile filaments due to the bollard's small diameter. The absence of standardized procedures for filament tensile testing has led to variations in gauge length and testing speed across studies.

The growing environmental consciousness surrounding AM has driven interest in sustainable materials, such as natural fibers for polymer composites. These fibers offer biodegradability, renewability, and superior tensile properties compared to synthetic fibers, which require more energy-intensive production processes. Plant fibers, including hemp and jute, demonstrate tensile strengths of up to 1830 MPa, providing a viable alternative for reinforcing polymers. Treatments such as alkali processing enhance their mechanical performance, as highlighted in recent reviews. Recycled PLA (rPLA) has emerged as a sustainable alternative due to its compatibility with PLA-based composites and environmental benefits [20]. While rPLA often results in reduced tensile strength and hardness, it maintains a stable elastic tensile modulus. Current research on rPLA typically restricts short fiber loadings to below 5 wt% due to challenges like inconsistent flow and nozzle blockages. These limitations underscore the need for further investigation into the tensile and compressive properties of fiber-reinforced thermoplastic composites produced through MEX [21].

The reinforcement of polymers with fibers, such as carbon fibers (CF), glass fibers (GF), and Kevlar fibers, has significantly improved the mechanical properties of AM components. For instance, Ferreira et al. [22] demonstrated a 220% increase in tensile modulus by incorporating short carbon fibers into PLA. Hanon et al. [23] observed superior elongation performance in polyethylene terephthalate glycol (PETG) compared to PLA. Other studies have shown enhanced tensile and yield strength in carbon fiber-reinforced PETG and PLA composites [24–27].

The AM technology has advanced to support the production of application-specific components with tailored properties. The multi-material approach in MEX allows for the fabrication of components with functional gradients and distinct mechanical characteristics. This capability holds promise in various industries, including biomedical implants, construction, automotive coatings, and drones [28–30]. The success of multi-material AM depends on factors such as material selection, cost, strength requirements, and, crucially, interfacial bonding between different polymers. Recent studies have explored multi-material fabrication techniques, such as single and multiple nozzle setups, to enhance the mechanical properties of additively manufactured components [31].

Despite extensive research into polymer and fiber-reinforced composites, limited attention has been given to the systematic investigation of reprocessed and reinforced recycled filaments, particularly PLA and PETG, with natural fibers like hemp and jute. Furthermore, a noticeable gap exists in the development of multi-material composites with functional gradients specifically designed for structural applications. This study addresses these gaps by advancing the understanding of how natural fiber reinforcements enhance the tensile strength of recycled filaments, providing a comprehensive evaluation of both tensile and compressive properties of additively manufactured fiber-reinforced specimens, and introducing functionally graded multi-material composites—a new direction in AM with potential applications in industries requiring tailored mechanical properties. Several commercially produced filaments (rPLA, CF–PLA, PLA, PETG, CF–PETG) were subjected to tensile tests to determine their Ultimate Tensile Strength (UTS) before printing. The filaments were re-processed in the laboratory using a single-screw extruder to evaluate UTS after recycling, and natural fibers were added to explore potential improvements. By examining the inter-layer adhesion and gradient arrangements, this research advances the understanding of how functional gradients optimize the properties of additively manufactured multi-material composites.

2. Materials and Methods

2.1. Material Specifications

Recycled polylactic acid (rPLA) and polyethylene terephthalate glycol-modified (PETG) filaments, along with carbon fiber-reinforced PETG (CF–PETG) and carbon fiber-reinforced PLA (CF–PLA), were utilized for filament tensile testing.

This study also incorporated re-manufactured filaments of PLA, rPLA, and PETG. To enhance the mechanical properties of the re-manufactured PLA and rPLA filaments, hemp fiber was used as reinforcement. Similarly, jute fiber was employed for the re-manufactured PETG filaments (see Figure 1a,b). Comprehensive tensile property testing was subsequently conducted on these filaments. Table 1 presents detailed information about filament manufacturers and the mechanical properties of the original filaments used in the printed specimens, as reported by the manufacturers. Additionally, the properties of the reinforcement fibers, as extracted from the material data sheets, are listed in Table 2.

(a)

(b)

Figure 1. *Cont.*

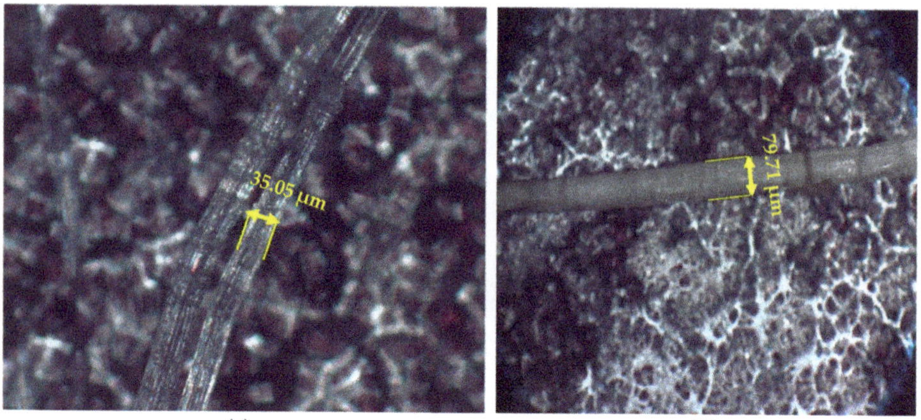

(c) (d)

Figure 1. Fibers used in the study along with their microstructure: Hemp fibers shown in (**a**,**c**), and Jute fibers in (**b**,**d**).

Table 1. Mechanical characteristics and manufacturers of the materials utilized.

Properties	Materials				
	PLA	rPLA	PETG	CF–PLA	CF–PETG
Density [g/cm^3]	1.25	1.24	1.29	1.3	1.32
Tensile strength [MPa]	51	63	61	65	45
Elastic Modulus [MPa]	2400	3251	3100	12,500	4250
Elongation at break [%]	3.5	4	4	0.5	4.9
Manufacturer	Creality	3D JAKE	Extruder	Spectrum	Spectrum

Table 2. Mechanical and physical properties of jute and hemp fiber [32].

Fibers	Length [mm]	Density [g/cm^3]	Tensile Strength [MPa]	Young's Modulus [GPa]
Jute	2–120	1.5	393–773	10–30
Hemp	5–55	1.4	550–1110	30–70

To ensure the resulting filament meets specific quality, performance, and application-related criteria, it is essential to determine the dimensions of small hemp pieces and the diameter of jute fibers before transforming them into filament form. Under 10×10 magnification, the microscope provided the following detailed measurements: the hemp fiber length was 35.05 µm, the width was 23.78 µm, and the jute fiber diameter measured 79.71 µm (see Figure 1c,d). Figure 2 presents a schematic roadmap that clarifies the distinction between the materials used, specifying which were purchased, which were employed in multi-material designs, and which were recycled, along with the tests performed on each group.

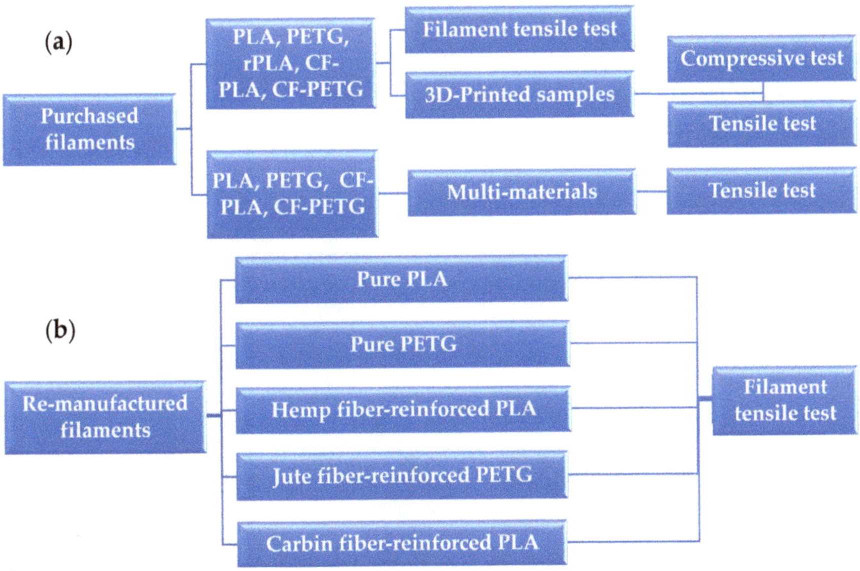

Figure 2. Schematic roadmap illustrating the categorization of materials used in the study: (**a**) purchased materials and (**b**) re-manufactured materials.

2.2. Filament Fabrication

The objective was to create new filaments by re-manufacturing the original commercially produced filaments to evaluate their ultimate tensile strength (UTS) and compare these results with those of the initial filaments. The preparation process began with chopping the original filament into small particles utilizing a filament cutting machine. These particles were then fed into a single-screw extruder to produce new filaments.

In the Materials Science laboratory, a filament-cutting machine was utilized. A segment of the commercially produced filament was sliced into pieces and placed into the machine's chamber. After securing the chamber to the apparatus, it was operated for seven seconds to generate the final particles (see Figure 3). Five materials—PLA, rPLA, CF–PLA, PETG, and CF–PETG—were cut separately and prepared for manufacturing new filaments. The filament was created using a filament maker machine after being ground with the WZ-1 grinder machine. The PLA and PETG particles were fed into the machine's barrel and heated to specific temperatures. The extrusion temperatures were set to 185 °C and 200 °C for PLA and PETG polymers, respectively. The extrusion speed was set to the maximum level on the machine to ensure material melting. Inside the barrel, the rotating screw advanced the molten material, simultaneously mixing and blending it to maintain uniform quality. After the material was fully melted and homogenized, it was extruded through a die to produce filaments with a specified diameter of 1.75 mm. The filament was then cooled with a fan and readied for subsequent testing.

The same process was repeated while mixing hemp fiber with PLA and rPLA filaments and jute fiber with PETG filaments. The selection of hemp with PLA and jute with PETG was based on preliminary trials that demonstrated better interfacial bonding and mechanical performance with these specific material combinations. Additionally, extruding hemp and jute composites with 7 wt% was achieved without difficulty. The percentages of fibers were chosen based on prior research [33] and practical considerations of printability and mechanical properties. However, attempts to re-manufacture CF–PETG were unsuccessful. According to previous research [34], recycling carbon fiber-reinforced thermoset composites has been particularly challenging, and similar difficulties may apply to carbon fiber-reinforced PETG. When natural fibers were added to CF–PETG and CF–PLA, the extrusion process became problematic. Extruding hemp with CF–PLA or jute fiber with CF–PETG was not feasible.

Due to circumstances that were difficult to control, varying filament diameters were obtained, which differed from commercially manufactured filament. These conditions included the inability of the mixing chamber in the extrusion machine to provide a homogeneous blend of the molten material, leading to inconsistent filament uniformity. Additionally, difficulties in calibrating the melting temperature and extrusion speed resulted in diameter variations and noticeable disturbances in the amount of extruded filament. Moreover, the lack of control over cooling, represented by the presence of a single-speed fan, exacerbated issues during the manufacturing process. Despite these challenges and difficulties, the required and sufficient samples were obtained for testing. Experimental adjustments were made to the temperature settings to align with the extrusion speed, and the available cooling system was adjusted to produce testable samples. The total number of the re-manufactured filament tensile test is 35.

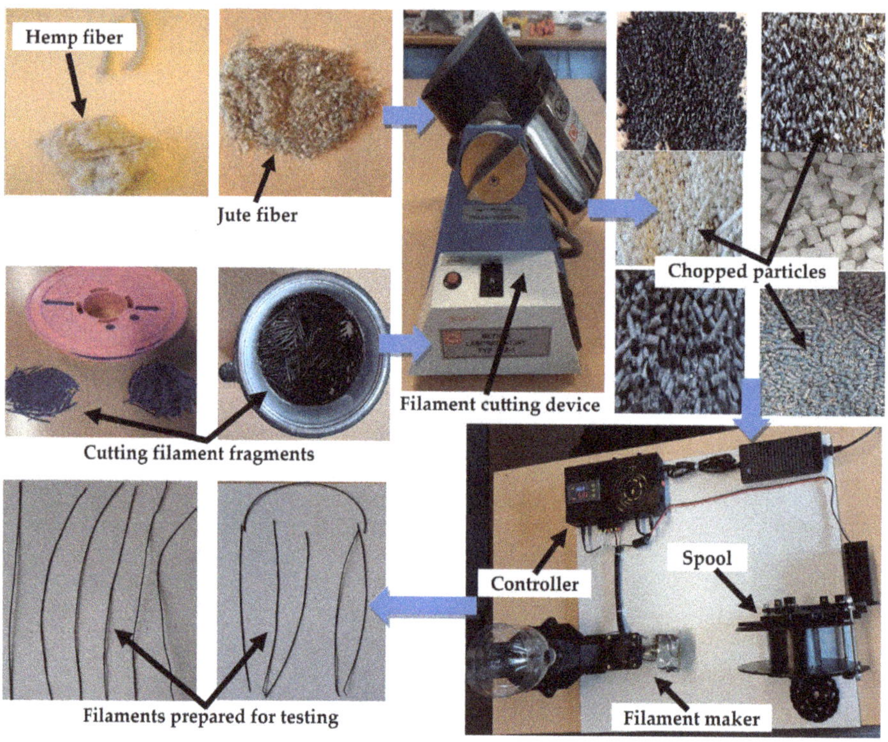

Figure 3. Filament cutting procedure.

The filament diameter varied for each material. Table 3 shows the different diameters of the materials after they were re-manufactured and reinforced with natural fibers.

The reduced filament diameter observed was likely due to challenges in controlling the fan speed and the rapid transition from the molten to solid state during extrusion. Additionally, the speed of the extruder played a role in this deviation, impacting the cooling rate and resulting in thinner filaments than expected.

Table 3. Diameters of the materials after re-manufacturing and reinforcing with natural fibers.

Specimens	Re-Manufactured Filament Diameters [mm]						
	PLA	rPLA	PETG	CF–PLA	PLA + Hemp	rPLA + hemp	PETG + Jute
1	1.1	1.2	1.4	1.2	1	1.3	1.3
2	1.2	1.2	1.5	1.2	0.9	1.3	1.3
3	1.2	1.2	1.5	1.2	0.7	1.2	1.3
4	1.1	1.1	1.5	1.1	0.7	0.9	1.2
5	1.1	1.1	1.4	1.2	0.9	1.2	1.2

2.3. Specimen Manufacturing

The experimental non-reinforced materials included PLA, recycled PLA (rPLA), and PETG filaments having a diameter of 1.75 mm. Additionally, CF–PLA and CF–PETG original filaments were enhanced with fibers containing 10 wt.% carbon fibers, each having a length of 100 µm. Tensile test specimens were designed according to ISO 527-2 standards [35], and compressive specimens were manufactured following ASTM D695 [36]. The CAD model of the samples was created using Solidworks 2024 software and saved as an STL file. The 3D specimens were produced using Material Extrusion (MEX) technology on a Geetech A20M 3D printer (Shenzhen, China), with printing operations managed through Ultimaker Cura 5.2.1 software (see Figure 4). Samples were produced in a flat orientation, with five samples produced for each original material using identical print settings. Examples of these samples are shown in Figure 5. The printing parameters outlined in Table 4, including the melting and platform temperatures, were selected based on the manufacturer's recommendations from the material data sheets. These parameters, such as temperatures, layer heights, and print speeds, were chosen to optimize adhesion, minimize warping, and ensure smooth surface finishes, consistent with best practices in the field. For instance, the nozzle temperatures were precisely matched to each filament's melting point to guarantee a consistent material flow during extrusion, while the selected layer height offered a balance between mechanical integrity and printing efficiency. The total number of the 3D manufactured specimens for the tensile and compressive test is 50.

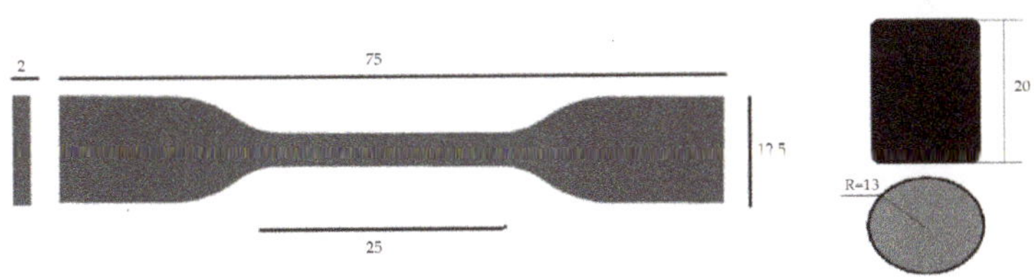

(**a**) (**b**)

Figure 4. Dimensions of samples designed using SolidWorks software: (**a**) tensile specimen, (**b**) compressive specimen.

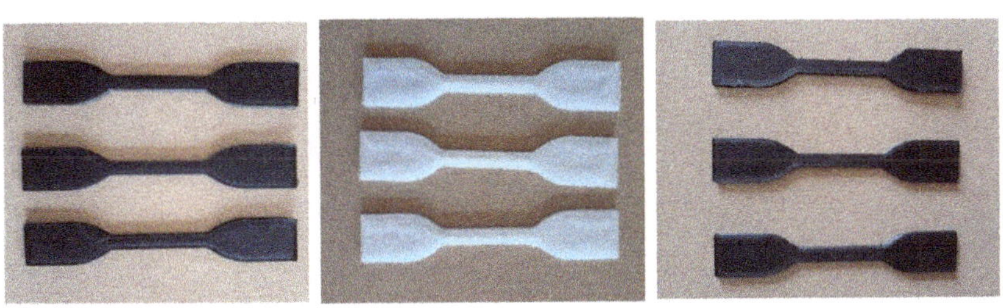

(**a**) (**b**) (**c**)

Figure 5. *Cont.*

(**d**) (**e**) (**f**)

Figure 5. Actual appearance of various additively manufactured tensile and compressive test specimens created with original filaments: (**a**) rPLA, (**b**) PLA, (**c**) PETG, (**d**) CF–PLA, and (**e**) CF–PETG. (**f**) Displays the compressive specimens for all materials.

Table 4. Printing parameters for additively manufactured samples produced with the Geetech A20M 3D printer.

Printing Parameters	Materials				
	rPLA	**PLA**	**PETG**	**CF–PLA**	**CF–PETG**
Layer thickness [mm]	0.2	0.2	0.2	0.2	0.2
Initial layer height [mm]	0.2	0.2	0.2	0.2	0.2
Print speed [mm/s]	50	50	50	50	50
Infill density [%]	100	100	100	100	100
Wall loops	2	2	2	2	2
Top/bottom layers	2	2	2	2	2
Infill pattern	Rectilinear	Rectilinear	Rectilinear	Rectilinear	Rectilinear
Platform temperature [°C]	60	60	70	45	75
Printing temperature [°C]	210	205	230	200	240
Nozzle diameter [mm]	0.4	0.4	0.4	0.4	0.4

2.4. Multi-Material Design

In this investigation, four distinct types of polymeric filaments tailored for additive manufacturing were extruded. These categories included pure PLA and PETG polymers, as well as PLA and PETG polymeric composite filaments reinforced with 10 wt% carbon fiber. According to the material provider, the average length of the carbon fibers was 100 μm. The samples were created using a Bambu Lab P1S 3D printer (Austin, TX, USA), which is capable of additive manufacturing with up to four different materials in a layer-by-layer manner. In our experimentation, all four feedstock materials were used, with each material corresponding to a dedicated feeding system. The STL files of the samples were sliced into standard G-code for printing.

Tensile specimens of single-gradient composites, denoted as PLA/CF–PLA and PETG/CF–PETG, consisted of 50% PLA and 50% CF–PLA, and 50% PETG and 50% CF–PETG, as well as 50% PLA and 50% PETG. Additionally, the multi-gradient composites were created using the following two specific combinations: PLA/CF–PLA (PCFP) and PETG/CF–PETG (GCFG). The PLA/CF–PLA setup featured 25% PLA on the outer layers and 50% CF–PLA in the central core (PCFPP). For the PETG/CF–PETG combination, 25% PETG was used for the outer layers with 50% CF–PETG as the core (GCFGG). Another variant, the PETG/PLA composite, included 25% PETG on the outer layers and 50% PLA in the inner core (GPG). For comparison, additively manufactured samples of pure neat PLA, PETG, CF–PLA, and CF–PETG were also prepared, and their properties were assessed alongside various gradient composites. The selection of the multi-material design samples was based on optimizing material properties for specific structural applications. The choice was influenced by prior studies that demonstrated the benefits of combining materials with different mechanical characteristics to enhance overall performance [37,38]. Figure 6 displays the schematic of the functional grading sequences for various PLA, PETG, and CF–PLA–CF–PETG composite layers. The samples were created with different gradient stacking arrangements, including both single- and multi-gradient setups with varying percentages of stacking layers.

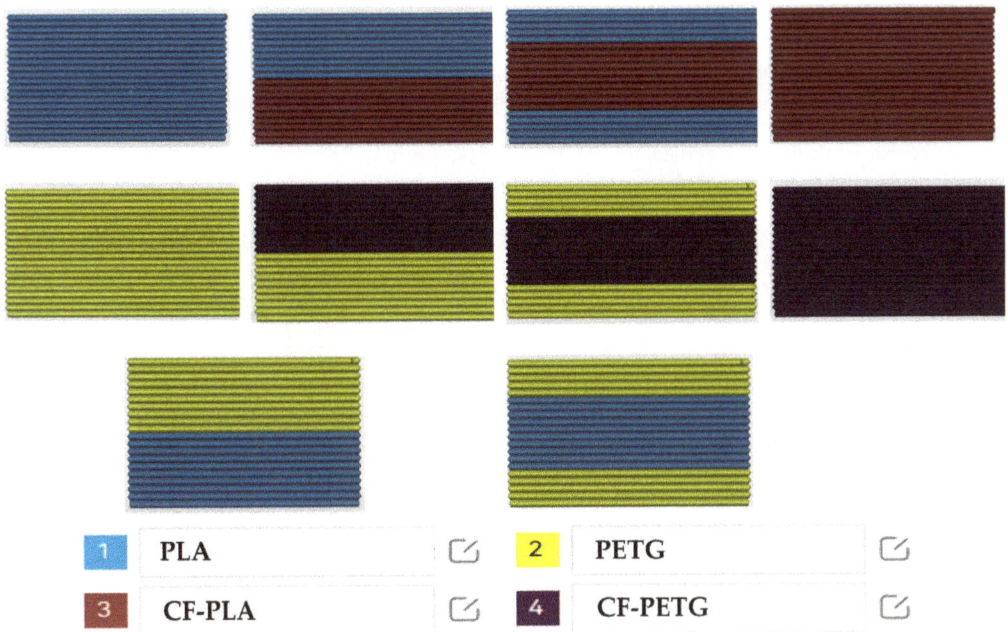

Figure 6. Schematic of functional grading sequences for various multi-material tensile specimens, produced using Bambu Studio v1.8.4.51 software.

The material properties of PLA and PETG produced under the specified printing parameters are outlined in Table 5. This study delved into the tensile strengths of the specimens, determined by evaluating functionally specific volume ratios. For printing with PLA and CF–PLA, an optimal printing temperature of 210 °C was employed, while printing with PETG and CF–PETG required nozzle temperatures of 220 °C and 240 °C, respectively. Although Table 5 suggests an ideal bed temperature of 45 °C for PLA, a bed temperature of 70 °C was used for all prints involving PETG.

Table 5. Printing settings employed while manufacturing multi-materials using Bambu 3D printer.

Printing Settings	PLA	PETG	CF–PLA	CF–PETG
Layer height [mm]	0.12	0.12	0.12	0.12
Initial layer height [mm]	0.2	0.2	0.2	0.2
Wall loops	2	2	2	2
Top/bottom layers	2	2	2	2
Infill density [%]	100	100	100	100
Infill pattern	Rectilinear	Rectilinear	Rectilinear	Rectilinear
Printing temperature [°C]	210	220	220	240
Platform temperature [°C]	45	70	45	70
Printing speed [mm/s]	50	50	50	50

2.5. Experimentation

2.5.1. Filament Fixation

In this study, the tensile properties of the original filaments (not printed specimens and 3D specimens) and the re-manufactured filaments (not printed into a specimen), such as ultimate tensile strength and fracture strength, were determined at a crosshead speed of 5 mm/min, with a minimum of five repetitions for each test. Filaments of 300 mm in length were clamped using Bollard-style grips (see Figure 7). The Bollard part of the fixture had a diameter of 120 mm, consisting of two cylindrical grips mounted on a Zwick Roell Z100 tensile testing machine (Ulm, Germany). Testing was performed at a displacement rate of 5 mm/min, with a distance of 150 mm between the grips. One grip was secured in place, while the other was adjustable. The adjustable grip was attached to the load cell of the testing machine, which recorded the force exerted on the specimen.

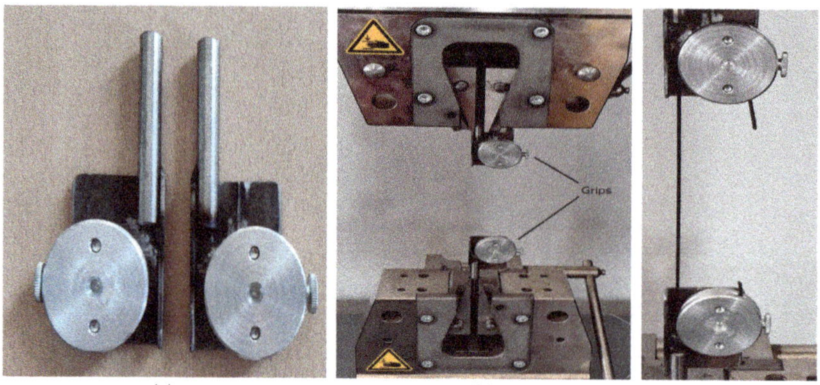

(a) (b) (c)

Figure 7. Tensile testing of filaments: (**a**) Bollard grips used, (**b**) grip positioning within the test machine, and (**c**) close-up view of the grips.

2.5.2. Tensile and Compressive Testing of Additively Manufactured Specimens

The tensile and compressive characteristics of the various configurations adhered to ISO 527-2 and ASTM D695 standards, respectively. For the tensile testing trial, specimens were manufactured with dimensions of 110 mm in length, 20 mm in width, and 4 mm in thickness. The different graded multi-material compressive samples had a diameter of 13 mm and a length of 20 mm. Thirty samples were prepared for both tensile and compressive tests, with each material configuration undergoing three repetitions. The experiments were conducted using a Zwick/Roell Z100 testing machine equipped with a 10 kN force transducer at a loading rate of 1 mm/min. Tensile properties, including tensile strength and failure mode, were evaluated under specific conditions, maintaining a consistent grip-to-grip distance of 78 mm. Additionally, compressive properties, including E modulus and failure mode, were computed for the samples. Figure 8 shows one of the samples during tensile and compressive tests conducted on the 3D-manufactured specimens.

(a) (b)

Figure 8. Experimental setup for mechanical testing of additively manufactured samples: (**a**) compressive test setup for PLA sample and (**b**) tensile test setup for PETG sample.

2.5.3. Tensile Properties of Multi-Material Composites

Tensile specimens of the various multi-material configurations, including those with gradient additives, were fabricated according to ISO 527-2 standards. These specimens were designed with dimensions of 150 mm in length, 20 mm in width, and 4 mm in thickness. The tensile strength and modulus of the prepared samples were assessed using the Universal Testing Machine Zwick/Roell Z100, applying a tensile loading rate of 1 mm/min. Each set of additively manufactured multi-material samples underwent five repetitions of testing. A representation of the tensile test samples is illustrated in Figure 9.

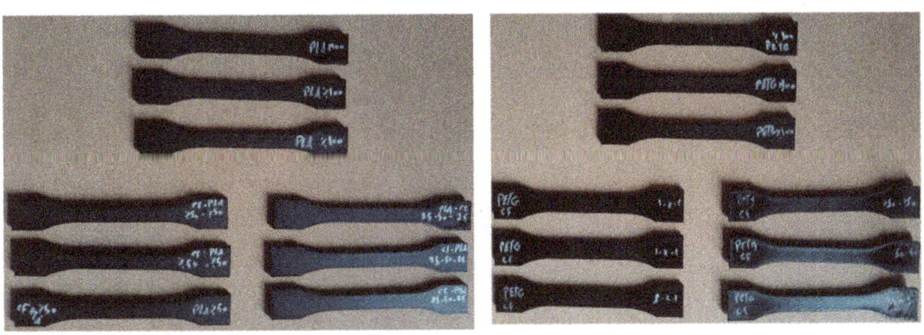

(a) (b)

Figure 9. Physical appearance of selected tensile specimens: (**a**) PLA and CF–PLA, and (**b**) PETG and CF–PETG.

3. Results and Discussion

3.1. Original Filament Tensile Test

Tensile testing was conducted on the original filament material, which had not undergone extrusion during the printing process. Figure 10 presents the stress–strain curves for the five different filament materials. The figure provides one representative result out of the three tests conducted for each material.

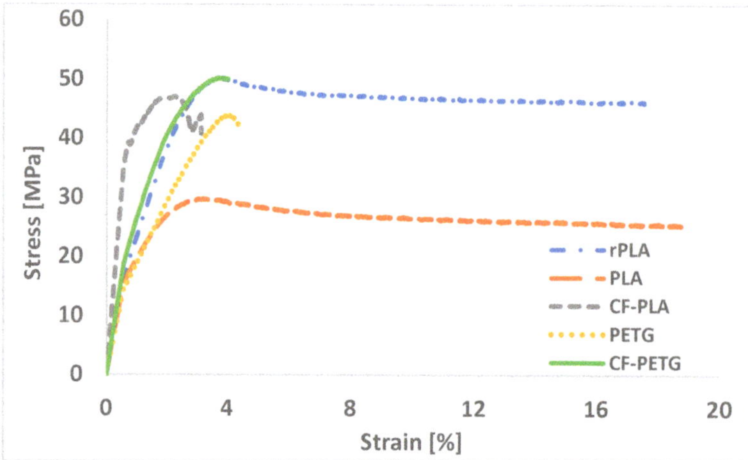

Figure 10. Stress–strain curves for the original (non-printed) materials.

The CF–PETG filament exhibited a maximum stress of 50.14 MPa, closely matching the stress of rPLA filaments. The stress of CF–PETG was higher than that of rPLA, PLA, CF–PLA, and PETG by 0.2%, 70%, 6.7%, and 14.4%, respectively. The CF–PETG filament demonstrated consistent stress resistance throughout the specified strain range. In contrast, the other filaments experienced a reduction in stress resistance after reaching their respective peak stress levels. Compared to the tensile strength values provided by the manufacturer, the tensile strength of PLA, rPLA, PETG, and CF–PLA decreased by 30%, 20%, 28%, and 27%, respectively. This reduction can be attributed to several factors, including variations in environmental conditions (such as humidity and temperature), differences in testing equipment or setup (e.g., grip type, alignment, or loading speed), and the storage conditions of the filament prior to testing (see Table 1 for manufacturer data).

3.2. *Additively Manufactured Specimen Tensile Testing*

Figure 11 shows the stress–strain curves obtained from the tensile tests of different original materials. The results indicate that rPLA and CF–PLA samples had the highest and nearly highest maximum stresses of 50 MPa and 46.5 MPa, respectively. This result highlights the enhancement that carbon fiber incorporation can bring to PLA's mechanical properties. Pure PLA exhibited a stress of 36 MPa, which was lower than that of rPLA and CF–PLA. This suggests that both recycling and reinforcement with carbon fibers can significantly enhance PLA's tensile strength. The PETG showed commendable performance at 45 MPa, indicating its good inherent mechanical properties, although it was still inferior to rPLA and CF–PLA. Interestingly, CF–PETG recorded a stress of 31 MPa, lower than pure PETG. This result could be due to factors such as inadequate bonding between the PETG matrix and carbon fibers or an unfavorable ratio of reinforcement. The rPLA exhibited 38% higher stress than pure PLA, while CF–PLA demonstrated a 50% increase in tensile stress compared to CF–PETG. The dissimilarities in chemical and mechanical properties between PLA and PETG may contribute to these variations. The varying compatibility of the added carbon fibers with each pure material might be influencing the overall performance observed in the tensile tests. Figure 12 illustrates the main failure scenarios observed during tensile tests. The fracture sites differed depending on the loading points, with pure PLA showing superior performance. While failure within the gauge zone is preferred for accurate strain measurement, the mechanical behavior of the material is still reliably reflected in the sample that failed outside the gauge zone. The test was conducted under standard conditions, with uniform stress distribution observed prior to failure. Therefore, despite the location of the fracture, the results remain valid and provide valuable insights into the material's performance.

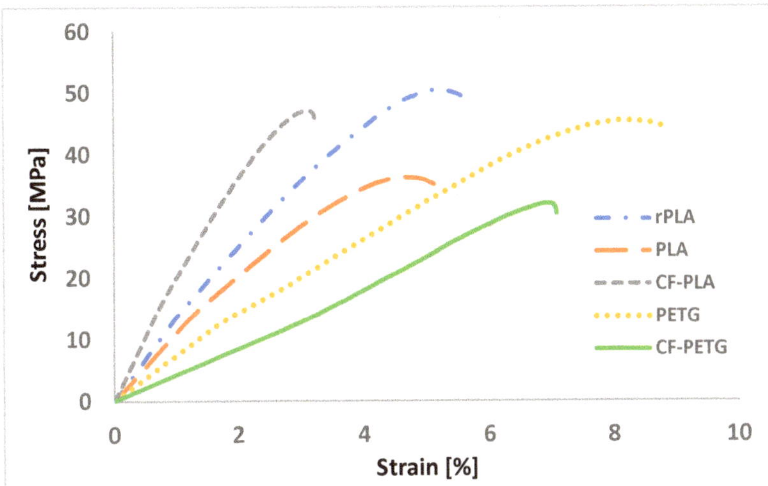

Figure 11. Stress–strain curves from tensile tests of additively manufactured specimens.

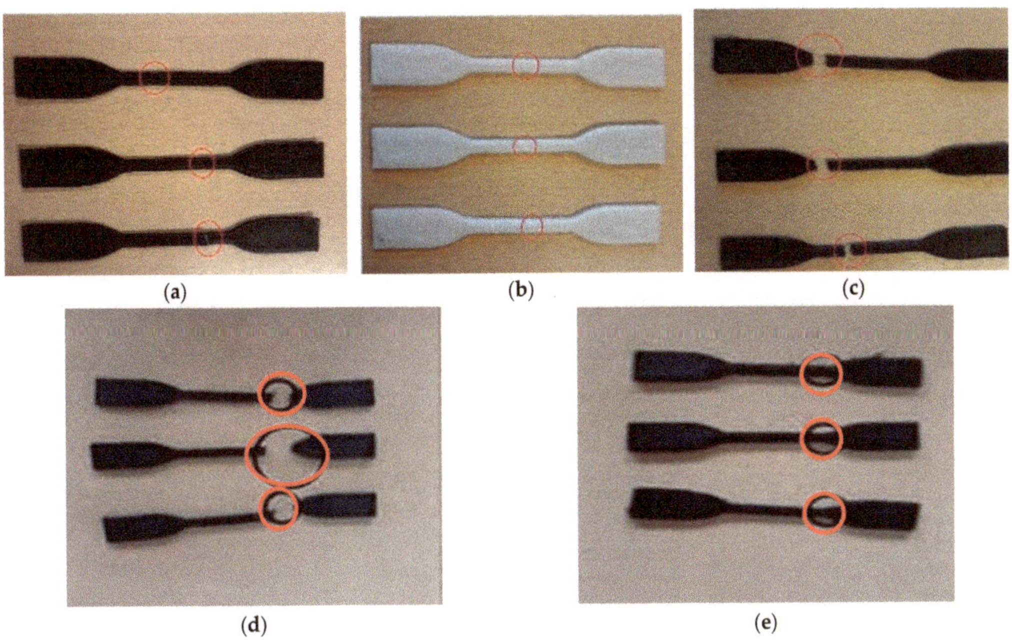

Figure 12. Specimens after tensile tests: (**a**) rPLA, (**b**) PLA, (**c**) CF–PLA, (**d**) PETG, and (**e**) CF–PETG. The red circles indicate the areas of failure.

3.3. Additively Manufactured Specimen Compressive Test

This section presents the results of compression tests conducted on additively manufactured specimens fabricated using the original filament materials. As depicted in Figure 13, the compression behavior during the tests varied among the different materials examined.

The stress–strain curves indicate that the highest maximum compressive stress of 81.7 MPa was recorded for rPLA. In contrast, the white PLA material exhibited the lowest stress at 42.8 MPa at a strain of 3.4%. The CF–PETG demonstrated a compressive stress of 48.6 MPa at a similar strain of 3.4%. Notably, CF–PLA specimens displayed a 61% higher compressive stress compared to CF–PETG. While pure PETG showed satisfactory mechanical properties, it was still inferior to CF–PLA. The compressive strength of PETG, measured at 53.7 MPa, aligns well with values reported in previous studies, which range from 41 to 98 MPa [39]. Both PETG and CF–PETG exhibited reduced compressive deformation, while rPLA exhibited the highest maximum deformation, consistent with the findings of Hsueh et al. [27]. Hsueh and colleagues concluded that, apart from thermal deformation, the mechanical properties of PLA surpassed those of PETG. In this study, adding 10 wt% carbon fiber to PETG decreased its compressive stress, which could be explained by factors such as the quality of dispersion and bonding with the PETG matrix. Numerous investigations have shown similar findings, highlighting the significance of fiber-matrix interactions in composite materials [27]. An optical microscope image of the compressed specimens is shown in Figure 14.

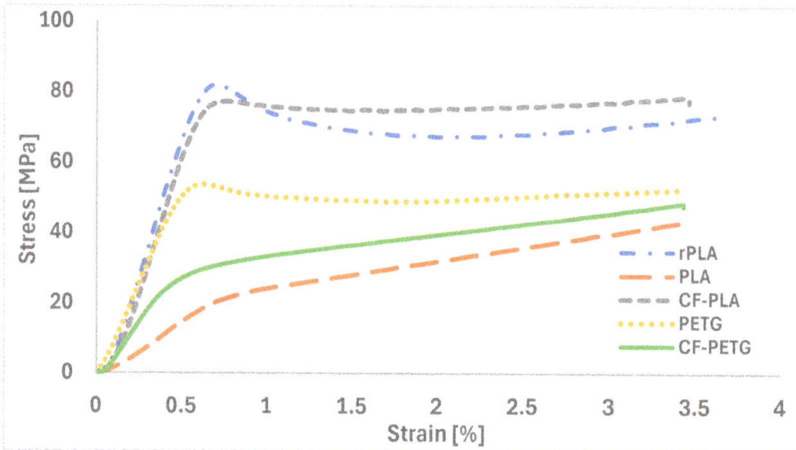

Figure 13. Stress–strain curves for compressive test of additively manufactured specimen.

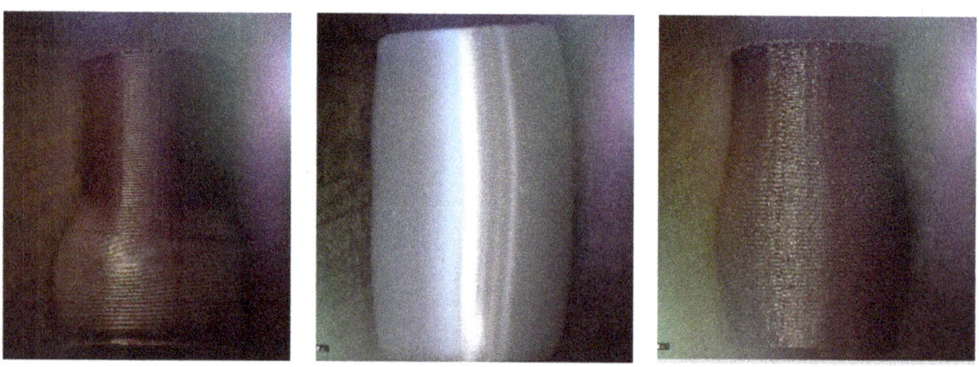

(a) (b) (c)

Figure 14. *Cont.*

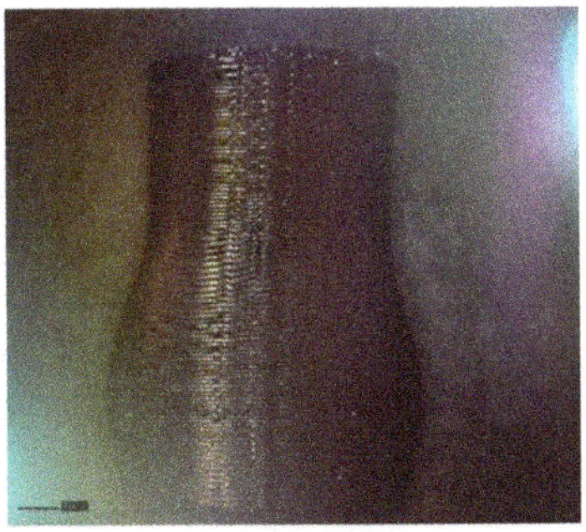

(d)

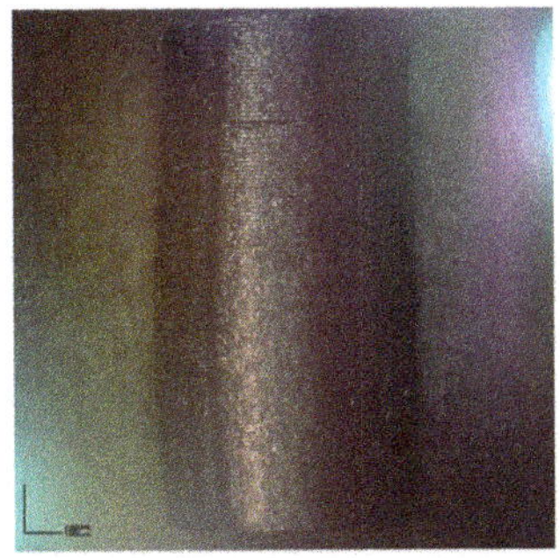

(e)

Figure 14. The morphology of the compressed samples following the compressive test: (**a**) rPLA, (**b**) PLA, (**c**) CF–PLA, (**d**) PETG, and (**e**) CF–PETG.

3.4. Re-Manufactured Filament Tensile Test

Tensile tests were conducted on re-manufactured filaments, encompassing a set of the following five specimens: rPLA, PLA, CF–PLA, PETG, and hemp fiber-reinforced PLA/rPLA filaments. Additionally, jute fiber-reinforced PETG underwent re-manufacturing and testing in this segment. Notably, the re-manufactured PLA and rPLA exhibited superior tensile strength compared to the original filaments, which can be attributed to the consistent layer bonding in the PLA samples. However, the recycled CF–PLA showed a 4% reduction in strength compared to the original filaments. This decrease in tensile strength is likely due to the formation of fiber clusters during recycling, weakening the adhesive bond between the PETG and carbon matrix. The clusters reduce the mechanical properties of the composite by impairing interlayer adhesion. Regarding PETG filament, the recycled PETG exhibited lower strength than the original filaments. On the contrary, incorporating 7 wt% hemp fiber into PLA and rPLA filaments and re-manufacturing them resulted in filaments with superior tensile strength compared to the original ones. In the case of white PLA with hemp filaments, the tensile strength surpassed that of the original filaments and pure recycled PLA by 29% and 26%, respectively. Furthermore, adding 7 wt% hemp fiber to rPLA increased its tensile strength by 7% compared to the original rPLA filament, though it resulted in a 6% decrease in tensile strength compared to re-rPLA. Moreover, adding 7 wt% jute fiber to PETG enhanced the tensile strength by 12% and 15% for the original and re-manufactured PETG filaments, respectively. Figure 15 shows the sample after the test. Additionally, a comparison between the maximum applied load and ultimate tensile strength (UTS) for both the original and re-manufactured filaments is shown in Table 6.

Table 6. Maximum applied load and tensile strength, values for the original, re-manufactured, and reinforced filaments.

Filaments	$F_{original\ filament}$	$F_{re\text{-}filament}$	$UTS_{re\text{-}filament}$	$UTS_{original\ filament}$
re-PLA	71.51 ± 2.7	18 ± 1.2	30.6 ± 1.8	29.7 ± 3.2
re-rPLA	120 ± 4.8	50 ± 1.7	58.6 ± 2.5	50 ± 3.7
re-PETG	105 ± 2.5	72 ± 3.1	42.43 ± 2.3	43.86 ± 2.6
re-CFPLA	113 ± 3.4	66.8 ± 2.6	42.4 ± 3.1	47 ± 3.6
re-PLA + Hemp	-	18 ± 1.1	38.5 ± 0.9	-
re-rPLA + Hemp	-	50 ± 2.8	53.5 ± 2.7	-
re-PETG + Jute	-	62.38 ± 3.2	49 ± 3.3	-

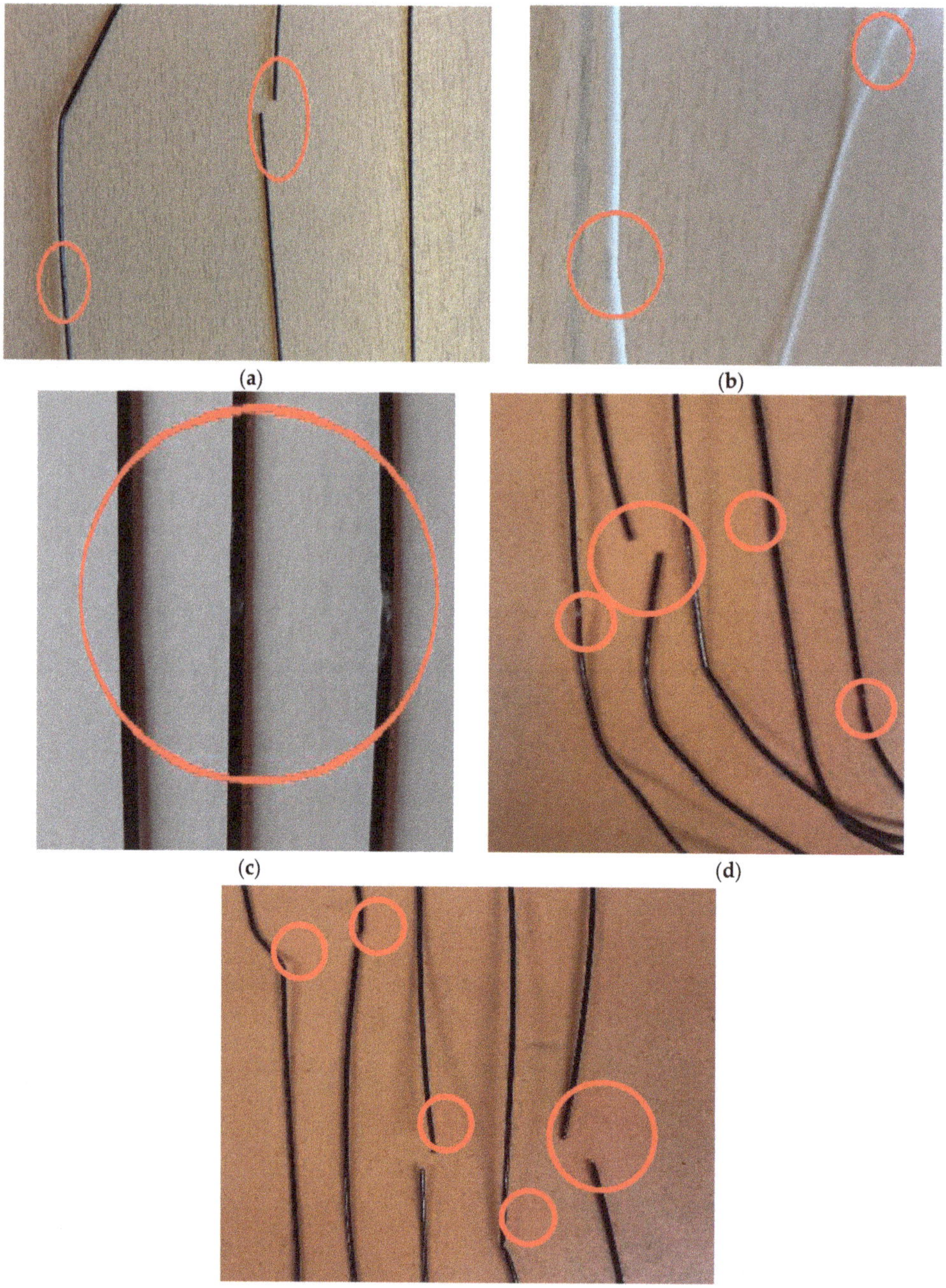

Figure 15. Fractured surfaces of re-manufactured filament samples: (**a**) re-rPLA, (**b**) re-PLA, (**c**) re-PLA + Hemp fiber, (**d**) re-PETG, and (**e**) re-PETG + Jute fiber. The red circles indicate the areas of failure.

3.5. Tensile Features of Multi-Gradient and Single PLA, PETG, CF–PLA, and CF–PETG Composites

Figure 16 presents the tensile properties of neat (100% PLA (P), 100% PETG (G), 100% CF–PLA (CFP), 100% CF–PETG (CFG)), single (50% PLA + 50% CF–PLA (PCFP), 50% PETG + 50% CF–PETG (GCFG), 50% PLA + 50% PETG (PG)), and multi-graded (25% PLA + 50% CF–PLA + 25% PLA (PCFPP), 25% PETG + 50% CF–PETG + 25% PETG (GCFGG), 25% PETG + 50% PLA + 25% PETG (GPG)) materials. Specifically, Figure 16a shows the tensile force-elongation curves for the neat, single, and functionally graded composites. Figure 16b,c displays the Young's modulus (E modulus) and tensile strength of different graded composites for the PLA, CF–PLA, PETG, and CF–PETG materials. In addition, Table 7 provides a detailed comparison of the tensile properties of the examined specimen materials. The results indicate that the tensile strengths of the neat materials are quite similar, with neat PLA exhibiting a slightly higher tensile strength than PETG by 0.03%. Generally, PLA is recognized for its higher tensile strength compared to PETG, making it a more rigid material with lower elongation at break. Conversely, PETG demonstrates higher elongation at break, making it more flexible, impact-resistant, and less brittle than PLA. For specimens composed of 100% layers, the tensile strength of CF–PLA was 0.04% higher than that of neat PLA. In contrast, the tensile strength of CF–PLA decreased by 65% compared to neat PETG. This reduction could be attributed to the introduction of carbon fibers, which potentially increases the occurrence of microvoids within the printed layers, thereby diminishing the overall tensile strength of the composite.

To compare the tensile strength of various additively manufactured specimens, the highest tensile strength, 47.4 MPa, was achieved by specimens composed of PCFPP. This result was close to the specimens manufactured by PCFP, which showed a tensile strength of 46.2 MPa. Specimens manufactured with multi-layers of CFPLA and PLA composite showed improved tensile strength, with a 4% increase for PCFP and 7% for PCFPP compared to neat PLA. Conversely, specimens manufactured with multi-material layers of PETG and CF–PETG experienced a reduction in tensile strength compared to neat PETG. The tensile strength decreased by 4% for GCFGG and by 5% for GCFG. These results demonstrate the distinct impacts of material compositions on tensile strength, showcasing the advantages of CFPLA in enhancing tensile properties while also highlighting the challenges of integrating carbon fiber with PETG.

Table 7. Mechanical qualities of the single, multi-graded, and neat materials.

Materials	Maximum Applied Load [N]	Elongation at Maximum Applied Load [mm]	Maximum Stress [MPa]	E Modulus [Mpa]
P	1920 ± 65.2	0.69 ± 0.07	44.4 ± 2.5	2768 ± 96
G	1759 ± 119.5	0.77 ± 0.09	42.9 ± 3.19	2185 ± 105
PCFP	1921.7 ± 82.1	0.63 ± 0.13	46.2 ± 3.56	3805 ± 173
GCFG	1675 ± 96.7	0.91 ± 0.05	40.6 ± 1.47	2214 ± 151
PCFPP	1926.2 ± 102.6	0.71 ± 0.04	47.4 ± 2.41	3626 ± 127
GCFGG	1736 ± 67.7	0.94 ± 0.16	41.2 ± 1.53	2138 ± 208
CFP	1919.8 ± 76.1	0.64 ± 0.15	46.3 ± 1.62	3914 ± 76
CFG	1032 ± 82.8	0.7 ± 0.06	26 ± 2.61	1318 ± 84
PG	1606 ± 99.7	0.69 ± 0.09	38.8 ± 2.2	2356 ± 169
GPG	1705 ± 52.9	0.88 ± 0.17	41.19 ± 1.84	2498 ± 197

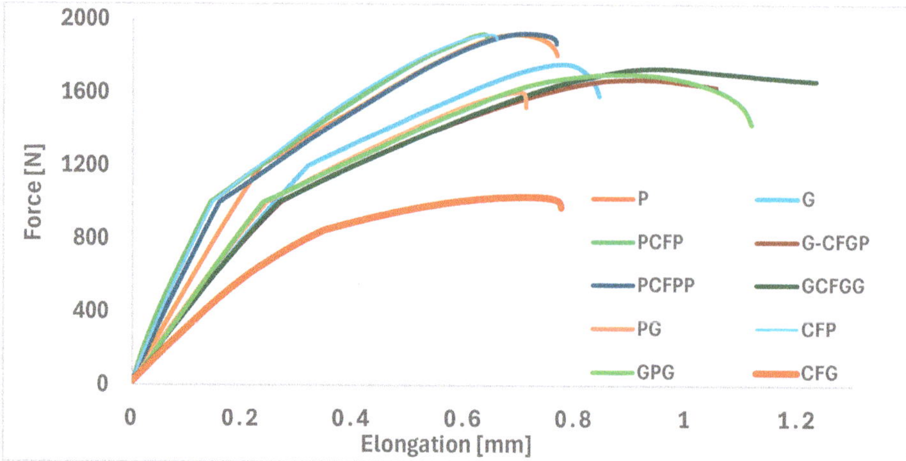

(a)

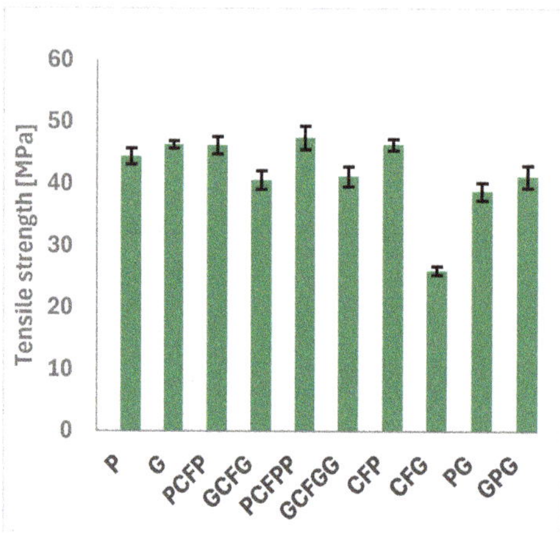

(b)

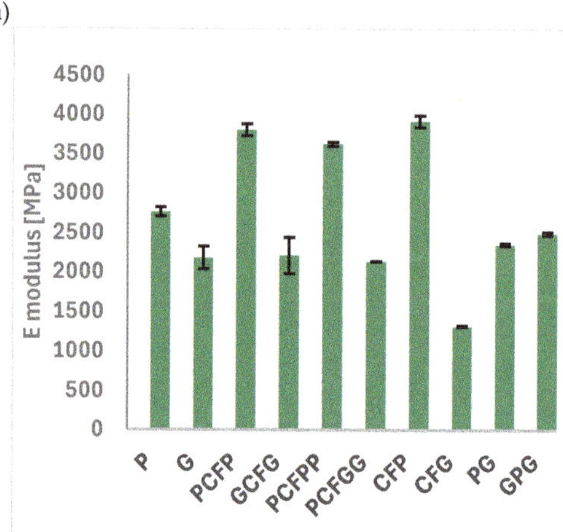

(c)

Figure 16. Tensile testing results: (**a**) Load-elongation curves, (**b**) tensile strength values, and (**c**) E-modulus measurements.

For the composite specimens of PLA and PETG, the highest tensile strength was observed in the GPG specimens, reaching 41.19 MPa. This value is lower than the tensile strength of specimens manufactured with 100% layers of neat PLA and neat PETG. The results suggest that while there are certain benefits to using a layered composite of PLA and PETG, the tensile strength of such composites does not always surpass that of using either material alone. This observation underscores the importance of optimizing the adhesion and intermolecular forces between different materials in composites to fully leverage their individual advantages. By enhancing the bonding quality between PLA and PETG, it is possible to create composites that better utilize the unique strengths of each material.

For the tensile modulus, specimens manufactured with 100% CF–PLA layers exhibited the highest tensile modulus at 3914 MPa. Among the single and multi-graded PLA + CFPLA multi-material composites, the specimens composed of 50% PLA + 50% CF–PLA showed a relatively high tensile modulus of 3805 MPa, outperforming other multi-layered composites. Conversely, the lowest tensile modulus was observed in the specimens manufactured with 100% CF–PETG, with a value of 1318 MPa. Furthermore, the multi-layered composites of PLA and PETG displayed reduced mechanical properties, which can be attributed to inadequate interfacial bonding between the two polymers. This poor adhesion likely stems from the differences in optimal printing temperatures and material characteristics of PLA and PETG, which hinder the formation of a strong, cohesive bond during the AM process.

Optical microscope images of the fractured surfaces of the neat, single, and multi-graded composite material specimens are presented in Figure 17. The fractured surfaces of PLA appeared brittle, as indicated by the clear and sharp fracture edges. The fracture process exhibited deformation and elongation of the PETG, highlighting an increase in toughness (Figure 17d). Moreover, the fracture of the PLA matrix within the CFPLA layers exhibited brittleness, and clear gaps between the PLA substrate and the carbon fibers suggested an inadequate bonding interface (Figure 17d). In CF–PETG, noticeable structural changes around the fibers can be observed, demonstrating how the stiffness of PETG interacts with the gritty texture of the carbon fibers (Figure 17h). The composite specimens manufactured by PLA mixed with PETG exhibited a clear, sharp boundary between the different parts. This difference in behavior between the materials can create some pressure, possibly causing the layers to separate (see Figure 17j).

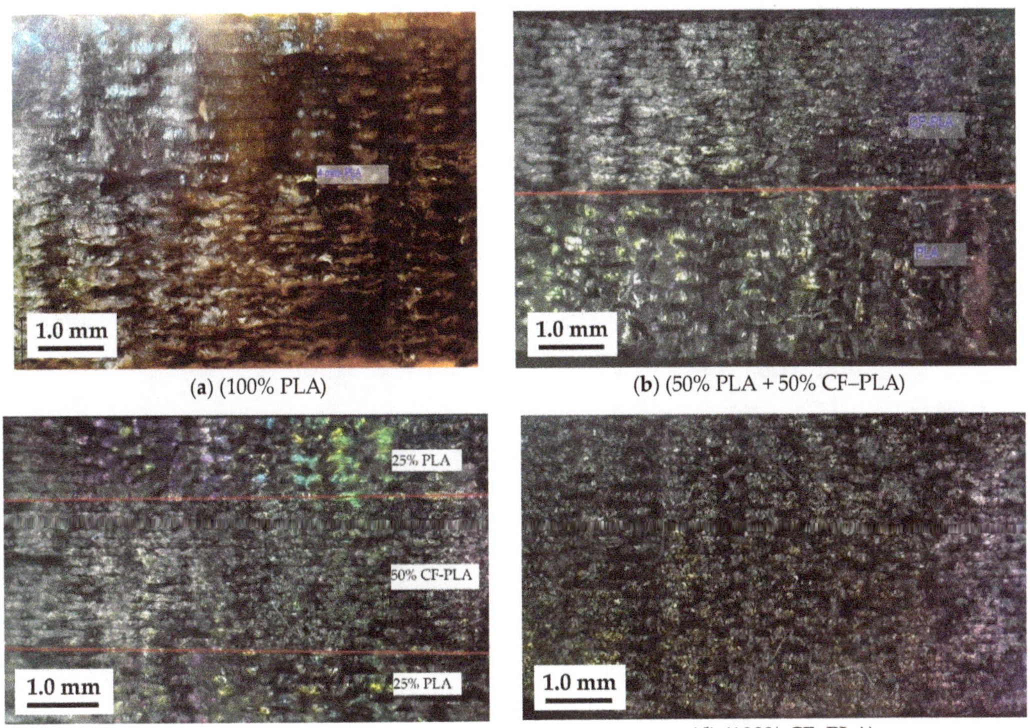

(**a**) (100% PLA)

(**b**) (50% PLA + 50% CF–PLA)

(**c**) (25% PLA + 50% CF–PLA + 25% PLA)

(**d**) (100% CF–PLA)

Figure 17. *Cont.*

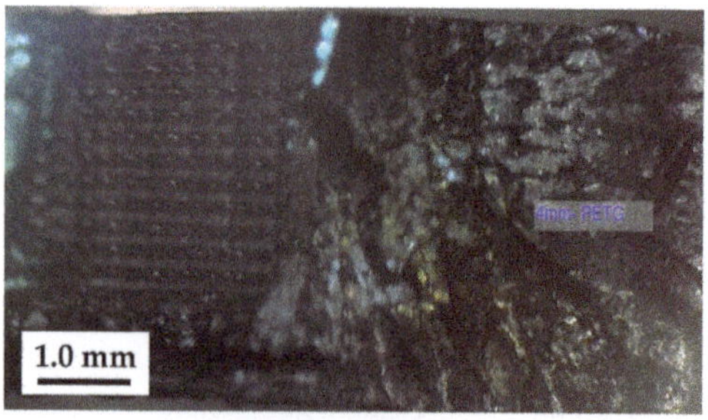

(e) (100% PETG)

(f) (50% PETG + 50% CF–PETG)

(g) (25%PETG + 50%CF–PETG + 25%PETG)

(h) (100% CF–PETG)

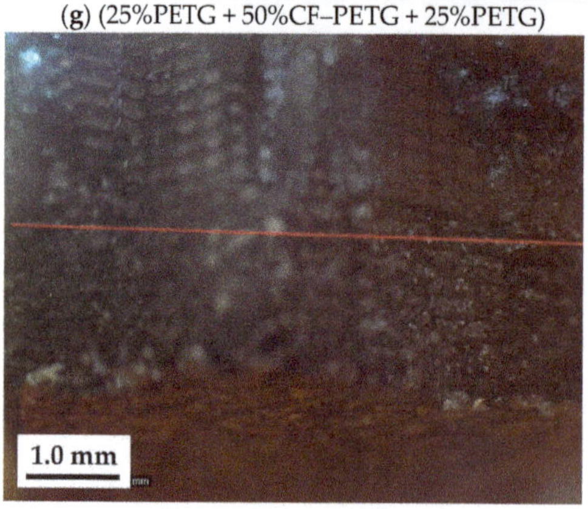

(i) (50% PLA+ 50% PETG)

(j) (25% PETG + 50% PLA+ 25% PETG)

Figure 17. Microscopy images of fractured zones in test specimens: (**a**) P, (**b**) PCFP, (**c**) PCFPP, (**d**) CFP, (**e**) G, (**f**) GCFG, (**g**) GCFGG, (**h**) CFG, (**i**) PG, and (**j**) GPG. The red lines show the border between the two materials in the case of composites of various materials.

The findings of this study have broader implications beyond the specific re-manufactured filaments tested. The results demonstrate that recycled and natural fiber-reinforced filaments can achieve mechanical properties comparable to or even better than original, commercially available filaments, offering sustainable and high-performance alternatives for industrial applications. Furthermore, the reinforcement techniques and multi-material composite designs explored in this research can be applied to other polymer-based materials, providing valuable insights for the additive manufacturing and materials science communities. These findings encourage further exploration into sustainable materials for real-world applications in industries like automotive, aerospace, and construction.

4. Conclusions

This research provides a thorough evaluation of the mechanical properties of various filaments and their additively manufactured equivalents, emphasizing the impact of material composition, reinforcement strategies, and processing techniques. Comprehensive tensile and compression testing revealed critical findings on the mechanical behavior of the specimens. The key conclusions of this study are summarized as follows:

- Among the tested original filaments, CF–PETG achieved the highest maximum stress at 50.14 MPa, outperforming rPLA, PLA, CF–PLA, and PETG.
- Re-manufactured filaments showed mixed results. Both CF–PLA and recycled PLA (rPLA) exhibited improved tensile strengths over their original versions, with rPLA's strength increasing by 29% with hemp fiber reinforcement. Recycled CF–PLA experienced a slight 4% reduction in strength compared to the original. Re-manufactured PETG displayed lower tensile strength than original PETG, but the addition of jute fiber increased the strength of both recycled and original PETG by 15% and 12%, respectively.
- The highest tensile strengths were found in additively manufactured specimens made from rPLA and CF–PLA, with values of 50 MPa and 46.5 MPa, respectively, indicating the reinforcing effect of carbon fiber. The PETG and pure PLA also performed well, although CF–PETG showed lower tensile strength (31 MPa), likely due to insufficient fiber-matrix bonding.
- The rPLA exhibited the highest compressive stress at 81.7 MPa, demonstrating strong resistance to compressive loads. The CF–PLA showed a 61% higher compressive stress compared to CF–PETG, while both PETG and CF–PETG displayed lower resistance to compressive deformation.
- Specimens made entirely of CF–PLA showed a tensile modulus of 3914 MPa, while a 50% PLA + 50% CF–PLA multi-material composite demonstrated effective reinforcement with a tensile modulus of 3805 MPa. However, 100% CF–PETG had the lowest tensile modulus at 1318 MPa, pointing to potential fiber-matrix interaction issues.
- Multi-layered PLA and PETG composites showed reduced mechanical performance, largely due to sharp interfaces between the materials causing stress at the transition points, which can lead to delamination. Ensuring strong adhesion between materials in multi-layered composites is essential to avoid this issue.

Author Contributions: Conceptualization, R.F.F. and Z.S.; methodology, R.F.F., Y.I. and M.N.B.; validation, R.F.F., Y.I. and M.N.B.; formal analysis, M.M.H. and N.D.S.; investigation, R.F.F. and M.M.H.; resources, Z.S. and I.O.; data curation, T.S.G. and N.D.S.; writing—original draft preparation, R.F.F.; writing—review and editing, R.F.F., M.M.H. and I.O.; visualization, R.F.F. and M.M.H.; supervision, Z.S. and I.O.; project administration, R.F.F.; funding acquisition, T.S.G. All authors have read and agreed to the published version of the manuscript.

Funding: This research received no external funding.

Data Availability Statement: The data supporting the reported results are available from the corresponding author upon reasonable request.

Conflicts of Interest: The authors declare no conflicts of interest.

Abbreviations

AM	Additive Manufacturing
MEX	Material Extrusion
PLA	Polylactic Acid
PETG	Polyethylene Terephthalate Glycol
CF	Carbon Fiber
rPLA	Recycled Polylactic Acid
CF-PLA	Carbon Fiber-Reinforced Polylactic Acid
CF-PETG	Carbon Fiber-Reinforced Polyethylene Terephthalate Glycol
re-PLA + Hemp	Re-manufactured PLA with adding 7 wt% hemp fiber
re-rPLA + hemp	Re- manufactured rPLA with adding 7 wt% hemp fiber
re-PETG + Jute	Re-manufactured PETG with adding 7 wt% jute fiber
P	100% PLA
G	100% PETG
PCFP	50% PLA + 50% CF-PLA
GCFG	50% PETG + 50% CF-PETG
PCFPP	25% PLA + 50% CF-PLA + 25% PLA
GCFGG	25% PETG + 50% CF-PETG + 25% PETG
CFP	100% CF-PLA
CFG	100% CF-PETG
PG	50% PLA+ 50% PETG
GPG	25% PETG + 50% PLA+ 25% PETG

References

1. Pajonk, A.; Prieto, A.; Blum, U.; Knaack, U. Multi-Material Additive Manufacturing in Architecture and Construction: A Review. *J. Build. Eng.* **2022**, *45*, 103603. [CrossRef]
2. Rudnik, M.; Hanon, M.M.; Szot, W.; Beck, K.; Gogolewski, D.; Zmarzły, P.; Kozior, T. Tribological Properties of Medical Material (MED610) Used in 3D Printing PJM Technology. *Teh. Vjesn.* **2022**, *29*, 1100–1108. [CrossRef]
3. Dizon, J.R.C.; Espera, A.H.; Chen, Q.; Advincula, R.C. Mechanical Characterization of 3D-Printed Polymers. *Addit. Manuf.* **2018**, *20*, 44–67. [CrossRef]
4. Hanon, M.M.; Zsidai, L. Sliding Surface Structure Comparison of 3D Printed Polymers Using FDM and DLP Technologies. *IOP Conf. Ser. Mater. Sci. Eng.* **2020**, *749*, 012015. [CrossRef]
5. Popescu, D.; Zapciu, A.; Amza, C.; Baciu, F.; Marinescu, R. FDM Process Parameters Influence over the Mechanical Properties of Polymer Specimens: A Review. *Polym. Test.* **2018**, *69*, 157–166. [CrossRef]
6. Dobos, J.; Hanon, M.M.; Oldal, I. Effect of Infill Density and Pattern on the Specific Load Capacity of FDM 3D-Printed PLA Multi-Layer Sandwich. *J. Polym. Eng.* **2022**, *42*, 118–128. [CrossRef]
7. Rasheed, R.K.; Mansoor, N.S.; Mohammed, N.H.; Qasim, S.S.B. Subtractive and Additive Technologies in Fixed Dental Restoration: A Systematic Review. *J. Tech.* **2023**, *5*, 162–167. [CrossRef]
8. Fonseca, J.; Ferreira, I.A.; de Moura, M.F.S.F.; Machado, M.; Alves, J.L. Study of the Interlaminar Fracture under Mode I Loading on FFF Printed Parts. *Compos. Struct.* **2019**, *214*, 316–324. [CrossRef]
9. Taylor, G.; Anandan, S.; Murphy, D.; Leu, M.; Chandrashekhara, K. Fracture Toughness of Additively Manufactured ULTEM 1010. *Virtual Phys. Prototyp.* **2019**, *14*, 277–283. [CrossRef]
10. Rashid, F.L.; Hashim, A.; Dulaimi, A.; Hadi, A.; Ibrahim, H.; Al-Obaidi, M.A.; Ameen, A. Enhancement of Polyacrylic Acid/Silicon Carbide Nanocomposites' Optical Properties for Potential Application in Renewable Energy. *J. Compos. Sci.* **2024**, *8*, 123. [CrossRef]
11. Sweeney, C.B.; Lackey, B.A.; Pospisil, M.J.; Achee, T.C.; Hicks, V.K.; Moran, A.G.; Teipel, B.R.; Saed, M.A.; Green, M.J. Welding of 3D-Printed Carbon Nanotube–Polymer Composites by Locally Induced Microwave Heating. *Sci. Adv.* **2017**, *3*, e1700262. [CrossRef] [PubMed]
12. Fatima Faidallah, R.; Hanon, M.M.; Szakál, Z.; Oldal, I. Study of the Mechanical Characteristics of Sandwich Structures FDM 3D-Printed. *Acta Polytech. Hung.* **2023**, *20*, 7–26. [CrossRef]
13. Delgado, G.F.; Pinho, A.C.; Piedade, A.P. 3D Printing for Cartilage Replacement: A Preliminary Study to Explore New Polymers. *Polymers* **2022**, *14*, 1044. [CrossRef] [PubMed]
14. Ansari, M.Q.; Redmann, A.; Osswald, T.A.; Bortner, M.J.; Baird, D.G. Application of Thermotropic Liquid Crystalline Polymer Reinforced Acrylonitrile Butadiene Styrene in Fused Filament Fabrication. *Addit. Manuf.* **2019**, *29*, 100813. [CrossRef]
15. Ferreira, I.; Vale, D.; Machado, M.; Lino, J. Additive Manufacturing of Polyethylene Terephthalate Glycol /Carbon Fiber Composites: An Experimental Study from Filament to Printed Parts. *Proc. Inst. Mech. Eng. Part L J. Mater. Des. Appl.* **2019**, *233*, 1866–1878. [CrossRef]

16. Santo, J.; Penumakala, P.K.; Adusumalli, R.B. Mechanical and Electrical Properties of Three-Dimensional Printed Polylactic Acid–Graphene–Carbon Nanofiber Composites. *Polym. Compos.* **2021**, *42*, 3231–3242. [CrossRef]
17. Vidakis, N.; Petousis, M.; Grammatikos, S.; Papadakis, V.; Korlos, A.; Mountakis, N. High Performance Polycarbonate Nanocomposites Mechanically Boosted with Titanium Carbide in Material Extrusion Additive Manufacturing. *Nanomaterials* **2022**, *12*, 1068. [CrossRef]
18. Cole, R.G.; Fayazbakhsh, K.; Avalos, A.; Nadeau, N.A. Improved Test Methods for Polymer Additive Manufacturing Interlayer Weld Strength and Filament Mechanical Properties. *ASTM Spec. Tech. Publ. STP* **2022**, *1637*, 325–338. [CrossRef]
19. Rahimizadeh, A.; Kalman, J.; Fayazbakhsh, K.; Lessard, L. Recycling of Fiberglass Wind Turbine Blades into Reinforced Filaments for Use in Additive Manufacturing. *Compos. Part B Eng.* **2019**, *175*, 107101. [CrossRef]
20. Liu, W.; Huang, H.; Zhu, L.; Liu, Z. Integrating Carbon Fiber Reclamation and Additive Manufacturing for Recycling CFRP Waste. *Compos. Part B Eng.* **2021**, *215*, 108808. [CrossRef]
21. Pierson, H.A.; Celik, E.; Abbott, A.; De Jarnette, H.; Sierra Gutierrez, L.; Johnson, K.; Koerner, H.; Baur, J.W. Mechanical Properties of Printed Epoxy-Carbon Fiber Composites. *Exp. Mech.* **2019**, *59*, 843–857. [CrossRef]
22. Ferreira, R.T.L.; Amatte, I.C.; Dutra, T.A.; Bürger, D. Experimental Characterization and Micrography of 3D Printed PLA and PLA Reinforced with Short Carbon Fibers. *Compos. Part B Eng.* **2017**, *124*, 88–100. [CrossRef]
23. Hanon, M.M.; Marczis, R.; Zsidai, L. Anisotropy Evaluation of Different Raster Directions, Spatial Orientations, and Fill Percentage of 3D Printed PETG Tensile Test Specimens. *Key Eng. Mater.* **2019**, *821*, 167–173. [CrossRef]
24. Kichloo, A.F.; Raina, A.; Haq, M.I.U.; Wani, M.S. Impact of Carbon Fiber Reinforcement on Mechanical and Tribological Behavior of 3D-Printed Polyethylene Terephthalate Glycol Polymer Composites—An Experimental Investigation. *J. Mater. Eng. Perform.* **2022**, *31*, 1021–1038. [CrossRef]
25. Singh, A.K.; Saltonstall, B.; Patil, B.; Hoffmann, N.; Doddamani, M.; Gupta, N. Additive Manufacturing of Syntactic Foams: Part 2: Specimen Printing and Mechanical Property Characterization. *JOM* **2018**, *70*, 310–314. [CrossRef]
26. Mansour, M.; Tsongas, K.; Tzetzis, D.; Antoniadis, A. Mechanical and Dynamic Behavior of Fused Filament Fabrication 3D Printed Polyethylene Terephthalate Glycol Reinforced with Carbon Fibers. *Polym.-Plast. Technol. Eng.* **2018**, *57*, 1715–1725. [CrossRef]
27. Hsueh, M.-H.; Lai, C.-J.; Wang, S.-H.; Zeng, Y.-S.; Hsieh, C.-H.; Pan, C.-Y.; Huang, W.-C. Effect of Printing Parameters on the Thermal and Mechanical Properties of 3D-Printed PLA and PETG, Using Fused Deposition Modeling. *Polymers* **2021**, *13*, 1758. [CrossRef] [PubMed]
28. Niu, C.; Luan, C.; Shen, H.; Song, X.; Fu, J.; Zhang, L.; Sun, Y.; Xu, G.; Ruan, Z. Tunable Soft–Stiff Hybridized Fiber-Reinforced Thermoplastic Composites Using Controllable Multimaterial Additive Manufacturing Technology. *Addit. Manuf.* **2022**, *55*, 102836. [CrossRef]
29. Faidallah, R.F.; Hanon, M.M.; Vashist, V.; Habib, A.; Szakál, Z.; Oldal, I. Effect of Different Standard Geometry Shapes on the Tensile Properties of 3D-Printed Polymer. *Polymers* **2023**, *15*, 3029. [CrossRef] [PubMed]
30. Salcedo, E.; Baek, D.; Berndt, A.; Ryu, J.E. Simulation and Validation of Three Dimension Functionally Graded Materials by Material Jetting. *Addit. Manuf.* **2018**, *22*, 351–359. [CrossRef]
31. Baca, D.; Ahmad, R. The Impact on the Mechanical Properties of Multi-Material Polymers Fabricated with a Single Mixing Nozzle and Multi-Nozzle Systems via Fused Deposition Modeling. *Int. J. Adv. Manuf. Technol.* **2020**, *106*, 4509–4520. [CrossRef]
32. Manaia, J.P.; Manaia, A.T.; Rodriges, L. Industrial Hemp Fibers: An Overview. *Fibers* **2019**, *7*, 106. [CrossRef]
33. Gupta, A.; Hasanov, S.; Fidan, I. Processing and Characterization of 3D-Printed Polymer Matrix Composites Reinforced with Discontinuous Fibers. In Proceedings of the 2019 International Solid Freeform Fabrication Symposium, Austin, TX, USA, 12–14 August 2019.
34. Zhang, J.; Chevali, V.S.; Wang, H.; Wang, C.H. Current Status of Carbon Fibre and Carbon Fibre Composites Recycling. *Compos. Part B Eng.* **2020**, *193*, 108053. [CrossRef]
35. *ISO 527-2:2012*; Plastics—Determination of Tensile Properties—Part 2: Test Conditions for Moulding and Extrusion Plastics. European Committee for Standardisation: Brussels, Belgium, 2012.
36. *ASTM D695-23*; Standard Test Method for Compressive Properties of Rigid Plastics. American Society for Testing and Materials: West Conshohocken, PA, USA, 2023. [CrossRef]
37. Palaniyappan, S.; Annamalai, G.; kumar Sivakumar, N.; Muthu, P. Development of Functional Gradient Multi-Material Composites Using Poly Lactic Acid and Walnut Shell Reinforced Poly Lactic Acid Filaments by Fused Filament Fabrication Technology. *J. Build. Eng.* **2023**, *65*, 105746. [CrossRef]
38. Yermurat, B.; Seçgin, Ö.; Taşdemir, V. Multi-Material Additive Manufacturing: Investigation of the Combined Use of ABS and PLA in the Same Structure. *Mater. Test.* **2023**, *65*, 1119–1126. [CrossRef]
39. Amza, C.G.; Zapciu, A.; Constantin, G.; Baciu, F.; Vasile, M.I. Enhancing Mechanical Properties of Polymer 3D Printed Parts. *Polymers* **2021**, *13*, 562. [CrossRef]

processes

MDPI

Article

Numerical Analysis of Optimising Liquid Desiccant Dehumidification for Sustainable Building Cooling: A Data-Driven Method Using Response Surface Methodology

Mohammed Azeez Hilal [1,*] and Saleem Jasim Abbas [2]

[1] Technical Institute of Baqubah, Middle Technical University, Baqubah 32001, Iraq
[2] Mechanical Power Technical Engineering Department, College of Engineering and Technologies, Al-Mustaqbal University, Babylon 51001, Iraq; saleem.jasim.abbas@uomus.edu.iq
* Correspondence: mohammed_azeez_hilal@mtu.edu.iq

Abstract: Leveraging data-driven methods such as Response Surface Methodology (RSM) has considerable potential for sustainable building cooling via mitigating energy consumption and environmental impacts. This research focuses on using the RSM to improve liquid desiccant dehumidification for sustainable building cooling performance using a D-optimal design. Specifically, the research intends to investigate the actual influence of the inlet air conditions and desiccant concentration on the performance of liquid desiccant dehumidification systems, i.e., the moisture removal rates and dehumidifier efficiency. To systematically conduct this research, a set of experimental data gathered from the open literature is utilised. This includes a specific set of inlet parameters of air temperature (27–34.5 °C), ratio of air humidity (20.5–25 g/kg), and solution temperature (27.5–38.5 °C) as the independent variables. Also, the feedback variables include the moisture removal rates (MRR) and efficacy (ϵ). The associated results of the analysis of variation indicate that the ratio of air humidity has the greatest influence on the moisture removal rate. However, the solution temperature and the ratio of air humidity have the most influence on efficacy. In the event of response optimisation, the result at MRR and (ϵ) are 0.54 g/s and 0.50, respectively, with a minimum desirability of 0.992 and 1.

Keywords: desiccant; dehumidification; RSM; D-optimal; desirability

1. Introduction

Liquid desiccant dehumidifiers are effective in removing moisture from the air by using a liquid solution to absorb water vapour. The liquid desiccant dehumidifier's performance of desiccant used can be determined via allocating the flow rate of the air and solution, and the temperature and humidity of the incoming air [1–3]. One advantage of liquid desiccant dehumidifiers is that they can operate at lower temperatures and humidity levels than other dehumidification technologies. They are also more energy-efficient than refrigerant-based dehumidifiers, which require a lot of energy to cool the air and remove moisture [4]. However, liquid desiccant dehumidifiers require a source of heat to regenerate the desiccant solution and release the captured moisture [5]. This heat source can be provided by a waste heat recovery system, solar collectors, or other sources of renewable energy. Liquid desiccant dehumidifiers can also be more complex and require more maintenance than other dehumidification technologies, as the liquid desiccant solution needs to be monitored and periodically replaced. Nevertheless, they are a good option for applications where low temperature and humidity levels are required, such as in the food and pharmaceutical industries, as well as for indoor air quality control in commercial and residential buildings.

Generally, the standard symmetrical designs cannot be used when the experimental domain is irregularly in form, or when the number of experiments appointed by a classical design is large. To sort out this issue, the principal notion of D-optimality can be used to select the proper design [6]. In hot and humid climates, the humid air is dehumidified with a strong liquid or solid dry, and the warm-dry air is cooled by evaporating water into it, which results in an acceptable final humidity and temperature. Dehumidification is a process that occurs when humid air interacts with a desiccant solution. The efficacy (ϵ) of the dehumidification process is measured using two indicators: the moisture removal rate (MRR) and efficacy of (ϵ) [7]. Several systems of liquid desiccant dehumidification for sustainable building cooling were constructed and improved during the last 30 years to improve the efficiency of classical air condition systems. According to Jain et al. [8], on account of the pressure vapour between the desiccant solution and air, the liquid desiccants absorb moisture from the air. The liquid desiccants available are used at slight temperatures, roughly between 50 °C and 80 °C [9]. Thus, the heat sources with a low temperature could be used to power the regeneration process at roughly 70 °C via using several techniques such as solar energy, waste heat, and geothermal power.

The aim of the current research is to study the effects of inlet air temperature (27–34.5 °C) and humidity (20.5–25 g/kg) on the moisture removal rate (MRR) and dehumidifier efficiency (ε), as well as to assess how different desiccant amounts impact energy efficiency and environmental performance. To systematically conduct this aim, the D-optimal method is used to reduce the number of experiments, besides developing a robust mathematical model using RSM to relate input variables (air temperature, humidity, solution temperature) to output responses (MRR and ε).

The outline of this article is as follows:

Section 2 introduces a review of the existing literature on dehumidifier performance, and categorizes modelling strategies into theoretical analysis, and experimental studies. In Section 3, the definition of Response Surface Methodology and a discussion of its role in optimizing system performance through systematic input testing are detailed. In Section 4, an analysis of inlet parameters affecting performance metrics together with assessing the model accuracy with key statistical indicators is provided. Section 5 focuses on introducing a mathematical model developed from experimental data to predict performance under variable conditions. In Section 6, an explanation of regression analysis techniques and deriving accurate formulas for variable relationships is provided. In Section 7, D-optimal design implementation for response prediction including the coding of variables at three levels is ascertained. Section 8 provides an overview of components and subsystems involved in dehumidification and regeneration processes, while Section 9 provides a description of using the Design-Expert's optimization tool to enhance system performance via a desirability function. Section 10 presents experimental findings, detailing operating conditions and D-optimal optimization results. Section 11 explores the complexities of interactions between operational parameters while highlighting the significance of optimizing liquid desiccant systems. Lastly, Section 12 summarizes the key findings on temperature and humidity effects, model accuracy, and optimal conditions in addition to emphasizing implications for energy savings and sustainable cooling policies.

2. Literature Review

The performance of the dehumidifier and regenerator was experimentally and theoretically studied by a number of researchers. Modelling strategies for guess dehumidifier and regenerator performance can be grouped into three categories based on past research: models for theoretical numerical analysis, experimental analysis, and artificial intelligence.

The researchers were commonly focused on estimating the overall performance of humidity, besides evaluating the desiccant solution inlet rate of flow, temperature, and concentration [10–12]. Moon et al. [13] demonstrated the new mass transfer performance information of the flow liquid desiccant dehumidifying system by developing new statistical correlations on dehumidifying efficacy.

Sanjeev et al. [14] conducted a set of experiments to investigate the efficiency of liquid desiccant dehumidification for a sustainable building cooling method under various operation parameters. The dehumidifier and regenerator with a cooling tower are the three essential components of the system. As desiccants in the system, lithium chloride and calcium chloride were employed. The results specified that the dehumidifier's efficiency can be ranged between (0.25 and 0.44), while the regenerator's efficiency reached between (0.07 and 0.8).

Liu et al. [15] evaluated the operation of the adiabatic dehumidifier units that were internally cooled. The internally cooled dehumidifier was established to perform better in mass transfer than the adiabatic unit. By utilising mathematical analysis of variance, the consequence of intake operation elements on the dehumidifier's condensation rate and the regenerator's water evaporation was investigated. The results indicated that the ratio of humidity of intake air, the mass flow rate of the intake air, and the temperature of the inlet solution have a straight effect on the water condensation rate with a significance ($p < 0.01$). Furthermore, the rate of water evaporation in the regeneration can also be affected by the intake air mass flow rate and the inlet solution temperature ($p < 0.01$). The mass transfer of aqueous solutions comprising lithium chloride and lithium bromide was also assessed. The results exposed that the (LiCL) can perform better than the (LiBr) in terms of dehumidification at the same rate of desiccant flow.

Mc-Donald et al. [16] and Sameer et al. [17,18] utilised the regression analysis to quantify data generated from packing the tower desiccant dehumidifier. Precisely, the regression formula was used to allocate the temperature of outlet air and the ratio outlet of humidity desiccant temperature and concentration.

Although previous studies have determined the impact of distinct elements on the performance of Liquid Desiccant Dehumidification, insufficient studies have evaluated the collective impacts of these characteristics quantitatively using experimental data obtained from various sources. Previous evaluations have predominantly concentrated on theoretical models or have been restricted to particular case studies, which may not be entirely reflective of wider operational circumstances. The main purpose of the present investigation is to appraise how the input parameters can impact the critical outlet air and desiccant qualities of Liquid Desiccant Dehumidification. The D-optimal design is engaged to improve the overall performance of liquid desiccant dehumidification for sustainable building cooling while assessing the relevance of independent factors on the moisture removal rate (MRR) and dehumidifier efficacy.

3. Response Surface Methodology (RSM)

The Response Surface RSM is a statistical and mathematical practice that finds the relationship between input factors and output responses to maximise the performance of a system or process Hern et al. [19]. It is regularly used in engineering, chemistry, and other fields where it is imperative to optimise a system's performance while reducing cost or obtaining the best out of efficiency. RSM demands planning a series of tests with numerous input variable combinations and calculating the suitable output answers. By analysing the data from these experiments, RSM can help in defining the best combination of input variables to produce the intended output response. RSM typically includes fitting a mathematical model to the experimental data, which can be used to guess the response for any combination of input variables considering the range of the experiments. This model can then be used to recognize the optimal combination of input variables that will result in the desired output response. RSM is utilised for both linear and nonlinear systems and can handle interactions between input variables. It can also be utilised to signify the relative importance of each input variable on the output response and can be utilised to optimise the performance of a liquid desiccant dehumidifier by recognizing the optimal values for the input variables that maximise the output variables. It can also help to reduce the number of experiments required to optimise the process, which can save time and resources. Overall, RSM is an influential tool for optimising systems and processes and therefore can help save time and resources by determining the optimal combination of input variables without the need for in-depth testing. It is predominantly useful for systems with multiple variables that affect performance.

4. Evaluation of RSM Model

Additionally, a thorough analysis that systematically examines the effects of multiple inlet parameters, namely, air temperature, humidity ratio, and solution temperature, on critical performance metrics like moisture removal rates (MRR) and dehumidifier efficiency (ϵ), is often lacking in previous studies that have examined various aspects of liquid desiccant systems. To assess the accuracy of the proposed RSM model, several statistical indicators were employed to compare predicted and actual values. Key indicators considered include Mean Square Error (MSE), coefficient of determination (R^2), and absolute relative error Khoshraftar et al. [20]. The calculation of MSE value is defined by Wu et al. [21]:

The MSE is calculated as follows:

$$\text{MSE} = \frac{1}{\text{N}}\sum_{i=1}^{\text{N}}(\text{P}_{\text{mea}} - \text{P}_{\text{for}})^2$$

P_{mea} represents the measured power and P_{for} stands for the forecasted or predicted power. The R^2 is calculated as follows:

$$\text{R}^2 = 1 - \frac{\sum_{i=1}^{m}(\text{P}_{\text{mea}} - \text{P}_{\text{for}})^2}{\sum_{i=1}^{m} P_{mea}^2 - \frac{\sum_{i=1}^{m} P_{for}^2}{m}}$$

5. Mathematical Model

By utilising the data from the experiments to figure out a mathematical model that clarifies the relationship between the variables and the performance of the dehumidifier, this model can be a simple linear relationship or a more complex polynomial equation. Utilising the mathematical model to guess the efficiency of the dehumidifier in different conditions can be achieved by inputting the values of the variables into the model and calculating the predicted performance. Adjusting the accuracy of the model is important for comparing the estimated performance to actual performance under the same conditions. The performance of the dehumidifier is predicted mathematically via computing the moisture dissipation average and its performance. Nelson and Goswami [22] and Oberg and Goswami [23] inspected the performance of dehumidifying with respect to the moisture removal rate. The moisture dissipation rate and dehumidifier effectiveness are mathematically estimated to determine its performance. The water carried from the air to the liquid desiccant was defined by Yin et al. [24], as represented below:

$$\Delta m_{con} = m_a \times (w_{in} - w_{out}) \tag{1}$$

m_a represents the rate of mass flow on the dehumidifier inlet, w_{in} the ratio of an inlet of air humidity, and w_{out} the ratio of outlet air humidity. The efficacy of the (dehumidifier/regenerator) is measured as the ratio of the change in actual humidity in the air to the greatest potential change in humidity feasible according to the following operating parameters:

$$\varepsilon = {w_{in} - w_{out}}/{w_{in} - w_{out-eq}} \tag{2}$$

ω_{out-eq} denotes the ratio of outlet air humidity in balance with the desiccant solution with the desiccant solution under the following operational conditions of Kinsara et al. [25]:

$$w_{out-eq} = 0.62185 \times {P_{wz}}/{(P - P_{wz})} \tag{3}$$

P signifies the total pressure of air on the solution and $P_{\omega z}$ is the partial pressure of water vapour in the solution.

The rate of moisture removal refers to the quantity of moisture extracted from the air per time unit, typically measured in grams per hour (g/h). This parameter is influenced by various factors such as the volume of air undergoing treatment, the moisture content of the incoming and outgoing air, and the duration of the process.

The moisture removal rate is commonly determined using the formula: Moisture Removal Rate = Q (W_{in} − W_{out}), where

- Q represents the volumetric flow rate of air (e.g., in cubic meters per hour);
- W_{in} indicates the moisture content of the incoming air (in grams of water per cubic meter);
- W_{out} signifies the moisture content of the outgoing air (in grams of water per cubic meter).

6. Regression Analysis

Regression analysis is a statistical approach that is used to analyse the association between one or more dependent variables (response) and independent variables (predictors) in order to predict the value of the response variable based on the values of the predictor variables. Finding the mathematical formula that most accurately captures the relationship between the variables is the aim of regression analysis. Three types of regression analysis are distinguished: polynomial, logistic, and linear regression. The most popular type of regression analysis is linear regression, which seeks to establish a linear relationship between predictor variables and response variables. In simple linear regression, there is only one predictor variable, while in multiple linear regression, there are multiple predictor variables. Logistic regression is used when the response variables is binary, which is significant as it only has two possible outcomes. Polynomial regression is used when the relationship between the predictor variables and the response variables is not a linear formula.

Regression analysis involves finding a mathematical correlation that best represents the association between predictor variables via a response variable. The equation depends on the type of regression analysis being used.

6.1. Simple Linear Regression

The simple linear regression correlation is:

$$Y = \beta_0 + \beta_x$$

Y is the independent variable (response), x is the independent variable (predictor), and β_x is the slope coefficient. β_x signifies the change in Y for a unit change in x, and the intercept (β_0) represents Y value when x is zero.

6.2. Multiple Linear Regression

The following correlation signifies the multiple linear regression with β_x predictor variables

$$Y = \beta_0 + \sum \beta_i x_i + \sum \beta_{ii} x_i^2 + \sum\sum \beta_{ij} x_i x_j$$

β_0 is the intercept or constant term. β_1, β_2, . . ., β_i are the slope coefficients; the slope coefficients represent the change in Y on a unit change in the corresponding predictor variable while keeping the other predictor variables constant.

6.3. Logistic Regression

The logistic regression equation is expressed in the following correlation

$$p = 1/(1 + exp\ (-(\beta_0 + \beta_1 x_1 + \beta_2 x_2 + \ldots + \beta_i x_i)))$$

p is the probability of the dependent variable (response) being in a certain category (usually 1 or 0), and β_1, β_2, . . ., β_i are the slope coefficients. The slope coefficients signify the change in the log odds of the dependent variable for a unit change in the corresponding predictor variable, holding all other predictor variables constant.

7. Statistical Analysis and Experimental Design (D-Optimal)

Every variable that required optimisation was coded at three different levels.: −1, 0, and +1. For predicting individual Y variables, a quadratic polynomial regression model was adopted. The following model is projected for each response of Y,

$$Y = \beta_0 + \Sigma\beta_i\, xi + \Sigma\beta_{ii}\, x^2 + \Sigma\Sigma\beta_{ij}\, x_i\, x_j \quad (4)$$

In this research, the "Mini-tab INC., version 14, Pa. USA", a statistical software package data, is used. This software is widely used for data analysis for multiple regression analysis (ANOVA) and analysis of ridge maximum of data in the response surface regression (RSREG) procedure [23]. The coefficient of determination R^2 is used to assess the model's quality of fit and its statistical significance is determined using the F-test. The reaction of the moisture removal rate and the performance of the dehumidifier are studied using a three-factor D-optimal design. The primary response criteria and their values are chosen after an initial screening stage. The variables X_1, X_2, and X_3 respectively stood for temperature of inlet air (°C), and the inlet ratio of air humidity (g/kg), with the inlet solution temperature (°C). The independent variables are included in Table 1, which includes the intake air temperatures of 27 and 34 °C, inlet air humidity ratios of 20 and 25, and inlet solution temperatures of 27 and 38 °C.

Table 1. D-Optimal design requires independent variables and their levels.

Independent Variables		Variables Levels		
		−1	0	+1
Temperature of Air (°C)	X_1	27	30	34
Humidity Ratio of Air (g/kg)	X_2	20	22	25
Solution Temperature (°C)	X_3	27	33	38

Moisture analysis plays an important role in evaluating and optimising the liquid desiccant dehumidification processes for sustainable building cooling. This analysis involves estimating the amount of moisture present in the desiccant that can affect the performance of the process. The efficiency of a liquid desiccant is determined by determining the volumetric flow rate of the liquid. This is specifically conducted using liquid flow meters, which work by sensing the flow and generating a signal proportional to the flow rate. There are several types of liquid flow meters, such as the following:

1. Differential pressure flow meters: Measure the pressure difference across a primary device such as an orifice plate or Venturi tube. This is suitable for dirty liquids.
2. Positive displacement flow meters: Contain moving components that recurrently fill and displace a known volume of liquid. Highly precise and suitable for highly viscous liquids.
3. Velocity flow meters: Determine the velocity of the liquid and convert it to a flow rate. These include turbine, magnetic, and electrical.
4. Mass flow meters: Determine the mass flow rate rather than the volumetric flow rate.

There are other ways to improve the efficiency of the liquid dryer, including the following:

- Enhancing the design: The performance of the liquid dryer can be enhanced by better designing the unit. The design should be optimised to attain an even distribution of the liquid and reduce losses.
- Using advanced materials: Use advanced materials that enhance heat transfer and reduce losses.
- Monitoring and regulating the process: The dryer process should be continually monitored and attuned according to the ambient conditions. Sensors can be utilised to measure temperature, pressure, and fluid flow.

- Enhancing thermal insulation: The performance of the liquid dryer can be enhanced by using high-quality thermal insulators to preserve the compulsory temperature inside the unit.
- Enhancing the control system: Use an advanced control system that can better alter the dryer process according to the ambient conditions.

The rate of moisture removal and efficacy of a liquid desiccant dehumidifier are investigated using a three-factor response surface methodology (RSM). Table 2 shows various responses at various experimental combinations for coding. For several experimental combinations, such as 0.259 to 0.577 for MRR and 0.318 to 0.532 for ε, there was a considerable range in all replies. The maximum and minimum values of these parameters are attained as a result of the test being pre-treated on experimental combinations (run order) of 5, 9, 12, and 17.

Table 2. D-Optimisation of optimal design of MRR and efficiency.

<table>
<tr><th rowspan="2">No. of Run</th><th colspan="3">Independent Variable Levels</th><th colspan="2">Responses</th></tr>
<tr><th>T_a (°C)</th><th>W_a (g/kg)</th><th>T_s (°C)</th><th>MRR (g s^{-1})</th><th>ε (%)</th></tr>
<tr><td>1</td><td>27.00</td><td>20.50</td><td rowspan="3">38.50</td><td>0.451</td><td>0.457</td></tr>
<tr><td>2</td><td rowspan="2">34.50</td><td>25.00</td><td>0.342</td><td>0.483</td></tr>
<tr><td>3</td><td>20.50</td><td>0.337</td><td>0.424</td></tr>
<tr><td>4</td><td rowspan="3">27.0</td><td rowspan="2">25</td><td>33.00</td><td>0.543</td><td>0.514</td></tr>
<tr><td>5</td><td>38.50</td><td>0.541</td><td>0.532</td></tr>
<tr><td>6</td><td rowspan="2">22.75</td><td>27.50</td><td>0.491</td><td>0.452</td></tr>
<tr><td>7</td><td>34.50</td><td>33.00</td><td>0.390</td><td>0.417</td></tr>
<tr><td>8</td><td rowspan="2">27.00</td><td rowspan="2">20.50
25.00</td><td rowspan="5">27.50</td><td>0.474</td><td>0.418</td></tr>
<tr><td>9</td><td>0.577</td><td>0.462</td></tr>
<tr><td>10</td><td rowspan="3">34.5</td><td>20.50</td><td>0.359</td><td>0.322</td></tr>
<tr><td>11</td><td>25.00</td><td>0.486</td><td>0.406</td></tr>
<tr><td>12</td><td>25.50</td><td>0.399</td><td>0.318</td></tr>
<tr><td>13</td><td>30.75</td><td>23.88</td><td rowspan="3">38.50</td><td>0.388</td><td>0.416</td></tr>
<tr><td>14</td><td>34.50</td><td>25.00</td><td>0.352</td><td>0.495</td></tr>
<tr><td>15</td><td>27.00</td><td rowspan="3">20.5</td><td>0.355</td><td>0.432</td></tr>
<tr><td>16</td><td>30.75</td><td>33.00</td><td>0.388</td><td>0.436</td></tr>
<tr><td>17</td><td>34.50</td><td rowspan="2">38.50</td><td>0.259</td><td>0.430</td></tr>
<tr><td>18</td><td>30.75</td><td>22.75</td><td>0.318</td><td>0.431</td></tr>
</table>

8. Explanation of the Operating System

The system and measured data are shown in Figure 1. Vapour compression (VCS), regeneration, and dehumidification are the three subsystems that make up the system. Before the solution is put into the dehumidifier and regenerator, it is first dehumidified and then regenerated using the VCS. Every subsystem for dehumidification and regeneration has installed a centrifugal fan, pump, valves, filter, eliminator, and flow meter.

The two pumps are utilised to move the solution back and forth between the regenerator and the condenser and the dehumidifier and evaporator, respectively. The ambient air at the bottom of the regenerator and the dehumidifier is moved around by two centrifugal fans.

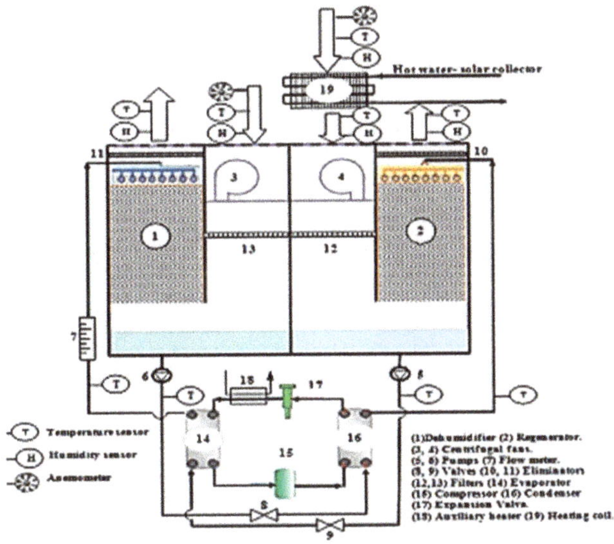

Figure 1. The diagram system of the liquid desiccant.

With an accuracy of ±0.1 °C, nine thermocouples were fitted to measure the temperature of the air and solution. The relative humidity at the input and outflow air streams was measured using five TR-RH2W humidity sensors, which have an accuracy of ±3% RH. To gather and send the recorded data to the PC system, each of these sensors is attached to a data acquisition logger. A flow meter was added to regulate the mass flow rate of the desiccant solution, with an accuracy of 2 L/h. Within the parameter ranges shown in Table 3, experimental experiments were conducted to assess the liquid adsorption dehumidification unit's performance.

Table 3. Characteristic measured data representing the monitoring period.

Air flow rate through the dehumidifier	0.125 kg/s
Air flow rate through the regenerator	0.125 kg/s
Dehumidifier/regenerator inlet desiccant flow	0.1 kg/s
Solar hot water flow rate through heating coil	0.25 kg/s
Dehumidifier air inlet temperature	27 to 34.5 °C
Regenerator air inlet temperature	39 to 57 °C
Humidity percentage at dehumidifier air inlet.	20.5 to 25 g/kg
Humidity percentage at regenerator air inlet.	18.9 to 21.8 g/kg
Dehumidifier inlet solution temperature.	27.5 to 38.5 °C
Regenerator solution inlet temperature	42 to 73 °C
Solar hot water inlet temperature	40 to 85 °C
Desiccant concentration	36.8%wt.

The simulation estimated the temperature, concentration, and mass flow rate at each state point, from P1 to P6, on the desiccant solution side of the liquid desiccant system. The air flow rate, dry bulb temperature, and humidity ratio for the process and regeneration air before and after the absorber and the regenerator were also calculated to evaluate the energy performance of both the reference and proposed systems.

9. Optimisation

Design-numerical Expert's optimisation is based on a desirability objective function. The overall desirability (D) is the multiplicative mean of all individual desirability (d_i) values between 0 and 1.

$$D = (d_1 \times d_2 \times \ldots .d_n)^{1/n} = \left(\prod_{i=1}^{n} d_i\right)^{1/n} \quad (5)$$

D is the optimal design optimisation of moisture removal rates and effectiveness. n is the number of responses, and the total function becomes 0 if any of the answers go outside of their desirable range.

Design Optimisation of MRR

Design optimisation of moisture removal rate and efficacy can be accomplished using D-optimal design, which is a statistical optimisation technique that reduces the variance of the estimated model parameters while maximising the accuracy of the predicted response. To optimise the moisture removal rates and effectiveness, the first step is to outline the parameters that affect moisture removal, such as temperature, humidity, airflow rate, and the type and size of the moisture removal system. Then, an experimental design matrix is formed based on these factors, and the moisture removal rates and efficacy are restrained for each combination of factors. The experimental data are then analysed using statistical software to fit a model that relates the moisture removal rates and efficiency to the factors. The model can be used to forecast the moisture removal rates and efficiency for any combination of factors. The D-optimal design method is used to recognize the optimal combination of factors that maximise the moisture removal rates and effectiveness. This is conducted by selecting the factor levels that reduce the variance of the estimated model parameters while maximising the exactness of the predicted response.

10. Results and Analysis

10.1. Operating Conditions

To assess the performance of the liquid dehumidifier, experiments are directed using the parameters listed in Table 3. When conducting a D-optimal design optimisation for MRR and efficiency, it is significant to carefully consider the specific parameters and operating conditions that are most relevant to the experiments being conducted. Generally, the operating conditions for D-optimal design optimisation of MRR and efficiency depend on a diversity of factors, and a wide-ranging analysis of these factors is mandatory to select the optimal operating conditions.

10.2. Analysis of Response Surface

Quadratic regression equations are improved to anticipate the moisture removal rate and validation under the designated experimental condition using *RSM*. The regression Equations (6) and (7) of the response properties are developed as a function of three input parameters examined in the experiment:

$$MRR = +0.38 - 0.057 \times X_1 + 0.043 \times X_2 - 0.039 \times X_3 + 0.052 \times X_1{}^2 + 0.028 \times X_2{}^2 - 0.022 \times X_3{}^2 + 0.009419 \times X_1 \times X_2 - 0.00364 \times X_1 \times X_3 - 0.00364 \times X_2 \times X_3 \quad (6)$$

$$\varepsilon = +0.43 - 0.029 \times X_1 + 0.032 \times X_2 + 0.033 \times X_3 + 0.016 \times X_1{}^2 + 0.016 \times X_2{}^2 - 0.027 \times X_3{}^2 + 0.001879 \times X_1 \times X_2 + 0.016 \times X_1 \times X_3 - 0.001883 \times X_2 \times X_3 \quad (7)$$

X_1 represents the air temperature (*Ta*), X_2 represents the air humidity (W_a), and X_3 represents the solution temperature (Ts). It should be noted that (Ta) and (Ts) have a negative effect, while (Wa) has a positive effect on (MRR). The ratio of air humidity to solution temperature has a positive effect and a negative effect on air temperature (ϵ). Based on the associated results of Figure 1 for the predicted and actual values, it can be ascertained that the predicted values are statistically similar to the actual measured values.

In this regard, Figure 2 illustrates that the predicted and actual values are approximately linear; i.e., the errors are evenly distributed.

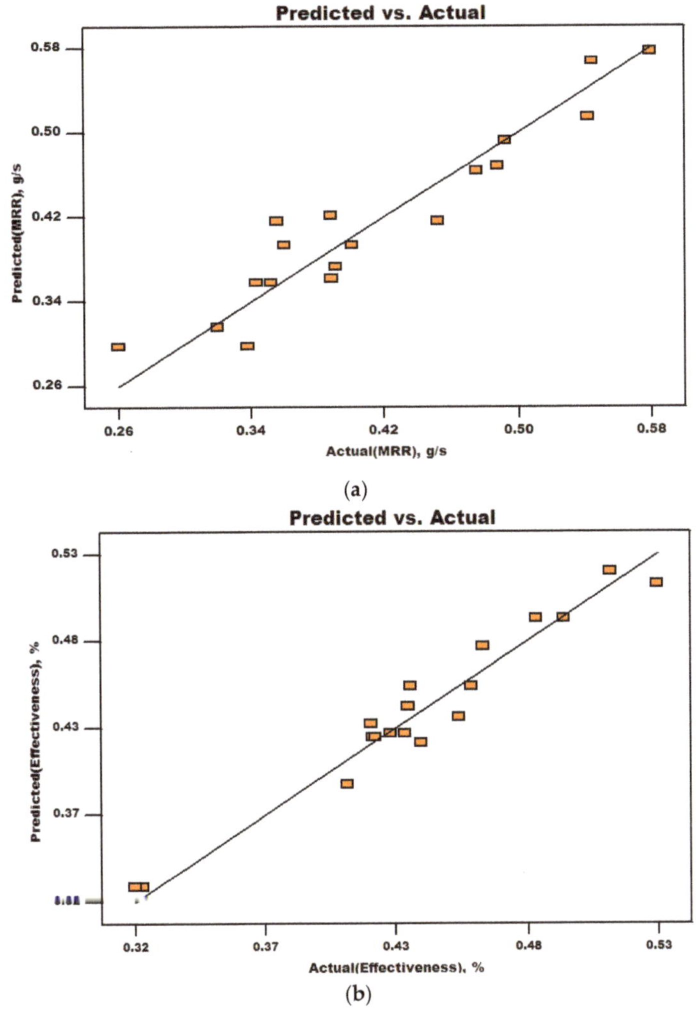

Figure 2. Plot of predicted response Y versus actual response: (**a**) MRR, (**b**) ϵ.

To assess the regression coefficients and statistical indication of type variables, the numerical analysis of variance was performed by fitting the experimental data (Equation (4)). The F-ratio was used to determine the importance of the model components at a probability (P) of 0.01. The adequacy model was assessed via the analysis model, the test for lack of fit, the coefficient of calculated R^2, the sum of the squares of the expected error, and the coefficient of variation (CV). The ANOVA findings for several runs of the experiment are shown in Tables 4 and 5.

Table 4. The anticipated quadratic polynomial model's regression coefficients for the response variables Y MRR and ϵ.

Predictors	Coefficients (β) MRR (g/s)	T	Prob.	Nota.	Coefficients (β) ε (%)	T	Prob.	Nota.
Intercept	+0.38	8.01	0.0037	*	+0.43	17.27	0.0003	*
Linear								
X_1	−0.058	27.58	0.0008	*	−0.029	36.69	0.0003	*
X_2	+0.044	14.39	0.0053	*	+0.032	40.49	0.0002	*
X_3	−0.040	12.07	0.0084	*	+0.033	45.32	0.0001	*
Quadratic								
X_{11}	+0.053	2.84	0.1309	**	+0.017	1.36	0.2796	**
X_{22}	+0.029	1.00	0.3478	**	+0.017	1.62	0.2405	**
X_{33}	−0.023	0.60	0.4595	**	−0.028	4.84	0.0592	**
Interaction								
X_{12}	+0.009419	0.66	0.4425	**	+0.001879	0.15	0.7218	**
X_{13}	−0.013	1.06	0.3360	**	+0.017	9.62	0.0148	**
X_{23}	−0.00365	0.090	0.7724	**	−0.001884	0.14	0.7322	**
R^2	0.91				0.96			

* $p < 0.01$; ** $p < 0.1$; Nota. = Notability.

Table 5. The ANOVA of models optimally MRR and ε.

Variable and Source	df	Sum of Squares	F-Value	*p*-Value
		MRR, g/s		
Model	9	0.13	8.02	0.0038
Residual	8	0.015		
Lack of fit	4	4.712×10^{-3}	1.56	0.7093
Pure error	4	8.502×10^{-3}		
R^2	0.90			
CV,%	9.82			
ε, %				
Model	9	0.048	17.28	0.0004
Residual	8	2.522×10^{-3}		
Lack of fit	4	2.130×10^{-3}	5.43	0.0654
Pure error	4	3.929×10^{-4}		
R^2	0.95			
CV,%	4.09			

For the model presented in Table 4, a probability > F less than 0.01 indicates that the model is significant in the linear model. This is desirable because it signifies that the parameters significantly affect the responses. In this case, the important terms in the model are X_1, X_2, and X_3.

The most significant linear order effect of air humidity ratio (X_2) is the most significant factor associated with MRR, while the most significant linear order effect of air humidity ratio (X_2) and solution temperature (X_3) is the most significant factor related to ϵ. It is clear from Table 5 that the discrepancy and impurity of MRR and ϵ are small with the sum of squared errors of 4.712×10^{-3}, 8.502×10^{-3}, 2.129×10^{-3}, and 3.928×10^{-4}, respectively. The residual is in the variance and the CV is relatively small. The most significant component was the linear order main effect of air humidity ratio (X_2) on MRR, followed by the linear order main effect of air humidity ratio and solution temperature X_3, which are the most significant aspect ratios with ϵ. The lack of fit and impurity of MRR and ϵ appear to be minimal in Table 5, with the sum of squared errors of 4.712×10^{-3} and 8.502×10^{-3}, 2.129×10^{-3}, and 3.928×10^{-4}, respectively, leading to residual contrast and relatively low coefficient of variation.

The obtained results of Table 4 indicate that our model is compatible with the data of the experiment on R^2 values greater than 0.91 for the MRR model and 0.96 for the ϵ model.

A regression model with RSM may be tested and its statistical significance evaluated via the ANOVA table displayed in Table 5. The suggested regression model is considered statistically significant with an F-value of 8.02 and a p-value of 0.0038, allowing for successful generation power prediction under the examined operational circumstances. In particular, for the relevant model terms X_1, X_2, X_3, X_1X_2, X_1X_3, and X_2X_3, the probability value (p-value) is less than 0.0004, which shows that the coefficients in Table 5 are significant. The model terms are not critical if the values are higher than 0.05. Model reduction may improve the overall design if many model terms (apart from those needed to sustain hierarchy) are deemed to be unimportant. The Adjusted R^2 of 0.95 and the Predicted R^2 of 0.90 diverge by less than 0.05, which is a reasonable alignment. The experimental design and findings are assessed by choosing the proper linear, quadratic, etc., models.

10.3. Influence of Operational Parameters on MRR and ε

The relationships between independent and dependent variables can be exhibited in the 3D exemplification as response surfaces. Curves with similar response values are constructed on a plane whose coordinates signify the levels of the contour plot's independent components. For a given set of factor levels, every contour appears with a surface value that is significantly higher than the stated level. Three-dimensional surface diagrams with contour graphs to the MRR verses ε are illustrated in Figures 3 and 4 and Figures 5–8, respectively.

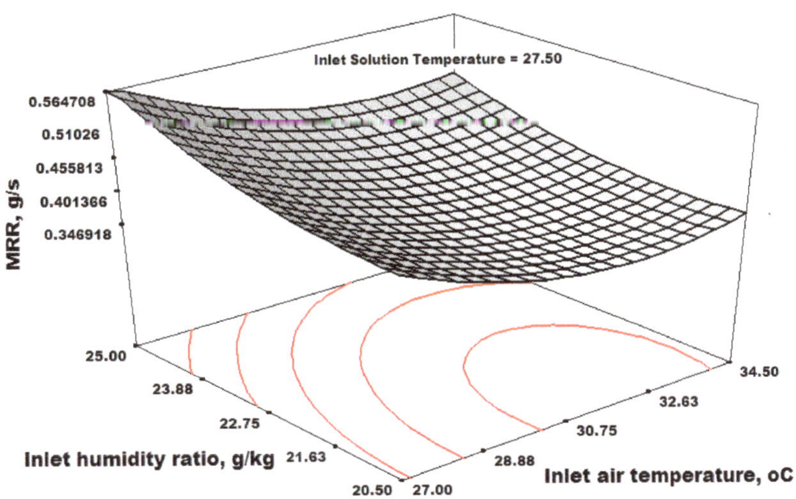

(a)

Figure 3. *Cont.*

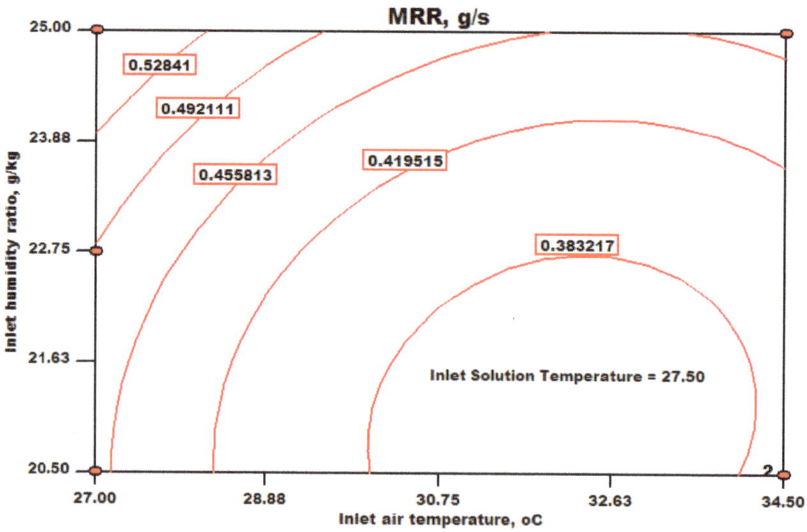

(b)

Figure 3. Effect of the inlet air temperature vs. ratio of humidity MRR: (**a**) response surfaces, (**b**) contour plot.

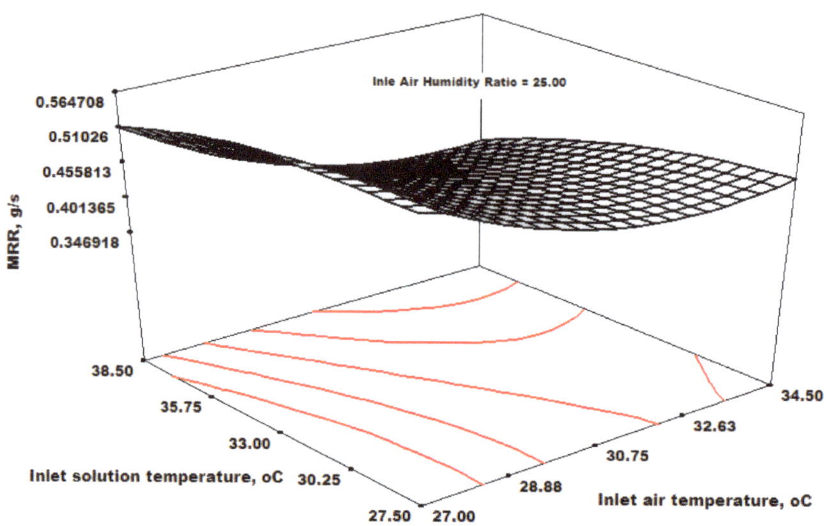

(a)

Figure 4. *Cont.*

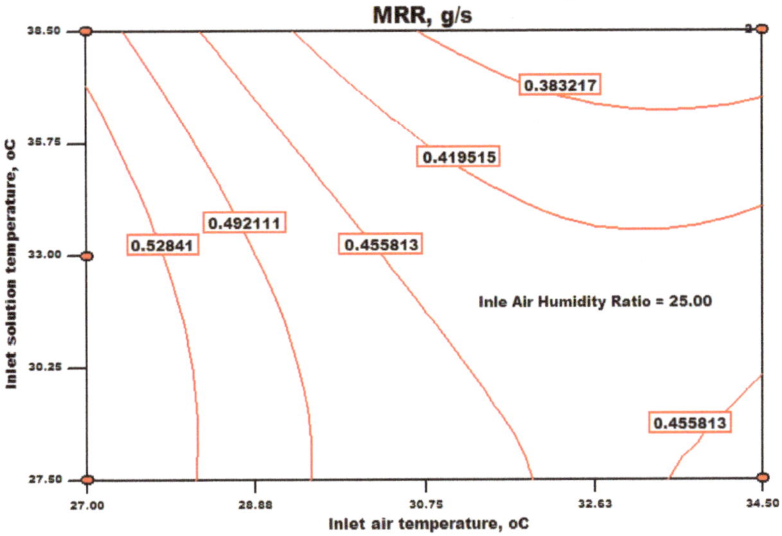

(**b**)

Figure 4. Influence of inlet air temperature vs. solution temperature on the MRR: (**a**) response surface, (**b**) contour plot.

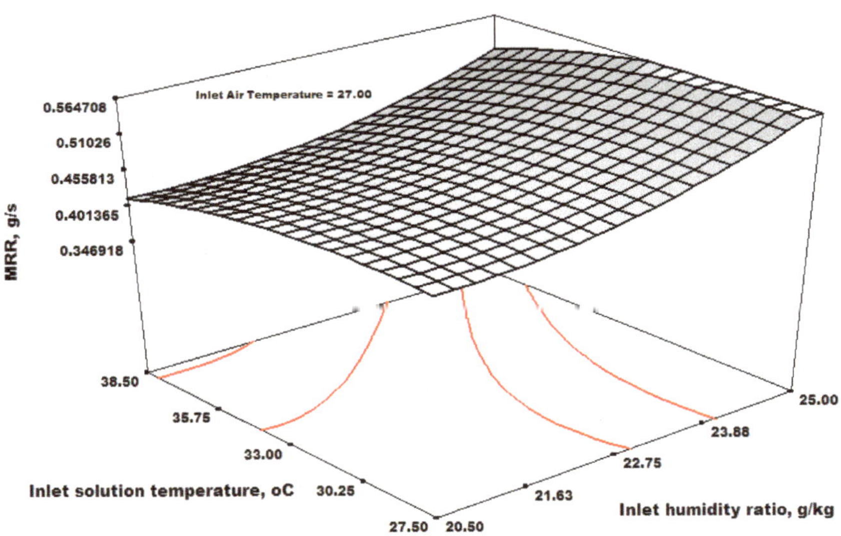

(a)

Figure 5. *Cont.*

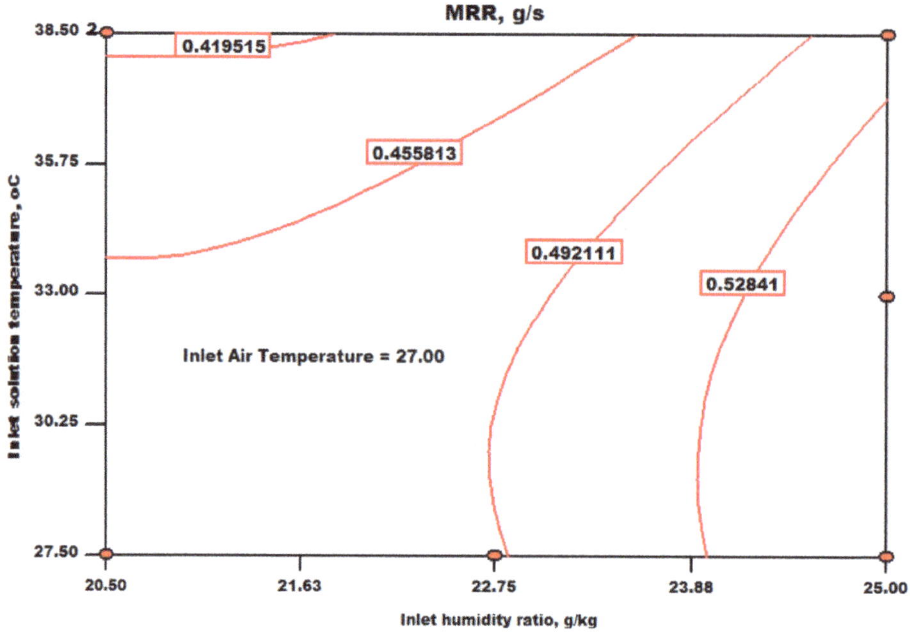

(b)

Figure 5. Influence ratio of air temperature vs. solution temperature on the MRR: (**a**) response surface, (**b**) contour plot.

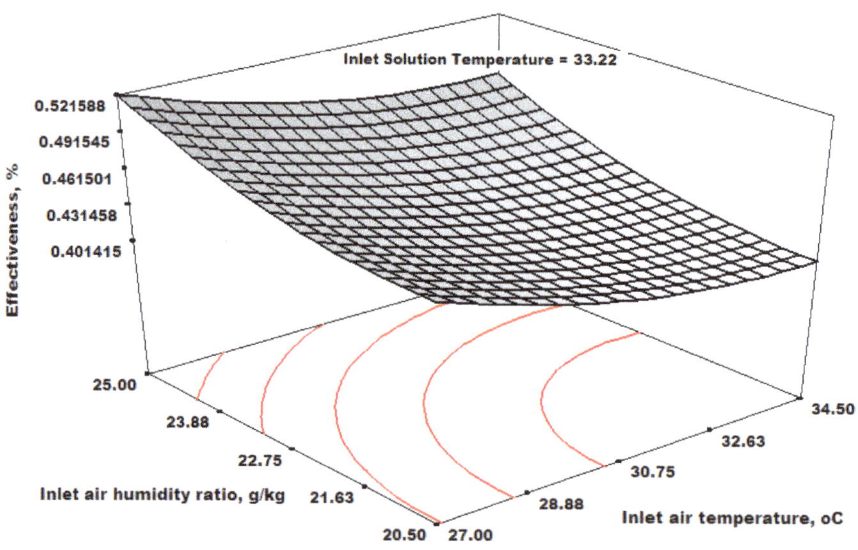

(a)

Figure 6. *Cont.*

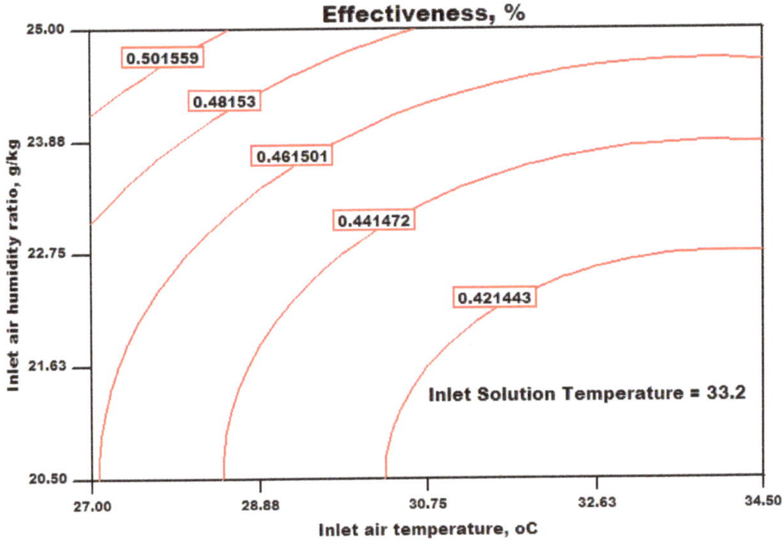

(b)

Figure 6. Effect of the inlet air temperature vs. the ratio of humidity on the ε: (**a**) response surface, (**b**) contour plot.

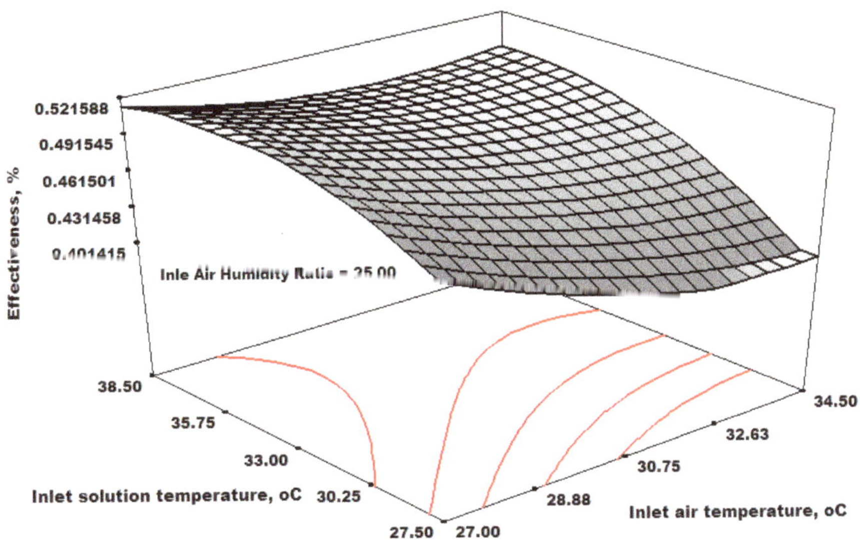

(a)

Figure 7. *Cont.*

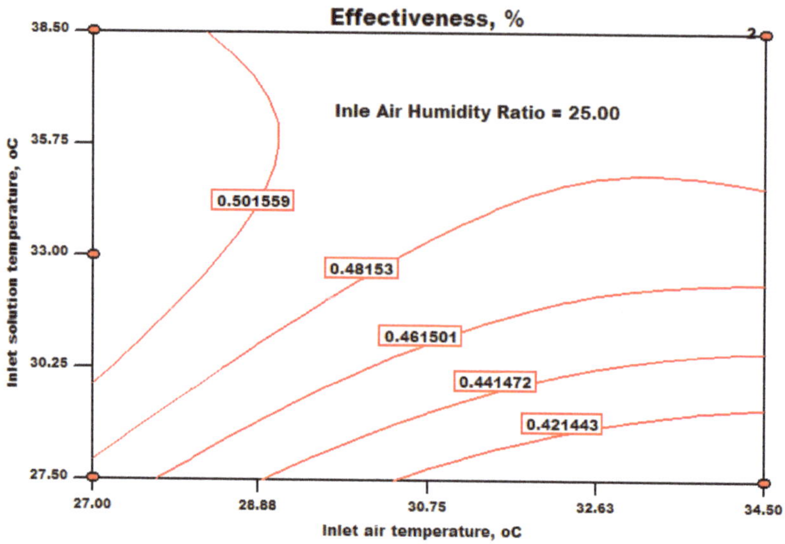

(b)

Figure 7. Influence between inlet air and solution temperature on the ε: (a) response surface, (b) contour plot.

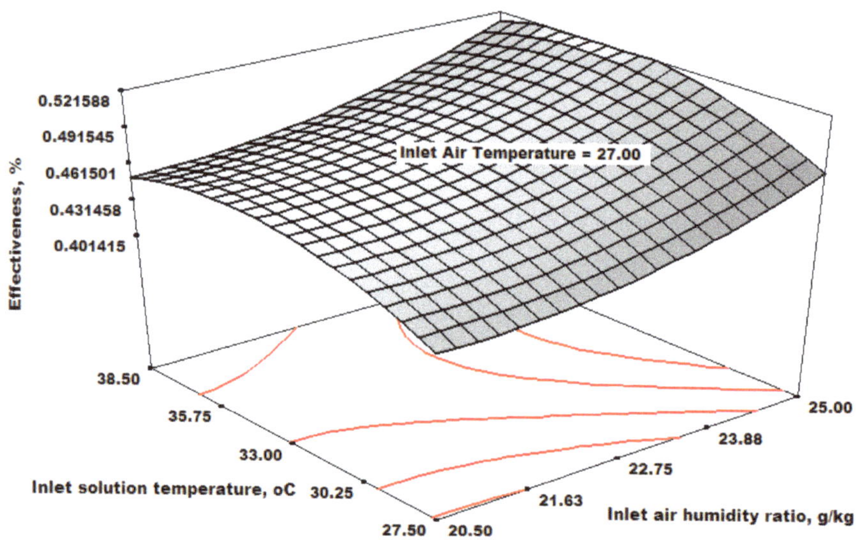

(a)

Figure 8. *Cont.*

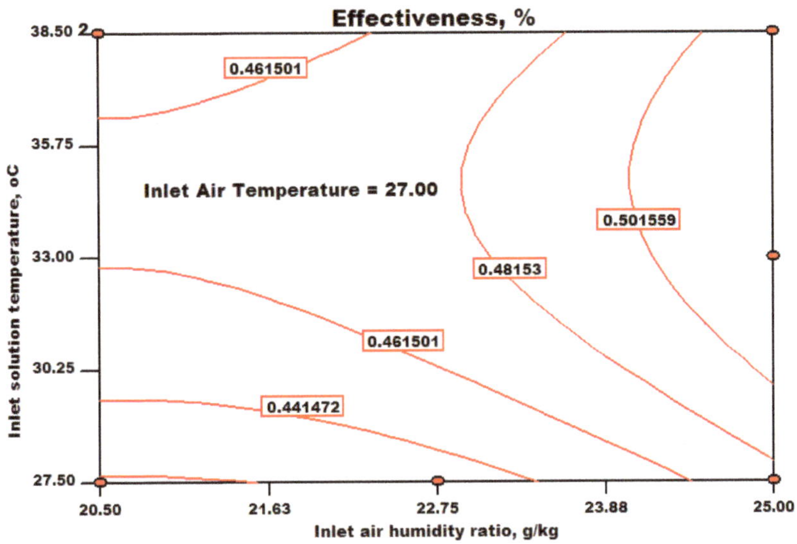

(b)

Figure 8. Influence of the inlet air humidity ratio and solution temperature on the ϵ: (**a**) response surface, (**b**) contour plot.

Response surfaces are 3D representations of the relationships between independents and dependents. Curves with the same response values are plotted on a plan whose coordinates indicate the levels for the independent components in a contour plot. Each contour indicates a distinct height from the surface over the plane formed by a combination of the factors' levels. Figures 2–4 and Figures 5–7, respectively, exhibit 3D surface graphs and contours for the MRR and ε.

The responses as MRR and ε are shown in the figures as a function of air temperature, the ratio of air humidity, and solution temperature. The fitted quadratic model accounts for the curved profile in the figure. MRR rises with decreasing air temperature to its minimum level (27 °C), as shown in Figures 2–4. Also, the input air temperature raises the solution temperature, which boosts the desiccant solution's partial pressure. As a result, the MRR of the dehumidifier is lowered. When the ratio is changed from (20 to 25 g/kg), the MRR is 0.577 gs^{-1} when the ratio of air humidity was 25 g/kg. This can be explained as when the humidity ratio rises, the partial water vapour pressure of the air rises, increasing the mass transfer potential. These results suggest that the air humidity ratio has a significant role in liquid desiccant dehumidification for sustainable building cooling. As the temperature of the solution rises, the MRR drops fast. The vapour pressure of the solution as the temperature rises lessons the mass transfer potential between the humid air and the liquid desiccant.

Figures 6–8 show that when the air temperature drops, the ε rises, indicating that raising the input temperature of air lowers the solution temperature, and this raises the partial pressure of the desiccant solution. As a result, the dehumidifier's efficacy is diminished. On the other hand, raising the inlet ratio of air humidity marginally reduces the efficacy. As the humidity ratio rises, the partial pressure of the water vapour of air rises, increasing the mass transfer potential. The inlet humidity ratio, furthermore, does not influence the dehumidifier's efficacy. As the solution increases in temperature, the equilibrium ratio of humidity at the solution surface and the humidity ratio of air outflow from the dehumidifier increase. The dehumidifier's efficacy increased marginally as a consequence of the offsetting effects between the numerator and denominator and denominator in (Equation (2)).

10.4. Optimisation of Operational Variables for MRR and ϵ

The mathematical optimisation approach gives the conditions of operation optimization whereby the optimum moisture removal rate and efficacy range from about 0 to 1. Each independent variable and response had a desired outcome indicated. The independent variables for both MRR and ϵ were kept within the range, while the response was set to maximum. Table 6 displays the optimum conditions for maximizing MRR obtained (27.4 °C, 24.5 g/kg) and (27.5 °C). Similarly, the maximized ϵ of the set of optimum conditions obtained was (27.4 °C, 24.8 g/kg) and (38.5 °C). Individual optimization yielded MR and ϵ responses of (0.54/s) and (0.5), respectively, with minimal desirability of (0.992) and (1). The optimal circumstances for maximum MRR achieved were (27.4 °C), (24.5 g/kg) (27.5 °C), and (27.5 °C), as indicated in Table 6. Similarly, the set of the best circumstances obtained to maximize the ε were (27.4 °C, 24.8 g/kg), and (38.5 °C). In the instance of individual optimisation, the MRR and ε responses were (0.54 g/s) and (0.5), respectively, with minimum desirability (0.992) and (1).

Table 6. Individual and multiple optimizations for maximum/minimum responsiveness.

Response Parameters	Coded Optimum Value			Uncoded Optimum Value			Responses	
	X_1	X_2	X_3	Ta	Wa	Ts	MRR, g/s	ε, %
MRR, g/s	−1	1	−1	27.4	24.5	27.5	0.54	0.48
ε, %	−1	1	1	27.4	24.8	38.5	0.52	0.50

11. Discussion

The results of this study highlight the complexity of interactions between operational parameters of dehumidification processes. The positive correlation between air humidity and the moisture removal rate (MRR) highlights that higher humidity levels promote mass transfer. By increasing humidity, water vapour is more likely to move from the air to the drying solution, improving dehumidification efficiency, especially when ambient humidity varies greatly.

Conversely, the negative impact of temperatures on MRR highlights the importance of optimal thermal conditions, as high temperatures can reduce the moisture absorption capacity of the dryer. This phenomenon illustrates the need for balanced temperature control to maintain system efficiency.

Statistical analysis, including analysis of variance, demonstrates the robustness of the regression model, with high R^2 values indicating a good explanation of the variance in the data. Specific interactions, particularly between air temperature and humidity, open the door to future research to refine these variables and optimize performance. Response surface methodology (RSM) provides a useful framework to visualize the complex interactions between variables, helping engineers make informed decisions when configuring systems. Moving forward, additional studies could evaluate the performance and durability of dehumidifiers under various environmental conditions, taking into account dust accumulation or corrosion. Exploring new materials or hybrid systems could also lead to significant improvements in performance and durability.

Overall, this research enriches the understanding of liquid desiccant systems and provides practical insights to optimize the design of energy-efficient refrigeration systems in the context of sustainable construction, paving the way for reduced energy consumption and improved indoor air quality.

12. Conclusions

The influence of temperature (air, ratio of air humidity and solution temperature) on the desiccant dehumidifier rate and performance was determined using response surface analysis. The most important conclusions can be made in the following:

1. With R^2 more than 0.90 for the moisture removal rate for the developed model and greater than 0.95 for the ϵ model, the model was able to reflect the experimental data.
2. Due to the examination of variation, the air humidity ratio has the greatest influence on the moisture removal rate. However, the ratio of air humidity to solution temperature has the greatest influence on efficacy.
3. Based on the table of D-optimal design optimisation, this model projected the highest moisture removal rate of 0.54 g/s. This is conducted within an optimal air temperature of 27.4 °C, humidity ratio of 25 g/kg, and solution temperature of 27.5 °C. Furthermore, the maximum effectiveness was approximated to be 0.50 below the optimum levels of air temperature of 27.4 °C, humidity ratio of 24.8 g/kg, and 38.5 °C of solution temperature.

Referring to the above conclusions, it is fair to ascertain that the optimisation of inlet parameters of liquid desiccant dehumidifiers via data analysis can attain substantial energy savings and lessen greenhouse gas emissions. Indeed, this would pave the way toward more sustainable building cooling policies.

Author Contributions: Methodology, M.A.H. and S.J.A.; Formal analysis, M.A.H. and S.J.A. All authors have read and agreed to the published version of the manuscript.

Funding: This research received no external funding.

Data Availability Statement: The original contributions presented in the study are included in the article; further inquiries can be directed to the corresponding author.

Conflicts of Interest: The authors declare no conflicts of interest.

References

1. Rafique, M.M.; Gandhidasan, P.; Bahaidarah, H.M. Liquid desiccant materials and dehumidifiers—A review. *Renew. Sustain. Energy Rev.* **2016**, *56*, 179–195. [CrossRef]
2. Chen, X.; Riffat, S.; Bai, H.; Zheng, X.; Reay, D. Recent progress in liquid desiccant dehumidification and air-conditioning: A review. *Energy Built Environ.* **2020**, *1*, 106–130. [CrossRef]
3. Qi, R.; Dong, C.; Zhang, L.Z. A review of liquid desiccant air dehumidification: From system to material manipulations. *Energy Build.* **2020**, *215*, 109897. [CrossRef]
4. Su, W.; Li, H.; Sun, B.; Li, S.; Zhang, X. Performance investigation on a frost-free air source heat pump system employing liquid desiccant dehumidification and compressor-assisted regeneration based on exergy and exergoeconomic analysis. *Energy Convers. Manag.* **2019**, *183*, 167–181. [CrossRef]
5. Jain, S.; Tripathi, S.; Das, R.S. Experimental performance of a liquid desiccant dehumidification system under tropical climates. *Energy Convers. Manag.* **2011**, *52*, 2461–2466. [CrossRef]
6. Triefenbach, F. Design of Experiments: The D-optimal Approach and Its Implementation as a Computer Algorithm. Bachelor's Thesis, South Westphalia University of Applied Sciences, Iserlohn, Germany, 2008.
7. Yang, Z.; Lian, Z.; Li, X.; Zhang, K. Concept of dehumidification perfectness and its potential applications. *Energy* **2015**, *91*, 176–191. [CrossRef]
8. Jain, S.; Dhar, P.L.; Kaushik, S.C. Experimental studies on dehumidifier and regenerator of a liquid desiccant cooling system. *Appl. Therm. Eng.* **2000**, *20*, 253–267. [CrossRef]
9. Patnaik, S.; Lenz, T.G.; Löf, G.O.G. Performance studies for an experimental solar open-cycle liquid desiccant air dehumidification system. *Sol. Energy* **1990**, *44*, 123–135. [CrossRef]
10. Chung, T.W.; GT, K.; Hines, A.L. Dehumidification of air by aqueous lithium chloride in a packed column. *Sep. Sci. Technol.* **1993**, *28*, 533–550. [CrossRef]
11. Kessling, W.; Laevemann, E.; Kapfhammer, C. Energy storage for desiccant cooling systems component development. *Sol. Energy* **1998**, *64*, 209–221. [CrossRef]
12. Lowensteing, L.A.; Gabruk, R.S. The effect of absorber design on the performance of a liquid desiccant air conditioner. *ASHRAE Trans.* **1992**, *106*, 712–720.
13. Moon, C.G.; Bansal, P.K.; Jainb, S. New mass transfer performance data of a cross-flow liquid desiccant dehumidification system. *Int. J. Refrig.* **2009**, *32*, 524–533. [CrossRef]
14. Liu, X.H.; Chang, X.M.; Xia, J.J.; Jiang, Y. Performance analysis on the internally cooled dehumidifier using liquid desiccant. *Build. Environ.* **2009**, *44*, 299–308. [CrossRef]
15. Liu, X.H.; Yi, X.Q.; Jiang, Y. Mass transfer performance comparison of two commonly used liquid desiccants: LiBr and LiCl aqueous solutions. *Energy Convers. Manag.* **2011**, *52*, 180–190. [CrossRef]

16. McDonald, B.; Wagaman, D.G.; Kettleborough, C.F. A statistical analysis of a packed tower dehumidifier. *Dry. Technol.* **1992**, *10*, 223–237. [CrossRef]
17. Ali, S.; Abdulateef, J.; Mohammad, A.T. Roughness Surface Methodology for Predicting the Performance of Heat and Mass Transfer Process in Liquid Desiccant Dehumidifier. In Proceedings of the International Conference on Mathematical Models & Computational Techniques in Science & Engineering, London, UK, 23–25 February 2019; Volume 1334, p. 012001.
18. Al-Obaidi, M.; Al-Nedawe, B.; Mohammad, A.; Mujtaba, I. Response surface methodology for predicting the dimethylphenol removal from wastewater via reverse osmosis process. *Chem. Prod. Process Model.* **2021**, *16*, 193–203. [CrossRef]
19. Hern, F.; Novi, V.; Alriols, M.G.; Labidi, J.; Erdocia, X. Optimisation of lignin liquefaction with polyethylene glycol/glycerol through response surface methodology modeling. *Ind. Crops Prod.* **2023**, *198*, 116729. [CrossRef]
20. Khoshraftar, Z.; Ghaemi, A. Evaluation of pistachio shells as solid wastes to produce activated carbon for CO_2 capture: Isotherm, response surface methodology (RSM) and artificial neural network (ANN) modeling. *Curr. Res. Green Sustain. Chem.* **2022**, *5*, 100342. [CrossRef]
21. Wu, M.; Ding, H.; Wang, S.; Xu, S. Optimization conditions for the purification of linoleic acid from sunflower oil by urea complex fractionation. *J. Am. Oil Chem. Soc.* **2008**, *85*, 677–684. [CrossRef]
22. Fumo, N.; Goswami, D.Y. Study of an aqueous lithium chloride desiccant system: Air dehumidification and desiccant regeneration. *Sol. Energy* **2002**, *72*, 351–361. [CrossRef]
23. Oberg, V.; Goswami, D.Y. Experimental study of the heat and mass transfer in a packed bed liquid desiccant air dehumidifier. *Sol. Energy Eng.* **1998**, *120*, 289–297. [CrossRef]
24. Yin, Y.; Zhang, X.; Chen, Z. Experimental study on dehumidifier and regenerator of liquid desiccant cooling air conditioning system. *Build. Environ.* **2007**, *42*, 2505–2511. [CrossRef]
25. Kinsara, A.A.; Elsayed, M.M.; Al-Rabghi, O.M. Proposed energy-efficient air- conditioning system using liquid desiccant. *Appl. Therm. Eng.* **1996**, *16*, 791–806. [CrossRef]

processes

MDPI

Article

Towards Sustainable Transportation: Adaptive Trajectory Tracking Control Strategies of a Four-Wheel-Steering Autonomous Vehicle for Improved Stability and Efficacy

Mazin I. Al-saedi [1,*] and Hiba Mohsin Abd Ali AL-bawi [2]

[1] Technical Instructors Training Institute, Middle Technical University, Baghdad 10071, Iraq
[2] Mechanical Power Technical Engineering Department, College of Engineering and Technologies, Al-Mustaqbal University, Babylon 51001, Iraq; hiba.mohsin.abd@uomus.edu.iq
* Correspondence: dr.mazin.ismael@mtu.edu.iq; Tel.: +964-07723704091

Abstract: The objective of continuous increase in the evolution of autonomous and intelligent vehicles is to attain a trustworthy, economical, and safe transportation system. Four-wheel steering (4WS) vehicles are favored over traditional front-wheel steering (FWS) vehicles because they have excellent dynamic characteristics. This paper exhibits the trajectory tracking task of a two degree of freedom (2DOF) underactuated 4WS Autonomous Vehicle (AV). Because the system is underactuated, MIMO, and has a nontriangular form, the traditional adaptive backstepping control scheme cannot be utilized to control it. For the purpose of rectifying this issue, two-state feedback-based methods grounded on the hierarchical steps of the block backstepping controller are proposed and compared in this paper. In the first strategy, a modified block backstepping is applied for the entire dynamic system. Global stability of the overall system is manifested by Lyapunov theory and Barbalat's Lemma. In the second strategy, a block backstepping controller has been applied after a reduction of the high-order model into various first-order subsystems, consisting of Lyapunov-based design and stability warranty. A trajectory tracking controller that can follow a double lane change path with high accuracy is designed, and then simulation experiments of the CarSim/Simulink connection are carried out against various vehicle longitudinal speeds and road surface roughness to demonstrate the effectiveness of the presented controllers. Furthermore, a PID driver model is introduced for comparison with the two proposed controllers. Simulation outcomes show that the proposed controllers can attain good response implementation and enhance the 4WS AV performance and stability. Indeed, enhancement of the stability and efficacy of 4WS autonomous vehicles would afford a sustainable transportation system by lessening fuel consumption and gas emissions.

Keywords: adaptive control; block backstepping; four-wheel steering; autonomous vehicle; electrical vehicle; lane changing; Lyapunov theory

1. Introduction

The crucial need for overcoming traffic congestion and accidents, environmental pollution, narrow places for operation [1], and poor fuel economy (due to the dissipation of vehicle kinetic energy because of braking through maneuvering [2]) stimulated the intelligent transportation systems. Recently, electric vehicles have been developed and prospered as a pattern of the most promising clean-energy vehicles, diminishing pollutant emissions, needing fewer maintenance requirements, and producing less noise [3–5].

To ensure a trustworthy and safe vehicle transmission even in severe driving circumstances, the development of Autonomous Vehicle (AV) has received a major attraction from engineers. Actually, intelligent systems can respond faster than a human driver, which will reduce the traffic incidents that usually result from the driver's faults [6].

The simple bicycle model of a vehicle is typically utilized for the study of a vehicle's maneuvering and handling stability. However, the desired handling performance cannot entirely be obtained because of the coupling of yaw rate and side slip angle and unsatisfactory control input by this bicycle model of a vehicle having front-wheel steering (FWS) [7].

If the cornering forces are produced at the rear wheels just like at the front wheels, rear wheels contribute to navigating the vehicle and changing the vehicle movement speedily. This perfects the transient response of the vehicle maneuvering in comparison with the traditional FWS vehicle, and as a result, it offers the freedom to preserve side slip and yaw rates at their required dynamics separately. This concept guided the principle of equipping a steering input at the rear wheels in addition to the front wheels, defined as the (4WS) system [7].

Four-wheel steering (4WS) (as well as named rear-wheel active steering) is an advanced control method that can evidently upgrade vehicle steering performance. In contrast to conventional front-wheel steering vehicles, 4WS vehicles can steer the front wheels and the rear wheels separately for cornering [8]. At low speeds, reducing the cornering radius can be achieved by orienting the rear wheels in the inverse sense to the front wheels. While, at medium and high vehicle speeds, reducing the cornering radius by steering the rear wheels in the identical direction to the front wheels, thus enhancing the performance and stability [9].

Essentially, a passenger-riding feeling and operating stability are insistently required for controlling driverless vehicles to ensure the safety and comfort of AVs [10].

A constant proportion between front wheels and rear wheels was suggested in [11,12], which can ensure a zero sideslip angle during steady steering. This procedure can be used easily, but during transient motion, it does not carry out well [13]. According to control theory, the lateral and yaw motion are impossible to control simultaneously employing only active rear steering or only active front steering. At a minimum, two independent control inputs are wanted. Hence, the incorporation of active front and rear wheel steering would be a MIMO multi-input multi-output system [9].

The main motivation for the current study is that no works deal with normal block backstopping control techniques for 4WS AV directly. In the current study, lateral dynamics of a 4WS AV for following up a desired trajectory is presented to be dominated by two strategies of MIMO-modified adaptive block-backstepping controller. The 2DOF 4WS AV model and lateral position, with fixed longitudinal speed, have been utilized. Simulation for both proposed controllers has been performed and then compared with the PID controller using CarSim-Matlab/Simulink. The reference adhering lane is composed of a double lane change movement. For further proof of the proposed strategies, tests for the 4WS AV under low and high-friction driving maneuvers have been done.

The remaining of the current publication is structured as follows: Section 2 reviews the literature on 4WS AD controllers and Adaptive Backstepping; Section 3 illustrates the 4WS AV model state space dynamics. A model depending on the first and the second block backstepping control strategies is designed with respect to Lyapunov's function in Section 4. Section 5 assesses the suggested controllers' efficiency in employing various conditions for 4WS AV. Lastly, conclusions and suggested future works are presented in Section 6.

2. Literature Review

Numerous control techniques have been carried out on vehicle dynamics, like PID control [14], fuzzy control FC [15], neural network control NNC [16,17], or a combination of optimized NNC and FC [18]. Despite these control techniques improving vehicle performance to a specific limit, NNC requires an enormous set of experimental data, while FC needs human expertise to build the fuzzy rules [19].

The LQR technique was presented to 4WS AV by Mashadi et al. [20] for the path-tracking controller. A generic interface design between the local path planning and path tracking systems assuming linear interpolation between consecutive waypoints is presented by [21]. However, using the path's heading and nonsmoothness in the derivatives of the

$$\begin{aligned}\dot{z}_2 &= (f_1 + g_{11}u_1 + g_{12}u_2) - (f_{1d} + g_{11d}u_{1d} + g_{12d}u_{2d}) + c_1\dot{z}_1 + \lambda_1\dot{\Gamma}_1 \\ &-k_1(\tfrac{\partial\Phi_1}{\partial e_1}\dot{e}_1 + \tfrac{\partial\Phi_1}{\partial e_2}\dot{e}_2 + \tfrac{\partial\Phi_1}{\partial e_3}\dot{e}_3 + \tfrac{\partial\Phi_1}{\partial e_4}\dot{e}_4 \\ &+\tfrac{\partial\Phi_1}{\partial q_{1d}}\dot{q}_{1d} + \tfrac{\partial\Phi_1}{\partial q_{2d}}\dot{q}_{2d} + \tfrac{\partial\Phi_1}{\partial p_{1d}}\dot{p}_{1d} + \tfrac{\partial\Phi_1}{\partial p_{2d}}\dot{p}_{2d})\end{aligned} \quad (22)$$

where:

$$\begin{gathered}\tfrac{\partial\Phi_1}{\partial e_1} = \tfrac{\partial\Phi_1}{\partial e_2} = \tfrac{\partial\Phi_1}{\partial e_3} = \tfrac{\partial\Phi_1}{\partial e_4} = 0,\ \tfrac{\partial\Phi_1}{\partial q_{1d}} = 0, \\ \tfrac{\partial\Phi_1}{\partial q_{2d}} = g_{12}A_2 - g_{11}B_2,\ \tfrac{\partial\Phi_1}{\partial p_{1d}} = g_{12}A_1 - g_{11}B_1,\ and\ \tfrac{\partial\Phi_1}{\partial p_{2d}} = g_{12}A_3 - g_{11}B_3\end{gathered} \quad (23)$$

As $g_{11d} = g_{11}$, $g_{12d} = g_{12}$, $g_{21d} = g_{21}$, and $g_{22d} = g_{22}$, also assuming that:

$$\delta_1 = u_1 - u_{1d} \text{ and } \delta_2 = u_2 - u_{2d} \quad (24)$$

Then, Equation (22) will be written as follows:

$$\dot{z}_2 = f_1 - f_{1d} + g_{11}\delta_1 + g_{12}\delta_2 + c_1\dot{z}_1 + \lambda_1 z_1 + \Omega_1 \quad (25)$$

where:

$$\Omega_1 = -k_1(\frac{\partial\Phi_1}{\partial q_{2d}}\dot{q}_{2d} + \frac{\partial\Phi_1}{\partial p_{1d}}\dot{p}_{1d} + \frac{\partial\Phi_1}{\partial p_{2d}}\dot{p}_{2d}) \quad (26a)$$

Step-5: The desired dynamics of z_2 is written in the form of the following:

$$\dot{z}_2 = -z_1 - c_2 z_2 \quad (26b)$$

Step-6: Similarly, motivated by Rudra et al. [49] and by trial and error a suitable new control variable z_3 has been specified as follows:

$$z_3 = e_2 - k_2(g_{22}e_3 - g_{21}e_4) \quad (27)$$

where k_2 is a design parameter. The time derivative of z_3 produces the following:

$$\begin{aligned}\dot{z}_3 &= e_4 - k_2(g_{22}\dot{e}_3 + \dot{g}_{22}e_3 - g_{21}\dot{e}_4 - \dot{g}_{21}e_4) \\ &= e_4 - k_2(g_{22}((f_1 + g_{11}u_1 + g_{12}u_2) - (f_{1d} + g_{11d}u_d + g_{12d}u_{2d})) + e_3 D(g) \\ &-g_{21}((f_2 + g_{21}u_1 + g_{22}u_2) - (f_{2d} + g_{21d}u_{1d} + g_{22d}u_{2d})) - e_4 D(g))\end{aligned} \quad (28)$$

where $D(g) = \sum(\partial g_{ij}/\partial e_{k1})\dot{e}_k$. Because of g = constant, hence $D(g) = 0$.

For simplification, let:

$$\dot{z}_3 = e_4 - k_2\Phi_2 \quad (29)$$

where:

$$\begin{aligned}\Phi_2 = g_{22}((f_1 + g_{11}u_1 + g_{12}u_2) - (f_{1d} + g_{11d}u_{1d} + g_{12d}u_{2d})) \\ -g_{21}((f_2 + g_{21}u_1 + g_{22}u_2) - (f_{2d} + g_{21d}u_{1d} + g_{22d}u_{2d}))\end{aligned} \quad (30)$$

Step-7: here, a stabilization function is selected as follows:

$$O_2 = -c_3 z_3 - \lambda_2\Gamma_2 + k_2\Phi_2 \quad (31)$$

where: $\Gamma_2 = \int_0^t z_3 dt$. λ_2 and c_3 are greater than zero constants.

Step-8: defining the difference between stabilizing function error as z_4, and writing it as follows:

$$z_4 = e_4 - O_2 \quad (32)$$

Substitution Equations (31) and (32) into Equation (29) produces:

$$\dot{z}_4 = e_4 - O_2 - c_3 z_3 - \lambda_2\Gamma_2 \quad (33)$$

or it can be written as:

$$\dot{z}_3 = z_4 - c_3 z_3 - \lambda_2 \Gamma_2 \tag{34}$$

The derivative of Equation (32) is as follows:

$$\dot{z}_4 = \dot{e}_4 - \dot{O}_2 \tag{35}$$

$$\dot{z}_4 = (f_2 + g_{21}u_1 + g_{22}u_2) - (f_{2d} + g_{21d}u_{1d} + g_{22d}u_{2d}) + c_3\dot{z}_3 + \lambda_2\dot{\Gamma}_2 - k_2\dot{\Phi}_2 \tag{36}$$

$$\begin{aligned}\dot{z}_4 = &(f_2 + g_{21}u_1 + g_{22}u_2) - (f_{2d} + g_{21d}u_{1d} + g_{22d}u_{2d}) + c_3\dot{z}_3 + \lambda_2\dot{\Gamma}_2 \\ &-k_2(\tfrac{\partial\Phi_2}{\partial e_1}\dot{e}_1 + \tfrac{\partial\Phi_2}{\partial e_2}\dot{e}_2 + \tfrac{\partial\Phi_2}{\partial e_3}\dot{e}_3 + \tfrac{\partial\Phi_2}{\partial e_4}\dot{e}_4 \\ &+\tfrac{\partial\Phi_2}{\partial q_{1d}}\dot{q}_{1d} + \tfrac{\partial\Phi_2}{\partial q_{2d}}\dot{q}_{2d} + \tfrac{\partial\Phi_2}{\partial p_{1d}}\dot{p}_{1d} + \tfrac{\partial\Phi_2}{\partial p_{2d}}\dot{p}_{2d})\end{aligned} \tag{37}$$

where:

$$\begin{aligned}&\tfrac{\partial\Phi_2}{\partial e_1} = \tfrac{\partial\Phi_2}{\partial e_2} = \tfrac{\partial\Phi_2}{\partial e_3} = \tfrac{\partial\Phi_2}{\partial e_4} = 0, \\ &\tfrac{\partial\Phi_2}{\partial q_{1d}} = 0, \tfrac{\partial\Phi_2}{\partial q_{2d}} = g_{22}A_2 - g_{21}B_2, \tfrac{\partial\Phi_2}{\partial p_{1d}} = g_{22}A_1 - g_{21}B_1,\ and\ \tfrac{\partial\Phi_2}{\partial p_{2d}} = g_{22}A_3 - g_{21}B_3\end{aligned} \tag{38}$$

As $g_{11d} = g_{11}, g_{12d} = g_{12}, g_{21d} = g_{21}$, and $g_{22d} = g_{22}$, also assuming that $\delta_1 = u_1 - u_{1d}$ and $\delta_2 = u_2 - u_{2d}$, then Equation (37) will be written as:

$$\dot{z}_4 = f_2 - f_{2d} + g_{21}\delta_1 + g_{22}\delta_2 + c_3\dot{z}_3 + \lambda_2 z_2 + \Omega_2 \tag{39}$$

where:

$$\Omega_2 = -k_2(\frac{\partial\Phi_2}{\partial q_{2d}}\dot{q}_{2d} + \frac{\partial\Phi_2}{\partial p_{1d}}\dot{p}_{1d} + \frac{\partial\Phi_2}{\partial p_{2d}}\dot{p}_{2d})$$

Step-9: The desired dynamics of z_4 is written in the form of the following:

$$\dot{z}_4 = -z_3 - c_4 z_4 \tag{40}$$

After simplification of Equations (25) and (39), the control inputs (i.e., steering angles) can be formulated as follows:

$$\delta_1 = \frac{1}{g_{11}g_{22} - g_{12}g_{21}}(g_{22}(\dot{z}_2 - (f_1 - f_{1d}) - c_1\dot{z}_1 - \lambda_1 z_1 - \Omega_1) - g_{12}(\dot{z}_4 - (f_2 - f_{2d}) - c_2\dot{z}_3 - \lambda_2 z_3 - \Omega_2)) \tag{41}$$

and:

$$\delta_2 = \frac{1}{g_{11}g_{22} - g_{12}g_{21}}(-g_{21}(\dot{z}_2 - (f_1 - f_{1d}) - c_1\dot{z}_1 - \lambda_1 z_1 - \Omega_1) + g_{11}(\dot{z}_4 - (f_2 - f_{2d}) - c_2\dot{z}_3 - \lambda_2 z_3 - \Omega_2)) \tag{42}$$

Stability Analysis

Lemma 1. *Utilizing the control law of Equations (40) and (41) can track the 2-DOF 4WS AV model along the continuous desired path.*

Proof. for the closed loop system composed of Equations (19), (26), (34) and (40) should be demonstrated is asymptotic stabilization. Suggesting the following Lyapunouv's function, this step can be proved, as [35,55]:

$$V = \frac{1}{2}\lambda_1\Gamma_1^2 + \frac{1}{2}\lambda_2\Gamma_2^2 + \frac{1}{2}z_1^2 + \frac{1}{2}z_2^2 + +\frac{1}{2}z_3^2 + \frac{1}{2}z_4^2 \tag{43}$$

desired path leads to performance deterioration. Based on SMC theory, Hiraoka et al. [22] presented a controller for a trajectory-following of a 4WS vehicle, and they concluded that the controller has a more accurate trajectory-following ability for 4WS with respect to a 2WS vehicle.

The sliding mode control was applied by Kim at el. [23] to measure the yaw moment desired for dynamic control of an in-wheel motor driving vehicle. The output torques of the motors have been arranged to obtain the aforesaid yaw moment value. The output torques were consistent to obtain the desired yaw moment. Ref. [24] presented an adaptive sliding mode controller for 4WS AV. However, the high-frequency switching of the sliding mode controller produces a "chattering" condition, which impacts the accuracy of vehicle control [25].

A mix of H2/H∞ controllers for 4WS vehicle dynamics was formulated by [19,26,27], and a mix of Takagi-Sugeno fuzzy control method/H∞ To suppress the system perturbation caused by the uncertainties of driver behavior was formulated by [28]. An approach of fuzzy controllers founded on state feedback and the SMC is suggested by Alfi and Farrokhi [29] for 4WS vehicles. A control based on disturbance observer (DOB) was utilized by [30,31]. Although the controller runs pretty well if the speed is constant and low, the control effect is reduced at high speed [25]. Yakub et al. [32] and Falcone et al. [33] presented the MPC approach for 4WS AGVs controlling. They proved that MPC could preserve stabilization throughout the required track. Despite that, this type of controller has some limitations, such as high computations, and the stability of the system has not been proved.

Modern Adaptive Backstepping control theories have established some efficient techniques to tackle the aforementioned limitations. Numerous fruitful modified backstepping approaches have been suggested recently to expand the beneficial merits of the backstepping approach, aiming to ensure efficient handling of the control tasks of several types of nonlinear systems and their stabilization requirement [34,35].

Because the non-linear system isn't considered a normal form, normal backstepping supervising means are not to be employed straightforwardly to track to get control of these systems. Ref. [36] suggested observer-based backstepping control techniques, starting by converging the path's lateral offset, for small curvature, as a straightway path. Later, the convergent lateral offset is used to find the minimized model in the form of strict feedback. As a result, the backstepping control is to be utilized in favor of the minimized model. Eventually, a reinforced observer was employed to evaluate unspecified parameters and external disturbances. The authors in [37] proposed an incorporation of sliding mode and adaptive backstepping. Concurrently, Ref. [38] proposed a cascaded backstepping control structure. Nevertheless, because of the facilitation of the dynamic steering model, there was a disturbance when the vehicle departed the curved road. On top of that, when the radius of the path was reduced, the nonlinearity of the vehicle increased.

Ref. [39] espoused the hierarchical control structure. In the main controller, the lane adhering error model has been transformed to the yaw rate pursuing the matter for tracking the path with high accuracy, utilizing the backstepping method. Next, the MPC has been implemented to carry out the steering angle and yaw moment as well. In the secondary controller, the idealized torque has been performed to supply it to every tire. However, the large number of short-time steps of prediction, required by MPC, produce computational costs [40]; therefore, it is not appropriate for real-time applications [41].

A method consisting of two sliding surfaces in [42] is presented. Since the mistaken dynamics are not considered a strict feedback system, a brake steer force input has been proposed by implementing a backstepping method. Regardless of the clarity and competence of the method, utilizing this method avoids wasting the energy supplied to the wheels as heat [43]; moreover, the vehicle would be decelerated due to the control action [44].

Ref. [45] designed the adaptive backstepping robust controller by using the filtered-based dynamic surface controller DSC method to overcome the complexity of the adaptive backstepping controller. However, they have not applied their technique for controlling 4WS vehicles.

Nevertheless, the path tracking complications of the non-triangular regular form underactuated mechanical systems are until now an open research challenge [46], because most of the presented control methods have limitations that could not be applied to real-time applications or have several facilitations. The block backstepping method is considered an effective backstepping relied upon technique [47–49]. For a wheeled mobile robot based on block-backstepping, a controller has been improved by Rudra et al. [50].

Nevertheless, the studied approaches cannot be utilized straightforwardly to follow up on the problem of mostly generalized—considered dynamic systems [35]. Hence, proceedings must be implemented for the dynamic system placed under control of the modified block backstepping as following stages of design [35]:

- Firstly, transforming the system dynamic equations into block-strict feedback.
- Thereafter, on every dynamic block, apply the backstepping procedure for achieving the equation of control input to the entire non-linear system.

Dong et al. [51] presented a backstepping control scheme to remedy the path-tracking task of a nonlinear 3DOF underactuated Unmanned Surface Vessel. Their results indicated that the controller can track trajectory with high accuracy. Krafes et al. [52] developed a Backstepping controller for underactuated and MIMO 5DOF Spherical Inverted Pendulum. However, the traditional Backstepping control design cannot be utilized to dominate it because the system has a nonsquare form. Therefore, they modified an approach based on the hierarchical steps of the Backstepping controller.

3. Dynamical Modelling of 4WS Vehicle

The equations of motion for a 2-DOF bicycle model for modeling 4WS autonomous vehicles (Figure 1) can be formulated in line with Newton's second law as follows [53]:

$$m\ddot{y} = 2F_f + 2F_r, \quad (1)$$

$$I\ddot{\psi} = 2l_f F_f + 2l_r F_r \quad (2)$$

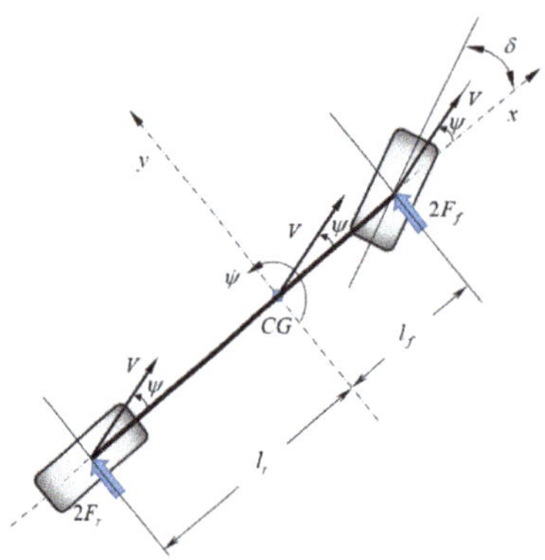

Figure 1. Dynamics of 2-DOF model of Autonomous Vehicle. F_f and F_r refer to the front and rear lateral tire forces.

The vehicle has mass m and moment of inertia I about a vertical axis through its center of gravity (CG). The front and rear wheel axles are placed at length l_f and l_r from CG, respectively. y and ψ refer to the lateral displacement and the yaw angle of the vehicle, respectively. The front and rear lateral tire forces F_f and F_r, respectively, which are reformulated as [53]:

$$F_f = K_f(\delta_f + \psi - \frac{1}{V_x}\dot{y} - \frac{l_f}{V_x}\dot{\psi}), \tag{3}$$

$$F_r = K_r(\delta_r + \psi - \frac{1}{V_x}\dot{y} + \frac{l_r}{V_x}\dot{\psi}) \tag{4}$$

where K_f and K_r denote the cornering stiffness of the front and rear tires, respectively. δ_f and δ_r are the front and rear steering angles, respectively, and are the control inputs in this paper. Assuming that V_x is a constant and the steering angles and tire slip angles of the left and right tires are the same, respectively. Substituting these variables into Equations (1) and (2) and reorganizing produces:

$$m\ddot{y} = -\frac{2(K_f + K_r)}{V_x}\dot{y} - \frac{2(l_f K_f - l_r K_r)}{V_x}\dot{\psi} + 2(K_f + K_r)\psi + 2K_f\delta_f + 2K_r\delta_r, \tag{5}$$

$$I\ddot{\psi} = -\frac{2(l_f K_f - l_r K_r)}{V_x}\dot{y} - \frac{2(l_f^2 K_f + l_r^2 K_r)}{V_x}\dot{\psi} + 2(l_f K_f - l_r K_r)\psi + 2l_f K_f\delta_f - 2l_r K_r\delta_r \tag{6}$$

In a matrix form, the dynamic model of the 2-DOF 4WS AV is [53]:

$$\frac{d}{dt}\begin{bmatrix} y \\ \dot{y} \\ \psi \\ \dot{\psi} \end{bmatrix} = \begin{bmatrix} 0 & 1 & 0 & 0 \\ 0 & -\frac{2(K_f+K_r)}{mV_x} & \frac{2(K_f+K_r)}{m} & -\frac{2(l_f K_f - l_r K_r)}{mV_x} \\ 0 & 0 & 0 & 1 \\ 0 & -\frac{2(l_f K_f - l_r K_r)}{IV_x} & \frac{2(l_f K_f - l_r K_r)}{I} & -\frac{2(l_f^2 K_f + l_r^2 K_r)}{IV_x} \end{bmatrix} \begin{bmatrix} y \\ \dot{y} \\ \psi \\ \dot{\psi} \end{bmatrix} + \begin{bmatrix} 0 & 0 \\ \frac{2K_f}{m} & \frac{2K_r}{m} \\ 0 & 0 \\ \frac{2l_f K_f}{I} & \frac{-2l_r K_r}{I} \end{bmatrix} \begin{bmatrix} \delta_f \\ \delta_r \end{bmatrix} \tag{7}$$

4. Block Backstepping Controller Design Strategies

Since the dynamic model is not a strict feedback form, the technique of integrator backstepping control in [54] cannot be applied directly. Accordingly, it should be transformed, at first, into a block-strict feedback form [49].

4.1. First Strategy for Block Backstepping Controller Design

For simplifications, rewriting the state model of 2DOF of 4WS AV with renaming the state variables such that; $q_1 = y$, $q_2 = \psi$, $p_1 = \dot{y}$, $p_2 = \dot{\psi}$, then the state model of Equation (7) is expressed in a reduced state as follows:

$$\begin{aligned} \dot{q}_1 &= p_1 \\ \dot{q}_2 &= p_2 \\ \dot{p}_1 &= f_1(q, p) + g_{11}(q)u_1 + g_{12}(q)u_2 \\ \dot{p}_2 &= f_2(q, p) + g_{21}(q)u_1 + g_{22}(q)u_2 \end{aligned} \tag{8}$$

where the state vector is: $X = \begin{bmatrix} q_1 & q_2 & p_1 & p_2 \end{bmatrix}^T$

Step-1: Defined the desired trajectory vector as: $X_d = \begin{bmatrix} q_{1d} & q_{2d} & p_{1d} & p_{2d} \end{bmatrix}^T$, and substituting it in Equation (8) gives the following:

$$\begin{aligned} \dot{q}_{1d} &= p_{1d1} \\ \dot{q}_{2d} &= p_{2d2} \\ \dot{p}_{1d} &= f_{1d}(q_d, p_d) + g_{11d}(q_d)u_{1d} + g_{12d}(q_d)u_{2d} \\ \dot{p}_{2d} &= f_{2d}(q_d, p_d) + g_{21d}(q_d)u_{1d} + g_{22d}(q_d)u_{2d} \end{aligned} \tag{9}$$

The tracking error vector is as follows:

$$e = X - X_d = \begin{bmatrix} e_1 & e_2 & e_3 & e_4 \end{bmatrix}^T \tag{10}$$

Hence, the control purpose for path tracking of 4WS AV is to formulate control law and to guarantee the tracking errors e which converge to an arbitrarily small neighborhood of zero as $t \to \infty$.

The time derivative of two sides of Equation (10) leads to the following:

$$\begin{array}{l} \dot{e}_1 = e_3 \\ \dot{e}_2 = e_4 \\ \dot{e}_3 = f_1(q,p) + g_{11}(q)u_1 + g_{12}(q)u_2 - (f_{d1}(q_d,p_d) + g_{d11}(q_d)u_{1d} + g_{12d}(q_d)u_{2d}) \\ \dot{e}_4 = f_2(q,p) + g_{21}(q)u_1 + g_{22}(q)u_2 - (f_{d2}(q_d,p_d) + g_{d21}(q_d)u_{1d} + g_{22d}(q_d)u_{2d}) \end{array} \quad (11)$$

Step-2: motivated by Rudra et al. [49] and by trial and error a suitable new control variable z_1 can be specified as follows:

$$z_1 = e_1 - k_1(g_{12}e_3 - g_{11}e_4) \quad (12)$$

where k_1 is a design parameter. The time derivative of z_1 produces the following:

$$\begin{array}{c} \dot{z}_1 = e_3 - k_1(g_{12}\dot{e}_3 + \dot{g}_{12}e_3 - g_{11}\dot{e}_4 - \dot{g}_{11}e_4) \\ = e_3 - k_1(g_{12}((f_1 + g_{11}u_1 + g_{12}u_2) - (f_{1d} + g_{11d}u_d + g_{12d}u_{2d})) + e_3D(g) \\ -g_{11}((f_2 + g_{21}u_1 + g_{22}u_2) - (f_{2d} + g_{21d}u_{1d} + g_{22d}u_{2d})) - e_4D(g)) \end{array} \quad (13)$$

where: $D(g) = \sum(\partial g_{ij}/\partial e_{k1})\dot{e}_k$. Because of g_{ij} = constant, hence $D(g) = 0$.

For simplification, let:

$$\dot{z}_1 = e_3 - k_1\Phi_1 \quad (14)$$

where:

$$\begin{array}{c} \Phi_1 = g_{12}((f_1 + g_{11}u_1 + g_{12}u_2) - (f_{1d} + g_{11d}u_{1d} + g_{12d}u_{2d})) \\ -g_{11}((f_2 + g_{21}u_1 + g_{22}u_2) - (f_{2d} + g_{21d}u_{1d} + g_{22d}u_{2d})) \end{array} \quad (15)$$

Step-3: here, a stabilization function is selected as follows:

$$O_1 = -c_1z_1 - \lambda_1\Gamma_1 + k_1\Phi_1 \quad (16)$$

where: $\Gamma_1 = \int_0^t z_1 dt$, and both λ_1 and c_1 are greater than zero constants.

Step-4: defining the difference between stabilizing function error as z_2, and writing it as:

$$z_2 = e_3 - O_1 \quad (17)$$

Substitution Equations (16) and (17) into Equation (14) gives the following:

$$\dot{z}_1 = e_3 - O_1 - c_1z_1 - \lambda_1\Gamma_1 \quad (18)$$

or it is expressed in the form of:

$$\dot{z}_1 = z_2 - c_1z_1 - \lambda_1\Gamma_1 \quad (19)$$

The derivative of Equation (17) is as follows:

$$\dot{z}_2 = \dot{e}_3 - \dot{O}_1 \quad (20)$$

$$\dot{z}_2 = (f_1 + g_{11}u_1 + g_{12}u_2) - (f_{1d} + g_{11d}u_{1d} + g_{12d}u_{2d}) + c_1\dot{z}_1 + \lambda_1\dot{\Gamma}_1 - k_1\dot{\Phi}_1 \quad (21)$$

Deriving Equation (43) and substituting Equations (19), (26), (34), and (40) into and simplifying, produces:

$$\begin{aligned}\dot{V} &= \lambda_1\Gamma_1 z_1 + \lambda_2\Gamma_2 z_2 + z_1\dot{z}_1 + z_2\dot{z}_2 + z_3\dot{z}_3 + z_4\dot{z}_4 \\ &= -c_1 z_1^T z_1 - c_2 z_2^T z_2 - c_3 z_3^T z_3 - c_4 z_4^T z_4 \leq 0\end{aligned} \tag{44}$$

Equation (44) shows that $V(t) \leq V(0)$ is confirmed, hence, $\Gamma_1, \Gamma_2, z_1, z_2, z_3$ and z_4 are bounded and deriving Equation (44) again leads to the following:

$$\ddot{V} = -2c_1 z_1^T \dot{z}_1 - 2c_2 z_2^T \dot{z}_2 - 2c_3 z_3^T \dot{z}_3 - 2c_4 z_4^T \dot{z}_4 \tag{45}$$

Thanks to that $z_1,\ z_2,\ z_3,\ z_4, \dot{z}_1,\ \dot{z}_2,\ \dot{z}_3$, and $\dot{z}_4$ are bounded, hence $\ddot{V}$ is bounded too. Applying Barbalat's Lemma, manifests that $z_1,\ z_2,\ z_3$ and z_4 are converging to zero as $t \to \infty$. □

Lemma 2. *The path tracking errors of Equation (15) with initial conditions* $\begin{bmatrix} q_1(0) & q_2(0) & p_1(0) & p_2(0) \end{bmatrix}^T$ *and under the employing of the control law, should asymptotically converge to zero.*

Proof. The object of this step is to verify that the tracking errors e_i, $i = 1, \ldots, 4$ explained in Equation (10) converge to zero. Taking into account z_1, z_2, z_3, and z_4 has been manifested it converges asymptotically to zero and the gain $k_1 > 0$, $k_2 > 0$, as a result, Equation (12) gives the following:

$$\lim_{t\to\infty} e_1 = \lim_{t\to\infty} [k_1(g_{12}e_3 - g_{11}e_4)] = 0 \tag{46}$$

Similarly, Equation (28) gives:

$$\lim_{t\to\infty} e_2 = \lim_{t\to\infty} [k_2(g_{22}e_3 - g_{21}e_4)] = 0 \tag{47}$$

And from Equations (16), (17), (31), and (32), the following can be achieved:

$$\lim_{t\to\infty} e_3 = \lim_{t\to\infty} \Omega_1 = 0 \tag{48}$$

$$\lim_{t\to\infty} e_4 = \lim_{t\to\infty} \Omega_2 = 0 \tag{49}$$

Substitute these results into Equations (46) and (47) produces:

$$k_1(g_{12}e_3 - g_{11}e_4) \to 0\ k_2(g_{22}e_3 - g_{21}e_4) \to 0 \tag{50}$$

which leads to the following:

$$\lim_{t\to\infty} e_1 = \lim_{t\to\infty} e_2 = 0 \tag{51}$$

For this reason, the four variables of the tracking error vector of Equation (10) converge asymptotically to zero. Accordingly, the global stabilization of the dynamic of the 2-DOF 4WS AV model can be certified by the proposed control law. □

4.2. Second Strategy for Block Backstepping Controller Design

For facilitating, rewriting the state variables as follows: $q_1 = y, p_1 = \dot{y}, q_2 = \psi, p_2 = \dot{\psi}$, the state model of 4WS AV can be written in another reduced form as follows [52]:

$$\begin{aligned}\dot{q}_1 &= p_1 \\ \dot{p}_1 &= f_1(q,p) + g_{11}(q)u_1 + g_{12}(q)u_2 \\ \dot{q}_2 &= p_2 \\ \dot{p}_2 &= f_2(q,p) + g_{21}(q)u_1 + g_{22}(q)u_2\end{aligned} \tag{52}$$

The state vector is described as: $X = \begin{bmatrix} q_1 & q_2 & p_1 & p_2 \end{bmatrix}^T$.

Step-1: Assuming the desired path as: $X_d = \begin{bmatrix} q_{1d} & p_{1d} & q_{2d} & p_{2d} \end{bmatrix}^T$ and substituting it into Equation (52) gives the following:

$$\begin{array}{l} \dot{q}_{1d} = p_{1d} \\ \dot{p}_{1d} = f_{1d}(q_d, p_d) + g_{d11}(q_d)u_{1d} + g_{12d}(q_d)u_{2d} \\ \dot{q}_{2d} = p_{2d} \\ \dot{p}_{2d} = f_{2d}(q_d, p_d) + g_{d21}(q_d)u_{1d} + g_{22d}(q_d)u_{2d} \end{array} \quad (53)$$

For simplification, let $x_1 = q_1$, $x_2 = p_1$, hence, the first subsystem considering the following:

$$\begin{array}{l} \dot{x}_1 = x_2 \\ \dot{x}_2 = A_1x_2 + A_2x_3 + A_3x_4 + A_4u_1 + A_5u_2 \end{array} \quad (54)$$

Step-2: Assuming that there is only one input, as a first step, and then calculate the control input utilizing the modified backstepping technique.

$$\begin{array}{l} \dot{x}_1 = x_1 \\ \dot{x}_2 = f_1(q, p) + g_{11}(q)u_1 \end{array} \quad (55)$$

Defining the error vector as the difference between the current state and the desired by:

$$e_i = x_i - x_{id} \quad (56)$$

The object is to design a virtual controller that produces $e_i = 0$, $i = 1, 2$. To perform that, a Lyapunov's function can be candidate as:

$$V_i = \frac{1}{2}e_i^2 \quad (57)$$

then:

$$\dot{V}_i = e_i\dot{e}_i = e_i(x_{i+1} - \dot{x}_{id}) \quad \text{i.e.,} \quad \dot{V}_1 = e_1\dot{e}_1 = e_1(x_2 - \dot{x}_{1d}) \quad (58)$$

Defining the derivative of the first virtual command $\dot{x}_{id} = x_{(i+1)d} + c_ie_i$, where $c_i > 0$. Then the second virtual command is described in the form of:

$$x_{(i+1)d} = -c_ie_i + \dot{x}_{id} \quad (59)$$

Assuming that the second error is as follows:

$$e_{i+1} = x_{i+1} - x_{(i+1)d} \quad (60)$$

Then Equation (58) becomes:

$$\dot{V}_i = e_i(e_{i+1} - c_ie_i)\dot{V}_1 = e_1(e_2 - c_1e_1) \quad (61)$$

The second Lyapunov function is specified in the form of the following:

$$V_{i+1} = V_i + \frac{1}{2}e_{i+1}^2 \quad (62)$$

So,

$$\begin{aligned} \dot{V}_{i+1} &= \dot{V}_i + e_{i+1}\dot{e}_{i+1} \\ &= e_i(e_{i+1} - c_ie_i) + e_{i+1}(c_ie_{i+1} + f_i(x) + g_{11}(x)u_1 - c_i^2e_i - \ddot{x}_{id}) \end{aligned} \quad (63)$$

$$\dot{V}_2 = e_1(e_2 - c_1e_1) + e_2(c_1e_2 + f_1(x) + g_{11}(x)u_1 - c_1^2e_1 - \ddot{x}_{1d}) \quad (64)$$

Using Equation (63) to achieve the first controller of the first subsystems following [52]:

$$u_i = -\frac{1}{g_{11}(x)}(e_i + c_{i+1}e_{i+1} + c_i e_{i+1} + f_i(x) - c_i^2 e_i - \ddot{x}_{id}) \tag{65}$$

$$u_1 = -\frac{1}{g_{11}(x)}(e_1 + c_2 e_2 + c_1 e_2 + f_1(x) - c_1^2 e_1 - \ddot{x}_{1d}) \tag{66}$$

Step-3: adding the second input u_2 to the subsystem and deriving the command using the modified backstepping technique similar to the first step [52]:

$$\begin{aligned} \dot{x}_1 &= x_1 \\ \dot{x}_2 &= f_1(q,p) + g_{11}(q)u_1 + g_{12}(q)u_2 \end{aligned} \tag{67}$$

And the second controller of the first subsystem can be written as follows [52]:

$$u_{i+1} = -\frac{1}{g_{12}(x)}(e_i + c_{i+1}e_{i+1} + c_i e_{i+1} + f_i(x) + g_{11}(x)u_1 - c_i^2 e_i - \ddot{x}_{id}) \tag{68}$$

$$u_2 = -\frac{1}{g_{12}(x)}(e_1 + c_2 e_2 + c_1 e_2 + f_1(x) + g_{11}(x)u_1 - c_1^2 e_1 - \ddot{x}_{1d}) \tag{69}$$

Now, for simplifications, and repeating the same procedure, let $x_3 = q_2$, $x_4 = p_2$, and considering the second subsystem as following and repeating the same procedure:

$$\begin{aligned} \dot{x}_3 &= x_4 \\ \dot{x}_4 &= B_1 x_2 + B_2 x_3 + B_3 x_4 + B_4 u_1 + B_5 u_2 \end{aligned} \tag{70}$$

Step-4: Again, assuming that there is only one input, as a first step, and then calculate the control input utilizing the modified backstepping technique [52]:

$$\begin{aligned} \dot{x}_3 &= x_4 \\ \dot{x}_4 &= f_2(q,p) + g_{21}(q)u_1 \end{aligned} \tag{71}$$

Defining the error vector as the difference between the current state and the desired by the following:

$$e_i = x_i - x_{id} \tag{72}$$

The object is to design a virtual controller that produces, $e_i = 0$ $i = 3, 4$. To perform that, a Lyapunov's function can be a candidate like:

$$V_i = \frac{1}{2}e_i^2 \tag{73}$$

and:

$$\dot{V}_i = e_i\dot{e}_i = e_i(x_{i+1} - \dot{x}_{id})\dot{V}_3 = e_3\dot{e}_3 = e_3(x_3 - \dot{x}_{3d}) \tag{74}$$

Defining the derivative of the first virtual command $\dot{x}_{id} = x_{(i+1)d} + c_i e_i$, where $c_i > 0$. Then, the second virtual command is described by the following:

$$x_{(i+1)d} = -c_i e_i + \dot{x}_{id} \tag{75}$$

Assuming that the second error as:

$$e_{i+1} = x_{i+1} - x_{(i+1)d} \tag{76}$$

Then, Equation (73) becomes:

$$\dot{V}_i = e_i(e_{i+1} - c_i e_i)\dot{V}_3 = e_3(e_4 - c_3 e_3) \tag{77}$$

The second Lyapunov's function is specified in the form of:

$$V_{i+1} = V_i + \frac{1}{2}e_{i+1}^2 \tag{78}$$

hence,

$$\begin{aligned}\dot{V}_{i+1} &= \dot{V}_i + e_{i+1}\dot{e}_{i+1} \\ &= e_i(e_{i+1} - c_i e_i) + e_{i+1}(c_i e_{i+1} + f_i(x) + g_{21}(x)u_1 - c_i^2 e_i - \ddot{x}_{id})\end{aligned} \tag{79}$$

$$\dot{V}_4 = e_3(e_4 - c_3 e_3) + e_4(c_3 e_4 + f_2(x) + g_{21}(x)u_1 - c_3^2 e_3 - \ddot{x}_{3d}) \tag{80}$$

Using Equation (79) to achieve the first controller of the first subsystems following:

$$u_i = -\frac{1}{g_{21}(x)}(e_i + c_{i+1}e_{i+1} + c_i e_{i+1} + f_i(x) - c_i^2 e_i - \ddot{x}_{id}) \tag{81}$$

$$u_1 = -\frac{1}{g_{21}(x)}(e_3 + c_4 e_4 + c_3 e_4 + f_2(x) - c_3^2 e_3 - \ddot{x}_{3d}) \tag{82}$$

Step-5: adding the second input u_2 to the second subsystem and concluding the second command utilizing the modified backstepping technique similar to the first step, gives:

$$\begin{aligned}\dot{x}_3 &= x_4 \\ \dot{x}_4 &= f_2(q,p) + g_{21}(q)u_1 + g_{22}(q)u_2\end{aligned} \tag{83}$$

Finally, the second controller of the first subsystem can be written as follows:

$$u_{i+1} = -\frac{1}{g_{22}(x)}(e_i + c_{i+1}e_{i+1} + c_i e_{i+1} + f_i(x) + g_{21}(x)u_1 - c_i^2 e_i - \ddot{x}_{id}) \tag{84}$$

$$u_2 = -\frac{1}{g_{22}(x)}(e_3 + c_4 e_4 + c_3 e_4 + f_2(x) + g_{21}(x)u_1 - c_3^2 e_3 - \ddot{x}_{3d}) \tag{85}$$

5. Simulations and Results

Based on the proposed two strategies of modified block backstepping controllers, numerical simulations have been conducted for validation using CarSim software to modeling 4WS AV, while the controllers were simulated by MATLAB/Simulink. The proposed schemes are applied to drive a 2DOF 4WS AV (Figure 2) along a desired trajectory of a fifth-degree polynomial function (quantic). Concurrently, both two suggested block backstepping control strategies (Figures 3 and 4) are compared with a PID controller for the same tracking path. The vehicle nominal parameters described in [56] are tabulated in Table 1.

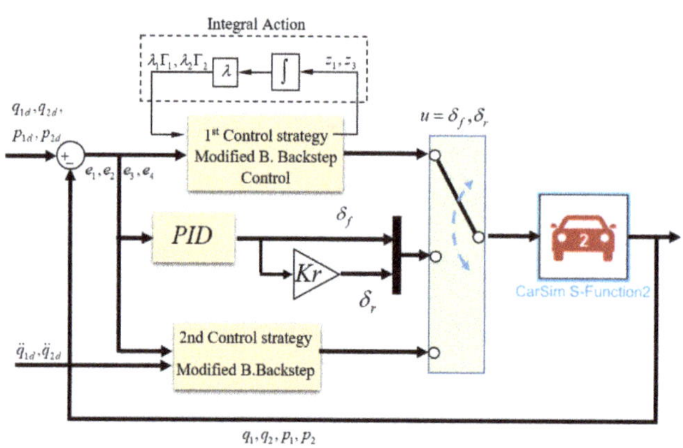

Figure 2. Scheme of the presented controllers: the first controller of modified block backstepping strategy, PID controller, and the second control strategy for 4WS AV.

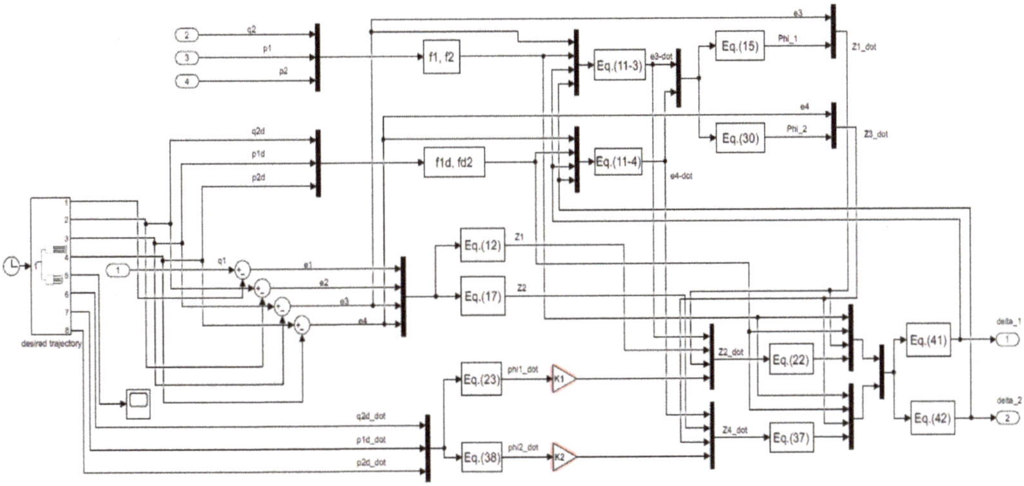

Figure 3. Scheme of the first presented modified block backstepping controller strategy for 4WS AV.

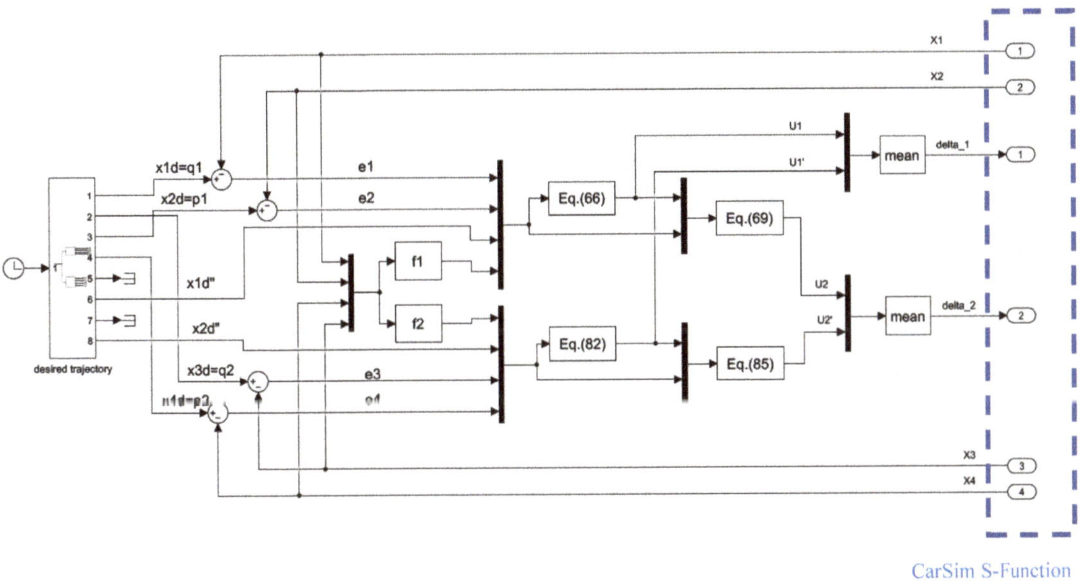

Figure 4. Scheme of the second presented modified block backstepping controller strategy for 4WS AV.

The design variables of the block backstepping controller have been chosen arbitrarily for each control strategy, as shown in Tables 2 and 3. To ensure system stability, the numerical amount of these design variables must be real and larger than zero.

Table 1. Design variables of 4WS AV implemented [56].

Symbol	Parameter	Value
m	Mass	1500 kg
I	Inertial moment about the z-axis	2500 kgm^2
l_f	length between C.O.G. and front-axle	1.1 m
l_r	length between C.O.G. and rear-axle	1.6 m
K_f	Front cornering stiffness	55,000 N/rad
K_r	Rear cornering stiffness	60,000 N/rad
$w_f = w_r$	Front and rear track	1.4 (m)
Kp	Proportional gain	0.15
Kd	Derivative gain	0.25

Table 2. Parameters of the first block backstepping control strategy which found manually.

μ	c_1	c_3	λ_1	k_1	λ_2	k_2
1	0.01	0.01	100	0.0001	100	0.0001
0.6	0.01	0.01	100	0.0001	100	0.0001

Table 3. Parameters of second block backstepping control strategy found manually.

μ	c_1	c_2	c_3	c_4
1	0.1	0.1	1	10
0.6	0.001	0.001	1	10

The transformation relations between the global coordinates (X_c, Y_c) and body coordinates (x_c, y_c) of the path's curvature center, according to Jazar [57], are as follows:

$$X_c = x - y'(x'^2 + y'^2)/(x'y'' - y'x'') \tag{86}$$

$$Y_c = y + x'(x'^2 + y'^2)/(x'y'' - y'x'') \tag{87}$$

$$x_c = (X_c - x)\cos(\psi) + (Y_c - y)\sin(\psi) \tag{88}$$

$$y_c = (Y_c - y)\cos(\psi) - (X_c - x)\sin(\psi) \tag{89}$$

where the angle ψ between the x-axis of a vehicle's body coordinate system and the global X-axis can be calculated as follows:

$$\psi = \tan^{-1}(y'/x') \tag{90}$$

The inner and outer steer angles of both front and rear wheels are computed, respectively, by:

$$\delta_{if} = \tan^{-1}(cc_1/(R - w_f/2)) \tag{91}$$

$$\delta_{of} = \tan^{-1}(cc_1/(R + w_f/2)) \tag{92}$$

$$\delta_{ir} = \tan^{-1}(cc_2/(R - w_r/2)) \tag{93}$$

$$\delta_{or} = \tan^{-1}(cc_2/(R + w_r/2)) \tag{94}$$

hence, the parameters cc_1, cc_2, and the radius of curvature R are as follows:

$$R = y_c \tag{95}$$

$$cc_2 = -l_r - x_c \tag{96}$$

$$cc_1 = l_f - x_c \tag{97}$$

Eventually, for the 4WS vehicle model, the required single steer angle is the mean of cotangent of the inner and outer steer angles:

$$\cot\delta_f = (\cot\delta_{of} + \cot\delta_{if})/2 \tag{98}$$

$$\cot\delta_r = (\cot\delta_{or} + \cot\delta_{ir})/2 \tag{99}$$

Then, the desired control input (u_d) can be calculated, wherein the desired control input u_1 and u_2 (i.e., the front and rear steering angles δ_f and δ_r) for the first strategy can be calculated by applying Equation (24). However, the required control input u_1 and u_2 for the second strategy can be calculated by applying Equations (66) and (69) and Equations (82) and (85), respectively.

The first simulations have been carried out under a constant longitudinal speed of V = 18 m/s, and the coefficient of friction of the road is μ = 1. These simulations, for tracking the 2DOF 4WS AV the desired trajectory path, have been done using both designed block backstepping control strategies. Simultaneously, for comparative analysis, a similar simulation for tracking the desired trajectory path by the 2DOF 4WS AV has been done utilizing the PID controller method. The maximum lateral displacement of the lane is 3.75 m.

For the sake of using the classical PID controller with two inputs, a constant velocity-dependent proportion of the rear wheel steering angle to the front wheel steering angle is suggested [12], which could be expressed as follows:

$$K_s = \frac{\delta_r}{\delta_f} = \frac{mu_0^2 l_f - l_r(l_r + l_f)K_r}{mu_0^2 l_r + l_f(l_r + l_f)K_f} \cdot \frac{K_f}{K_r} \tag{100}$$

Substituting the design parameters of 4WS AV used in this study for a velocity of u_o = Vx = 18 m/s in Equation (100) gives K_s = 0.2683.

The outputs of the first and second block backstepping strategies and the PID controller are displayed in Figure 5a–e. It can be concluded that the PID controller produces smaller front and rear wheel steering angles than those of the first and second block backstepping control strategies. However, their yaw angle curves exhibit the identical changing tendency and values. One can conclude that the amount of controller outputs will affect the precision of trajectory tracking remarkably.

Figure 5b,c display the lateral velocities and yaw rates of 4WS AV. The yaw rates by the PID controller are much larger than those of the first and second block backstepping strategies, and there is a considerable difference in lateral velocity.

The trajectory tracking achievement by the three controllers is shown in Figure 5c,e. It can be noticed that both the yaw angle and the lateral displacement achieve faster responses with lower overshoots and lower steady-state errors utilizing the first and second block backstepping strategies than utilizing the PID controller, which reveals that the trajectory tracking performance of the second block backstepping strategy controller is better than that of the PID controller for the same model. The percentage of maximum vehicle lateral drift by PID is (0.25 m/3.75 m) = 6.66%, while it is 3.33% by the first suggested control.

Later, a double-lane change activity by the two proposed controllers is evaluated in the following simulation cases. The 4WS AV is presumed to perform the activity prone to various longitudinal velocities and road surface friction.

To test both designed control strategies, different longitudinal velocities are considered in the procedure. Figures 6–8 display the simulation outcomes for double-lane change movement under various longitudinal speeds (Vx = 5 m/s, Vx = 15 m/s, Vx = 25 m/s).

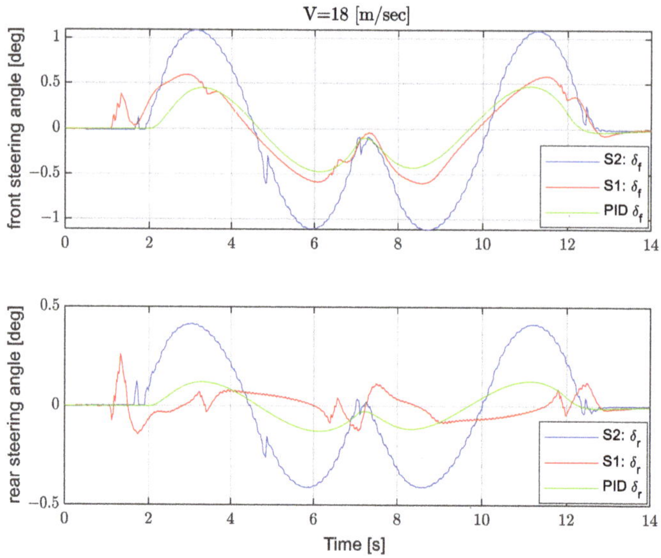

(**a**)

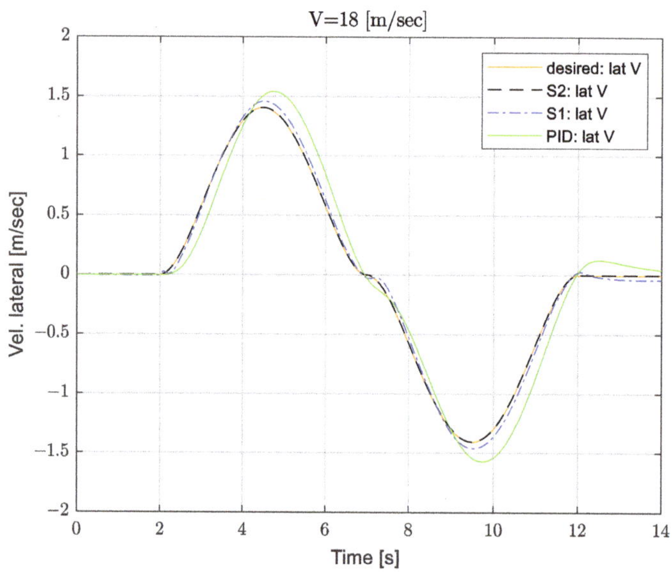

(**b**)

Figure 5. *Cont.*

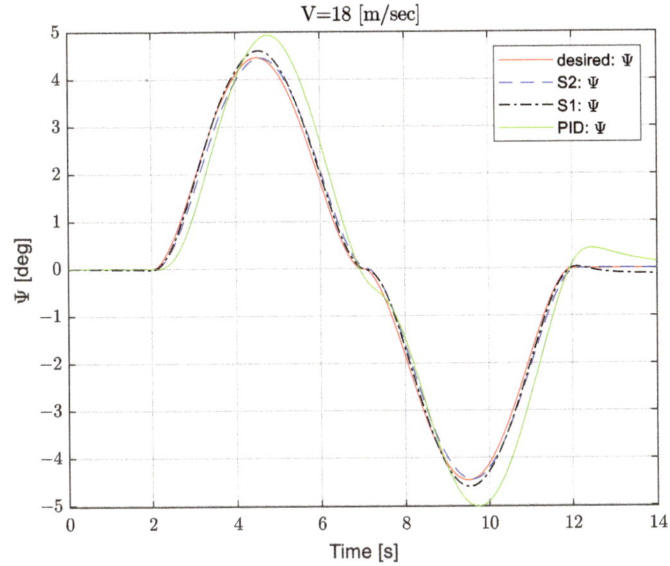

(**c**)

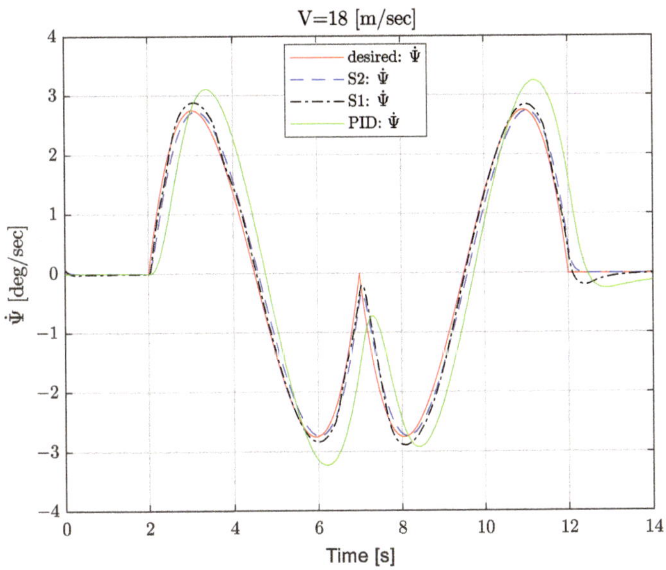

(**d**)

Figure 5. *Cont.*

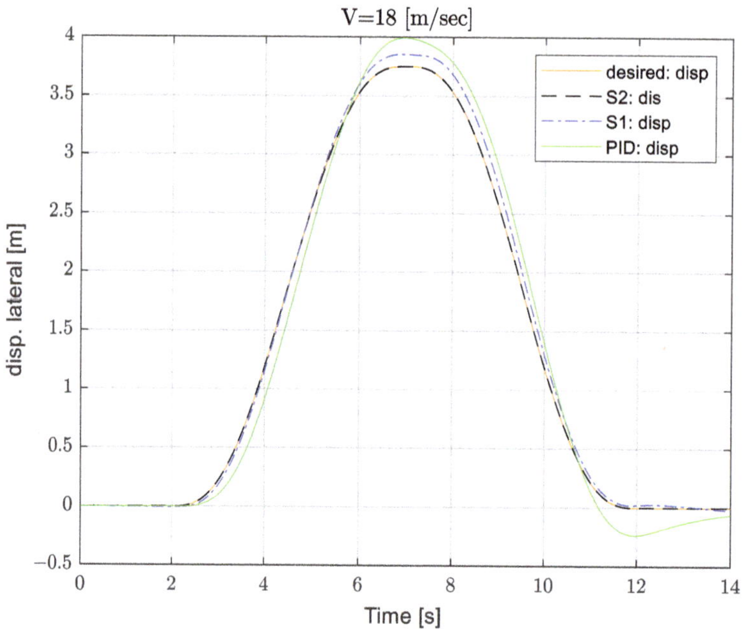

(e)

Figure 5. Simulation results for different control strategies. "S1: first Strategy, S2: second Strategy" and PID controller: (**a**) steering angles; (**b**) lateral velocity; (**c**) the yaw angle; (**d**) the yaw rate; (**e**) the lateral displacement.

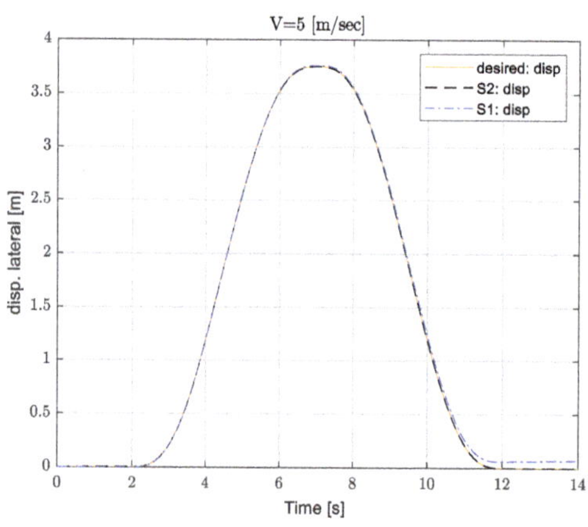

(a)

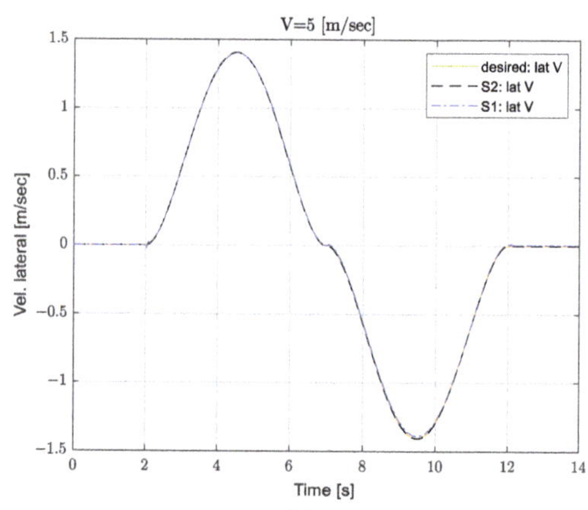

(**b**)

Figure 6. *Cont.*

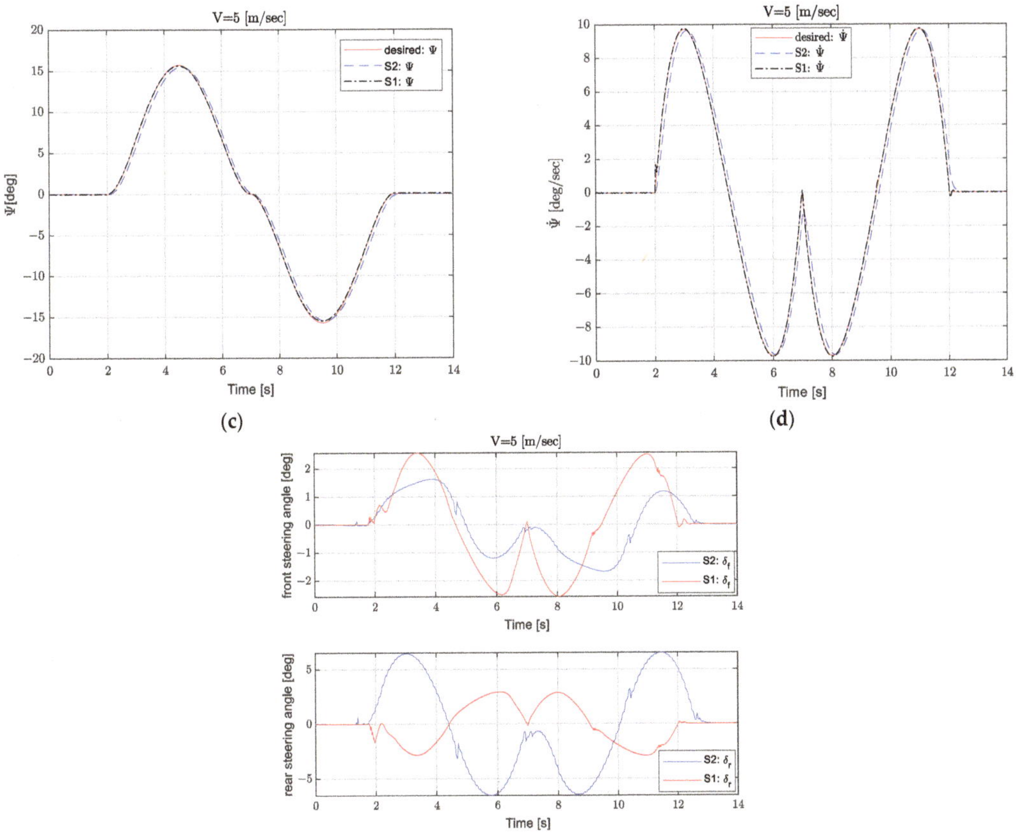

(c) (d) (e)

Figure 6. Simulation results for low velocity (5 m/s = 18 km/h) by two control strategies."S1: first Strategy, S2: second Strategy": (**a**) the lateral displacement; (**b**) lateral velocity; (**c**) the yaw angle; (**d**) the yaw rate; (**e**) steering angles.

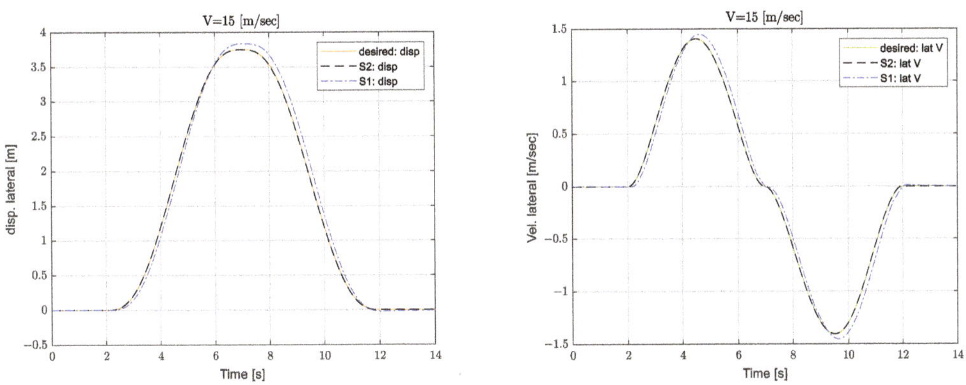

(a) (b)

Figure 7. *Cont.*

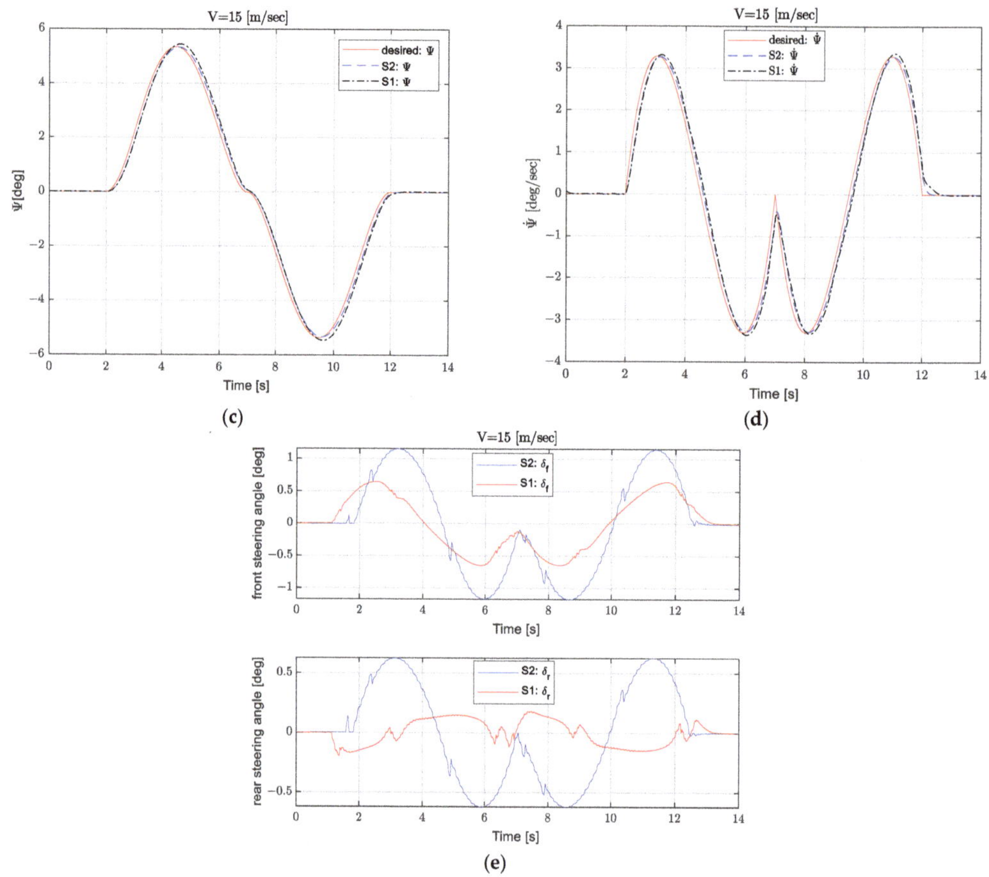

Figure 7. Simulation results for moderate velocity (15 m/s = 54 km/h) by two control strategies. "S1: first Strategy, S2: second Strategy": (**a**) the lateral displacement; (**b**) lateral velocity; (**c**) the yaw angle; (**d**) the yaw rate; (**e**) steering angles.

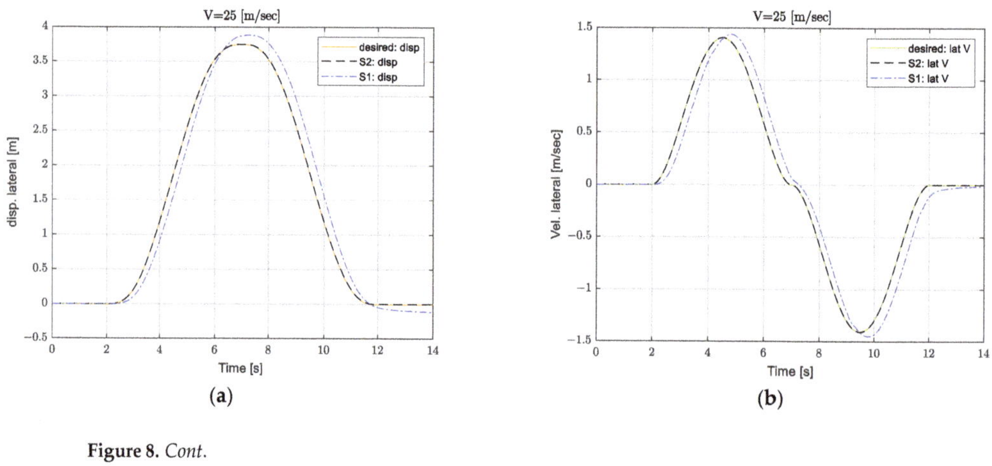

Figure 8. *Cont.*

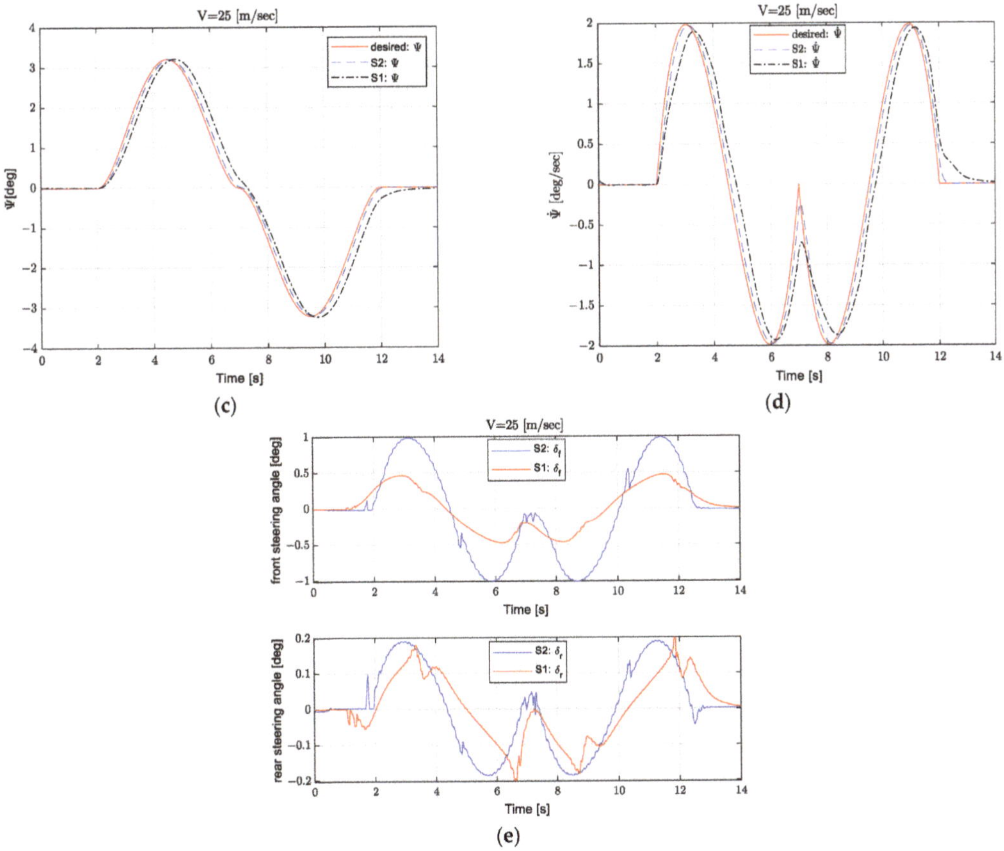

Figure 8. Simulation results for high velocity (25 m/s = 90 km/h) by two control strategies."S1: first Strategy, S2: second Strategy": (**a**) the lateral displacement; (**b**) lateral velocity; (**c**) the yaw angle; (**d**) the yaw rate; (**e**) steering angles.

As shown in Figures 6, 7 and 8a–e, double-lane change road simulation results exposed to different longitudinal velocities are compared. Figures 6, 7 and 8a–d display the lateral position, lateral velocity, yaw angle, and yaw rate, which pointed out that the errors increase according to the increase in longitudinal speed. Although the vehicle longitudinal velocity is subjected to changes, the path-tracking activity does not break down violently. Examining the lateral displacements by the first strategy in three cases shows that the maximum vehicle lateral drift is 0 m, 0.12 m, and 0.18 m, respectively, which means that the vehicle tracks the lane perfectly with decreasing speed. It can be seen from Figures 6, 7 and 8d that 4WS AV with the first and second block backstepping control strategies can well track the reference yaw rate, but at the end of the tracking path vehicle with the first control strategy can't quickly return at high lateral speed. AV vehicle by the second block backstepping control strategy can well track yaw rate. Figures 6, 7 and 8e depict the front and rear wheel angles. They are totally different with two kinds of controllers. Where, at low velocity, the first strategy constrained the rear wheels to follow the front wheels but with different angle values, while the rear wheels at the second strategy were in the inverse orientation to the front wheels, as explained in Figure 9. At moderate speed, the first strategy also constrained the rear wheels to follow the front wheels but with different angle values, while the rear wheels at the second strategy approximately are at zero angle. Moreover, at high speed, the first strategy still constrained the rear wheels to follow the front wheels, but with different angle values, while the rear wheels at the second strategy moved in the same direction as the front wheels. When the vehicle completes the desired trajectory movement and comes back to its horizontal lane, a slight fluctuation at the front and rear steering angles will occur, which increases with speed.

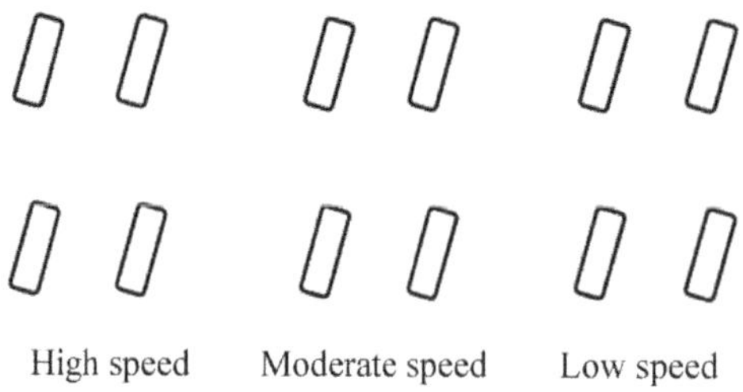

(a)

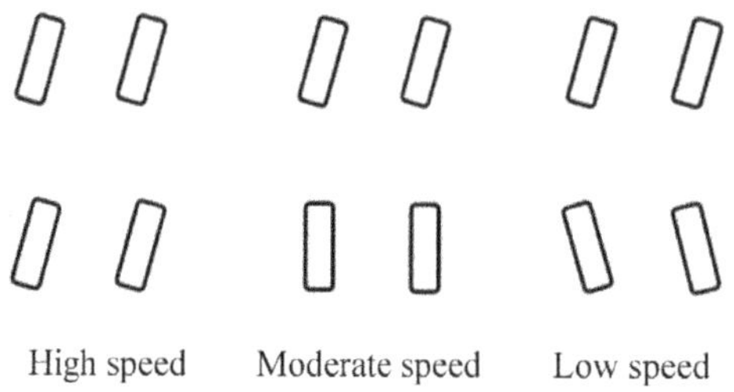

(b)

Figure 9. The orientation of the front and rear wheels by: (**a**) first control strategy, and (**b**) second control strategy.

However, it can be seen from Figures 6, 7 and 8a that the path tracking output of 4WS AV with the second block backstepping control strategy excels that with the first control strategy. The simulation outcomes indicate that the maneuverability of 4WS AV exposed to high speed with the second block backstepping control strategy is better than that with the first control strategy.

Aiming to prove the tracking performance for the reference model of both suggested controllers and the vehicle performance of the research subject, double lane change simulation tests in which 4WS AV moves on a high coefficient of friction (μ = 1) road and moves on a coefficient of friction (μ = 0.6) road at a velocity of 18 m/s (65 Km/h) are presented. Furthermore, simulation results with the first and second block backstepping control strategies are compared to inspect the performance of controllers.

Figure 10 exhibits the simulation outputs of double-lane change movement on dissimilar road coefficients of friction (wet road μ = 0.6, dry road μ = 1). The peak values of the controller outputs (steering angles) governed by high friction coefficient are lower than that prone to low friction coefficient as can be shown obviously in Figure 10e, and the fluctuations prone to low friction coefficient are noticeable. Another observation, the peak values of the controller outputs (steering angles) exposed to the first strategy are lower than those under the second strategy.

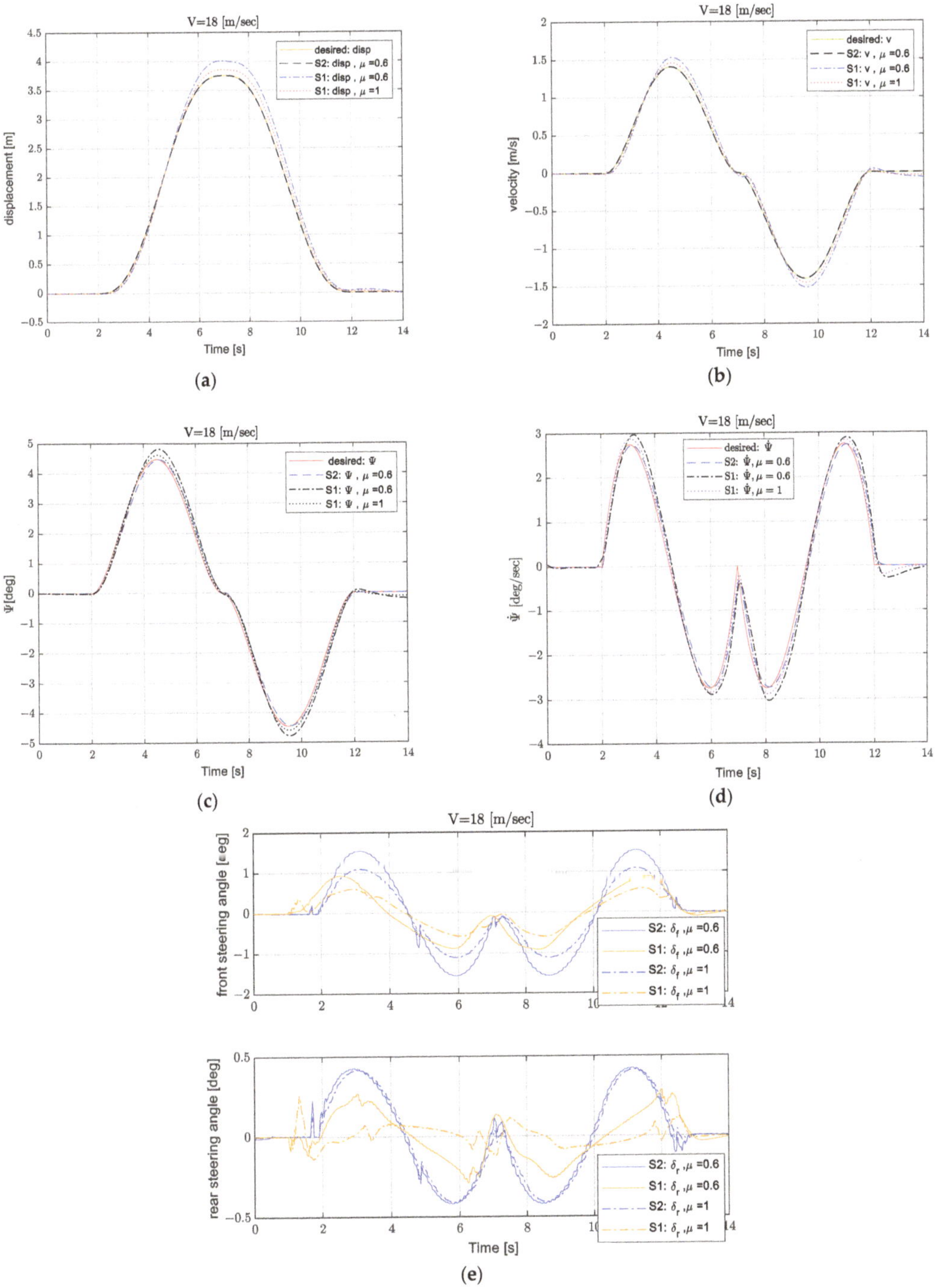

Figure 10. Simulation results for different friction values (μ = 0.6 & 1) by two control strategies. "S1: first Strategy, S2: second Strategy": (**a**) the lateral displacement; (**b**) lateral velocity; (**c**) the yaw angle; (**d**) the yaw rate; (**e**) steering angles.

In Figure 10b,d, the lateral velocities and yaw rates are seen when the vehicle is subjected to different road frictions, respectively. These figures indicate that when road roughness increases, it leads to a reduction of the amplitudes of both lateral velocity and yaw rate. Figure 10a,c exhibit the path-tracking of yaw angle and lateral position, where obviously the path-tracking errors in wet road conditions are larger than that in high roughness road (the maximum vehicle lateral drift by the first strategy is 0.25 m, while it is 0.12 m by the 2nd strategy). It is clear that the 4WS AV has better path-tracking performance under high-roughness roads. Moreover, the path-tracking errors are insignificant despite existing in the low-roughness road, which indicates that the designed controllers have excellent stability and good activity versus various road roughness, especially by using the second control strategy.

Overall, a comparison of the first and second block backstepping control strategy for enhancing path-tracking maneuver indicates that the second block backstepping control strategy has better path-tracking potential relative to the first block backstepping control strategy. Moreover, the robust performance of the second block backstepping control strategy is validated even in the existence of variation of longitudinal speed and road surface roughness.

6. Conclusions and Future Works

In this research, two adaptive schemes model-based backstepping state feedback controllers were designed and illustrated for trajectory tracking of underactuated 2DOF 4WS AV (MIMO) systems. In the first strategy, block backstepping was applied to the entire system. To enhance the tracking performance, an integral influence was inserted into this backstepping controller. Global stability was demonstrated by Lyapunov theory and Barbalat's Lemma. In the second strategy, block backstepping was applied after a turnover of the high-order model into multiple first-order subsystems, including Lyapunov-based design and stability guarantee.

To show the efficiency of our proposed methods, double-lane change maneuvers were performed at three different longitudinal speeds and different road conditions in tracking the ideal path. Beyond that, the double-lane change maneuver was performed by the PID controller for comparison with both proposed strategies. Depending on simulations in CarSim/Simulink and theoretical analysis, the outcomes of this research were addressed by the following points:

i. According to the model following control strategies, the transient response of 4WS AV, lateral displacement, yaw angle, lateral velocity, and yaw rate can trace the reference model exactly. Therefore, the vehicle performance and stability are enhanced greatly.
ii. Unlike presented in some control approaches, the rear wheels haven't steered as a fixed proportion of the front steering angle because of vehicle yaw rate and lateral acceleration response.
iii. From the simulation outcomes, it can be concluded that the proposed strategies achieve the effectiveness of desired performance. The second strategy controller is more accurate, simpler structured, and easy to implement.

Finally, this research has specifically performed the test supposing at a constant speed and constant road coefficient of friction without confirming the two proposed block backstepping control strategies under the condition of time-varying speed and changing road coefficient of friction. More importantly, the detailed trajectory tracking control of 4WS autonomous vehicles has enabled optimized path planning, which minimizes energy usage and environmental impact and sustainability. In this aspect, an optimization algorithm for calculating the excellent numerical amounts of the design constants for the block backstepping controllers will be conducted in future research instead of utilizing the trial and error method.

In the future, we intend to apply these strategies to the real world to explore practical aspects, including hardware limitations, sensor inaccuracies, or latency in real-time implementation. Furthermore, these proposed control strategies improve vehicle performance using simple algebraic equations without the need for enormous sets of experimental data as in NNC nor the need for human expertise as in FC to build the fuzzy rules, and without high computations such as in MPC, comparisons with other advanced controllers, in future work, would provide more comprehensive validation of these proposed methods.

Author Contributions: Conceptualization, M.I.A.-s.; methodology, M.I.A.-s.; software, M.I.A.-s.; validation, M.I.A.-s. and H.M.A.A.A.-b.; formal analysis, M.I.A.-s. and H.M.A.A.A.-b.; investigation, M.I.A.-s. and H.M.A.A.A.-b.; resources, M.I.A.-s. and H.M.A.A.A.-b.; data curation, M.I.A.-s. and H.M.A.A.A.-b.; writing—original draft preparation, M.I.A.-s. and H.M.A.A.A.-b.; writing—review and editing, M.I.A.-s. and H.M.A.A.A.-b.; visualization, M.I.A.-s. and H.M.A.A.A.-b.; supervision, M.I.A.-s. and H.M.A.A.A.-b.; project administration, M.I.A.-s. and H.M.A.A.A.-b.; funding acquisition, M.I.A.-s. and H.M.A.A.A.-b. All authors have read and agreed to the published version of the manuscript.

Funding: This research received no external funding.

Data Availability Statement: Data are contained within the article.

Conflicts of Interest: The author declares no conflicts of interest.

References

1. Fijalkowski, B. Steer-By-Wire (SBW) Four-Wheel Steering (4WS) Conversion Mechatronic Control Systems. In *Automotive Mechatronics: Operational and Practical Issues*; Springer: Berlin/Heidelberg, Germany, 2011; pp. 73–115.
2. Goodarzi, A.; Mohammadi, M. Stability enhancement and fuel economy of the 4-wheel-drive hybrid electric vehicles by optimal tyre force distribution. *Veh. Syst. Dyn.* **2014**, *52*, 539–561. [CrossRef]
3. Lipu, M.S.H.; Mamun, A.A.; Ansari, S.; Miah, M.S.; Hasan, K.; Meraj, S.T.; Abdolrasol, M.G.; Rahman, T.; Maruf, M.H.; Sarker, M.R. Battery management, key technologies, methods, issues, and future trends of electric vehicles: A pathway toward achieving sustainable development goals. *Batteries* **2022**, *8*, 119. [CrossRef]
4. Deb, N.; Alam, M.; Rahman, T.; Jami, M.; Bt Mansor, M.; Tajuddin, H. Design and Analysis of a Fuel Cell and Batteries in Energy Production for Electric Vehicle. *Iran. J. Energy Environ.* **2023**, *14*, 301–313. [CrossRef]
5. Shaker, F.N.; Obed, A.A.; Abid, A.J.; Saleh, A.L.; Hassoon, R.J. Energy Management Strategy for PV PSO MPPT/Fuel Cell/Battery Hybrid System with Hydrogen Production and Storage. *J. Tech.* **2023**, *5*, 52–60. [CrossRef]
6. Chebly, A.; Talj, R.; Charara, A. Coupled longitudinal and lateral control for an autonomous vehicle dynamics modeled using a robotics formalism. *IFAC-PapersOnLine* **2017**, *50*, 12526–12532. [CrossRef]
7. Chaudhari, S.; Shendge, P.D.; Phadke, S.B. Disturbance observer based controller under noisy measurement for tracking of nDOF uncertain mismatched nonlinear interconnected systems. *IEEE/ASME Trans. Mechatron.* **2020**, *25*, 1600–1611. [CrossRef]
8. Zhang, J.; Zheng, H.; Zhao, M. *Analysis of Vehicle Steering Stability of Nonlinear Four Wheel Steering Based on Sliding Mode Control*; SAE Technical Paper; SAE International: Warrendale, PA, USA, 2018; ISSN 0148-7191.
9. Li, B.; Yu, F. Optimal model following control of four-wheel active steering vehicle. In Proceedings of the 2009 International Conference on Information and Automation, Zhuhai/Macau, China, 22–24 June 2009; IEEE: New York, NY, USA, 2009; pp. 881–886.
10. Wang, R.; Yin, G.; Zhuang, J.; Zhang, N.; Chen, J. The path tracking of four-wheel steering autonomous vehicles via sliding mode control. In Proceedings of the 2016 IEEE Vehicle Power and Propulsion Conference (VPPC), Hangzhou, China, 17–20 October 2016; IEEE: New York, NY, USA, 2016; pp. 1–6.
11. Sano, S.; Furukawa, Y.; Shiraishi, S. Four wheel steering system with rear wheel steer angle controlled as a function of sterring wheel angle. *SAE Trans.* **1986**, *95*, 880–893.
12. Ulsoy, A.G.; Peng, H.; Çakmakci, M. *Automotive Control Systems*; Cambridge University Press: Cambridge, UK, 2012.
13. Yu, S.; Wang, J.; Wang, Y.; Chen, H. Disturbance observer based control for four wheel steering vehicles with model reference. *IEEE/CAA J. Autom. Sin.* **2016**, *5*, 1121–1127. [CrossRef]
14. Haytham, A.; Elhalwagy, Y.; Wassal, A.; Darwish, N. Modeling and simulation of four-wheel steering unmanned ground vehicles using a PID controller. In Proceedings of the 2014 International Conference on Engineering and Technology (ICET), Cairo, Egypt, 19–20 April 2014; IEEE: New York, NY, USA, 2014; pp. 1–8.
15. Krishna, S.; Narayanan, S.; Ashok, S.D. Fuzzy logic based yaw stability control for active front steering of a vehicle. *J. Mech. Sci. Technol.* **2014**, *28*, 5169–5174. [CrossRef]
16. Gao, L.; Jin, L.; Zheng, Y.; Li, K.-Y. Design of radial basis function neural network compound controller for four-wheel steering vehicle. *J. Jilin Univ. Eng. Technol. Ed.* **2016**, *46*, 366–372.
17. Napolitano Dell'Annunziata, G.; Ruffini, M.; Stefanelli, R.; Adiletta, G.; Fichera, G.; Timpone, F. Four-Wheeled Vehicle Sideslip Angle Estimation: A Machine Learning-Based Technique for Real-Time Virtual Sensor Development. *Appl. Sci.* **2024**, *14*, 1036. [CrossRef]
18. Noubissie Tientcheu, S.I.; Du, S.; Djouani, K.; Liu, Q. Data-Driven Controller for Drivers' Steering-Wheel Operating Behaviour in Haptic Assistive Driving System. *Electronics* **2024**, *13*, 1157. [CrossRef]

19. Xu, F.-X.; Liu, X.-H.; Chen, W.; Zhou, C.; Cao, B.-W. Improving handling stability performance of four-wheel steering vehicle based on the H2/H∞ robust control. *Appl. Sci.* **2019**, *9*, 857. [CrossRef]
20. Mashadi, B.; Ahmadizadeh, P.; Majidi, M. *Integrated Controller Design for Path Following in Autonomous Vehicles*; SAE Technical Paper; SAE International: Warrendale, PA, USA, 2011; ISSN 0148-7191.
21. Rumetshofer, J.; Stolz, M.; Watzenig, D. A generic interface enabling combinations of state-of-the-art path planning and tracking algorithms. *Electronics* **2021**, *10*, 788. [CrossRef]
22. Hiraoka, T.; Nishihara, O.; Kumamoto, H. Automatic path-tracking controller of a four-wheel steering vehicle. *Veh. Syst. Dyn.* **2009**, *47*, 1205–1227. [CrossRef]
23. Kim, J.; Park, C.; Hwang, S.; Hori, Y.; Kim, H. Control algorithm for an independent motor-drive vehicle. *IEEE Trans. Veh. Technol.* **2010**, *59*, 3213–3222. [CrossRef]
24. Norouzi, A.; Adibi-Asl, H.; Kazemi, R.; Hafshejani, P.F. Adaptive sliding mode control of a four-wheel-steering autonomous vehicle with uncertainty using parallel orientation and position control. *Int. J. Heavy Veh. Syst.* **2020**, *27*, 499–518. [CrossRef]
25. Huang, B.; Wu, S.; Huang, S.; Fu, X. Lateral stability control of four-wheel independent drive electric vehicles based on model predictive control. *Math. Probl. Eng.* **2018**, *2018*, 6080763. [CrossRef]
26. Lin, C.; Peng, C.L. Mixed H∞/H2 Output Feedback Stability Control for Dual-Motor Independent Drive Electric Vehicle. In *Advanced Materials Research*; Trans Tech Publications, Ltd.: Zurich, Switzerland, 2013; pp. 602–608. [CrossRef]
27. You, S.-S.; Chai, Y.-H. Multi-objective control synthesis: An application to 4WS passenger vehicles. *Mechatronics* **1999**, *9*, 363–390. [CrossRef]
28. Zhang, C.; Liu, H.; Dang, M. Robust Shared Control for Four-Wheel Steering Considering Driving Comfort and Vehicle Stability. *World Electr. Veh. J.* **2023**, *14*, 283. [CrossRef]
29. Alfi, A.; Farrokhi, M. Hybrid state-feedback sliding-mode controller using fuzzy logic for four-wheel-steering vehicles. *Veh. Syst. Dyn.* **2009**, *47*, 265–284. [CrossRef]
30. Fujimoto, H.; Tsumasaka, A.; Noguchi, T. Vehicle stability control of small electric vehicle on snowy road. *Rev. Automot. Eng.* **2006**, *27*, 279–286.
31. Kawashima, K.; Uchida, T.; Hori, Y. Rolling stability control of in-wheel electric vehicle based on two-degree-of-freedom control. In Proceedings of the 2008 10th IEEE International Workshop on Advanced Motion Control, Trento, Italy, 26–28 March 2008; IEEE: New York, NY, USA, 2008; pp. 751–756.
32. Yakub, F.; Abu, A.; Sarip, S.; Mori, Y. Study of model predictive control for path-following autonomous ground vehicle control under crosswind effect. *J. Control. Sci. Eng.* **2016**, *2016*, 6752671. [CrossRef]
33. Falcone, P.; Eric Tseng, H.; Borrelli, F.; Asgari, J.; Hrovat, D. MPC-based yaw and lateral stabilisation via active front steering and braking. *Veh. Syst. Dyn.* **2008**, *46* (Suppl. S1), S611–S628. [CrossRef]
34. Hao, Y.; Wang, J.; Chepinskiy, S.A.; Krasnov, A.J.; Liu, S. Backstepping based trajectory tracking control for a four-wheel mobile robot with differential-drive steering. In Proceedings of the 2017 36th Chinese Control Conference (CCC), Dalian, China, 26–28 July 2017; IEEE: New York, NY, USA, 2017; pp. 4918–4923.
35. Rudra, S.; Barai, R.K.; Maitra, M. Nonlinear state feedback controller design for underactuated mechanical system: A modified block backstepping approach. *ISA Trans.* **2014**, *53*, 317–326. [CrossRef]
36. Kang, C.M.; Kim, W.; Chung, C.C. Observer-based backstepping control method using reduced lateral dynamics for autonomous lane-keeping system. *ISA Trans.* **2018**, *83*, 214–226. [CrossRef]
37. Kim, W.; Kang, C.M.; Son, Y.S.; Chung, C.C. Nonlinear backstepping control design for coupled nonlinear systems under external disturbances. *Complexity* **2019**, *2019*, 7941302. [CrossRef]
38. Kang, C.M.; Kim, W.; Baek, H. Cascade backstepping control with augmented observer for lateral control of vehicle. *IEEE Access* **2021**, *9*, 45367–45376. [CrossRef]
39. Wang, C.; He, R.; Jing, Z.; Chen, S. Coordinated Path Following Control of 4WID-EV Based on Backstepping and Model Predictive Control. *Energies* **2022**, *15*, 5728. [CrossRef]
40. Kim, M.; Lee, D.; Ahn, J.; Kim, M.; Park, J. Model predictive control method for autonomous vehicles using time-varying and non-uniformly spaced horizon. *IEEE Access* **2021**, *9*, 86475–86487. [CrossRef]
41. Li, X.; Xie, N.; Wang, J. A variable weight adaptive cruise control strategy based on lane change recognition of leading vehicle. *Automatika* **2022**, *63*, 555–571. [CrossRef]
42. Son, Y.S.; Kim, W. Nonlinear Differential Braking Control for Collision Avoidance during Lane Change. *Mathematics* **2021**, *9*, 1699. [CrossRef]
43. Azad, N.L. Dynamic Modelling and Stability Controller Development for Articulated Steer Vehicles. Ph.D. Thesis, University of Waterloo, Waterlo, ON, Canada, January 2007.
44. Termous, H.; Shraïm, H.; Talj, R.; Francis, C.; Charara, A. Coordinated control strategies for active steering, differential braking and active suspension for vehicle stability, handling and safety improvement. *Veh. Syst. Dyn.* **2019**, *57*, 1494–1529. [CrossRef]
45. Pang, H.; Yao, R.; Wang, P.; Xu, Z. Adaptive backstepping robust tracking control for stabilizing lateral dynamics of electric vehicles with uncertain parameters and external disturbances. *Control. Eng. Pract.* **2021**, *110*, 104781. [CrossRef]
46. Moghanni-Bavil-Olyaei, M.-R.; Ghanbari, A.; Keighobadi, J. Trajectory tracking control of a class of underactuated mechanical systems with nontriangular normal form based on block backstepping approach. *J. Intell. Robot. Syst.* **2019**, *96*, 209–221. [CrossRef]

47. Akgün, O.; Türker, T. An Adaptive Block Backstepping Controller For A Quadrotor. In Proceedings of the 2018 6th International Conference on Control Engineering & Information Technology (CEIT), Istanbul, Turkey, 25–27 October 2018; IEEE: New York, NY, USA, 2018; pp. 1–6.
48. Chang, Y.; Cheng, C.-C. Block backstepping control of multi-input nonlinear systems with mismatched perturbations for asymptotic stability. *Int. J. Control* **2010**, *83*, 2028–2039. [CrossRef]
49. Rudra, S.; Barai, R.K.; Maitra, M. *Block Backstepping Design of Nonlinear State Feedback Control Law for Underactuated Mechanical Systems*; Springer: Berlin/Heidelberg, Germany, 2017.
50. Rudra, S.; Barai, R.K.; Maitra, M. Design and implementation of a block-backstepping based tracking control for nonholonomic wheeled mobile robot. *Int. J. Robust Nonlinear Control* **2016**, *26*, 3018–3035. [CrossRef]
51. Dong, Z.; Wan, L.; Li, Y.; Liu, T.; Zhang, G. Trajectory tracking control of underactuated USV based on modified backstepping approach. *Int. J. Nav. Archit. Ocean. Eng.* **2015**, *7*, 817–832. [CrossRef]
52. Krafes, S.; Chalh, Z.; Saka, A. Backstepping Controller Design for a 5 DOF Spherical Inverted Pendulum. *Int. J. Eng. Res. Afr.* **2019**, *41*, 37–50. [CrossRef]
53. Ackermann, J.; Bartlett, A.; Kaesbauer, D.; Sienel, W.; Steinhauser, R. *Robust Control: Systems with Uncertain Physical Parameters*; Springer: Berlin/Heidelberg, Germany, 1993.
54. Krstic, M.; Kokotovic, P.V.; Kanellakopoulos, I. *Nonlinear and Adaptive Control Design*; John Wiley & Sons, Inc.: Hoboken, NJ, USA, 1995.
55. Khalil, H.K. *Nonlinear Systems*, 3rd ed.; Prentice Hall: Saddle River, NJ, USA, 2002; pp. 322–329.
56. Al-Saedi, M.I. A Modified Block Backstepping Controller for Autonomous Vehicle Lane Changing. *Int. J. Intell. Eng. Syst.* **2023**, *16*, 240–254.
57. Jazar, R.N. Road Dynamics. In *Advanced Vehicle Dynamics*; Springer: Berlin/Heidelberg, Germany, 2019; pp. 297–350.

processes

MDPI

Review

Bubble Dynamics in Sustainable Technologies: A Review of Growth, Collapse, and Heat Transfer

Farhan Lafta Rashid [1], Mudhar A. Al-Obaidi [2,3,*], Ahmed Kadhim Hussein [4], Shabbir Ahmad [5], Mahmood Shaker Albdeiri [6] and Iqbal M. Mujtaba [7]

1 Petroleum Engineering Department, College of Engineering, University of Kerbala, Karbala 56001, Iraq; farhan.lefta@uokerbal.edu.iq
2 Technical Institute of Baquba, Middle Technical University, Baghdad 10074, Iraq
3 Technical Instructor Training Institute, Middle Technical University, Baghdad 10074, Iraq
4 Mechanical Engineering Department, College of Engineering, University of Babylon, Babylon 51002, Iraq; ahmedkadhim7474@gmail.com
5 Institute of Geophysics and Geomatics, China University of Geosciences, Wuhan 430074, China; shabbiraleem@cug.edu.cn
6 Mechanical Power Technical Engineering Department, College of Engineering and Technology, Al-Mustaqbal University, Babylon 51001, Iraq; mahmoodshaker@uomus.edu.iq
7 Department of Chemical Engineering, Faculty of Engineering and Informatics, University of Bradford, Bradford BD7 1DP, UK; i.m.mujtaba@bradford.ac.uk
* Correspondence: dr.mudhar.alaubedy@mtu.edu.iq; Tel.: +964-771-506-1121

Abstract: The study of bubble growth and collapse is of great significance in the context of sustainability due to its influence on numerous energy-related processes and technologies. Understanding the dynamics of bubble behavior is vital for optimising heat transfer efficiency, which has an energetic role in improving the performance of sustainable systems such as nuclear reactors, thermal inkjet printing, and nucleate boiling. Indeed, researchers can progress strategies to enhance the efficiency of these technologies by analysing the parameters influencing bubble growth and collapse, which can lead to reduced energy consumption and environmental impact. Although several theoretical models and experimental investigations have been achieved in the past to inspect bubble growth and collapse, a thorough review and critical assessment of the studies conducted have not yet been achieved. This review aims to provide a comprehensive understanding of the relationship between bubble dynamics and sustainability, highlighting the potential for further research and development in this area. Specifically, the scope and limitations of past research on bubble growth and collapse is conducted to fill this gap in the open literature. The review covers both numerical and experimental studies of bubble growth and collapse in a wide set of innovative industrial applications including nuclear reactors, thermal inkjet printing, nucleate boiling, hydrodynamic erosion, and ultrasonic and medicinal therapy. The current review also attempts to illustrate and evaluate the numerical methods used and underlines the most relevant results from the studies that were looked at in order to provide researchers with a clear picture of the growth and collapse of bubbles in different applications. The results give a precise understanding of the dynamics of bubble growth and collapse and the related temperature change and cumulative heat transmission from the thermal boundary layer. Additionally, it has been demonstrated that simulation-based models can effectively predict transport coefficients. However, the review observes a number of limitations of the past research on bubble growth and collapse. Due to numerical instability, very little work with respect to dynamic modelling has been carried out on the mechanisms of bubble collapse. Accordingly, a number of recommendations are made for the improvement of heat transmission during bubble growth and collapse. Specifically, future criteria for the highest heat transmission will demand more precise experimental and numerical approaches.

Keywords: bubble growth; collapse; cavitation; hydrodynamics; bubble dynamics; phase change

1. Introduction

The growth and collapse of bubbles in fluids can have significant impacts on a variety of physical processes and engineering applications, including subcooled nucleate boiling, bubble condensation, and liquid cavitation. In particular, as the liquid is being boiled, the bubbles will rise and promote the mixing of the liquid's molecules and heat transmission. The procedures are intrinsically complicated due to the occurrence of phase-change heat transfer in a moving interface, which is characterised by the variable radius of the bubble. This makes the operations more difficult [1].

It is crucial to have a thorough appreciation of the bubble dynamics that take place throughout nucleate boiling in order to accomplish operative and reliable heat transfer in several industrial processes, such as distillation columns and refrigeration [2]. Specifically, bubbles are created on a surface immersed in a liquid that is below its boiling point during the subcooled nucleate boiling process. Due to the heat transfer from the surface, the liquid boils at the bubble site with the expansion of the bubbles. The bubble separates from the surface as the bubble reaches a sufficient size, rises to the surface, and then bursts. High-speed liquid jets may be created when the bubble bursts, which could recover heat transmission from the surface [3]. Throughout the boiling process, it is well acknowledged that changes in the shape of the bubbles and the dynamics of the contact lines can have a noteworthy influence on heat transmission. Due to the controlling role that they play in heat transmission and hydrodynamics, the bubble departure frequency, departure diameter, and growth rate are the subjects of a significant amount of research [4].

As the phenomenon of bubble condensation happens, a bubble may condense and collapse when it touches the boundary between a liquid's surface and a gas phase. A fast-moving liquid jet is created when a bubble bursts, which can erode or damage surrounding surfaces. The collapse of bubbles is required in some applications, such as inkjet printing, since it can facilitate the production of tiny liquid droplets that are extruded from nozzles [5].

Cavitation, a physical phenomenon in liquids, is the act of rupturing a liquid by lessening the pressure at a relatively fixed liquid temperature. As the liquid pressure drops below its vapor pressure, bubbles start to develop and eventually collapse, which is known as liquid cavitation. These bubbles have the potential to expand and then burst, sending high velocity liquid jets and shockwaves toward nearby objects. Pumps, turbines, and propellers may experience cavitation in engineering applications, resulting in decreased performance or even failure [6,7]. Cavitation methods are specifically ultrasonic or ultrasound cavitation techniques. Under high-intensity sound wave irradiation (in industrial processes between 20 and 50 kHz), acoustic cavitation happens in liquids. Yet, the passageway of a liquid through a constriction, such as an orifice plate, throttling valve, venturi, etc., produces hydrodynamic cavitation [8]. The cavitation process is very attractive because it produces large amounts of local energy in the bulk liquid at a low cost, has a great degree of flexibility, and minimises environmental damage [9]. In nucleate boiling, bubble condensation, and liquid cavitation processes, the growth and collapse of bubbles play a substantial role in the overall behavior of systems. Understanding the dynamics of bubble growth and collapse can help engineers to design more efficient and reliable systems and improve our understanding of fundamental physical processes.

Both numerical and experimental studies were carried out concerning the growth and collapse of bubbles in different industrial applications. The ability of numerical modelling to forecast dynamic behavior and comprehend the mechanisms of heat transfer involved in nucleate boiling heat transfer has proven to be very promising. This potential can be ascribed to the rapid advancement of computational resources and the enhancement of various multi-phase models.

Both direct and indirect measurements of the bubble-induced pressure pulses have resulted in the development of a number of different methods over the years. Acoustic transients, such as shockwaves, micro-jets, turbulence, shear forces, etc., released during the collapse of a bubble are collected utilising a hydrophone at a distance from the centre of the collapse. These transients are then utilised to estimate the pressure that was exerted during the collapse [10,11]. During the collapsing bubble-driven pumping, the following effects are caused by a rise in the viscosity of the fluid: (1) an upsurge in the stability of the bubble oscillations, which restricts the creation of a jet that is vital in the operation of the pump; and (2) an upsurge in the viscous pressure head that has to be confronted in order to push the liquid through the hole [12].

The authors believe that the body of published research does not offer a thorough analysis of the advancements made in bubble growth and collapse. In this area of research, the drive of this review is to judgmentally evaluate the conducted studies of bubble growth and collapse illustrating the scope and limitations, innovative configurations, numerical methods, in addition to drawing some of the foremost findings from the existing open literature and identifying a number of unresolved questions for future studies. The scope of this review encompasses a broad diversity of subject areas, one of which is the configuration of experimental or numerical studies on bubble growth and collapse. This article goes into considerable detail about technology strategies for bubble growth and collapse, covering research, application, and development techniques. The consequences of this review may deliver a map for further investigations as they would make it easier for scientists to understand the numerous advancements made in bubble growth and collapse. On top of this, this review attempts to offer a comprehensive understanding of the link between bubble dynamics and sustainability, emphasizing the potential for further research and development in this region.

2. Review Method

The current review is presented to deliver an outline of research that has examined the scope and limitations of recent innovations and developments concerning the growth and collapse of bubbles. This overview facilitates a comprehensive understanding of the topic. The review is further augmented by a detailed examination of existing published studies. Different databases, including ScienceDirect, SpringerLink, Scopus, ResearchGate, and Google Scholar, were used to find related studies and peer reviewed content. The keywords designed to filter and catch related papers included: bubble growth, collapse, cavitation, hydrodynamics, bubble dynamics, and phase change. Although there were no restrictions on the publication dates of the articles, priority was given to the inclusion of the most successful current published studies between 2004 and 2023 to discourse the most innovative investigations.

3. Bubble Growth Studies

Indeed, it is essential to have a proper knowledge of the bubble dynamics that occur during boiling to achieve successful heat transfer in a wide set of applications. It is widely known that during boiling operations, variations in the bubble form and the dynamics of the contact lines may significantly affect the transport of bubbles [13]. Ebullition factors, including the bubble departure diameter, bubble departure frequency, and growth rate are the focus of a plethora of studies because of the controlling impact they have in heat transmission and hydrodynamics. In most situations where heterogeneous bubble nucleation happens on a solid surface, it is well known that the wettability of the surface will control the bubble shape during the whole ebullition cycle. This section focuses on presenting the associated studies of bubble growth and their influences on heat transmission within a wide set of applications with a critical evaluation. The next sections discuss the relative bubble growth studies that are classified as the liquid–gas and solid–gas studies.

3.1. Liquid–Gas Studies

Agarwal et al. [14] utilised a Volume of Fluid (VOF) based interface approach to model the film boiling and bubble production in water at 219 bars and 373 °C, while satisfying an isothermal horizontal surface. Together with an interface mass transfer model, the Navier–Stokes and thermal energy correlations were resolved. The models account for the impacts of interface mass transfer, surface tension, and conforming latent heat as well as the impact of temperature on the conveyed thermal characteristics (specific heat and thermal conductivity) of vapor. The findings provided a clear understanding of film boiling and yielded quantitative data on irregular periodic bubble release forms as well as on temporal and spatial variable film thickness. In comparison to simulations with constant fluid characteristics, the calculations also accurately anticipated the transport coefficients on the horizontal surface, which were strongly affected by the fluctuations in fluid characteristics.

Mukherjee and Dhir [15] numerically examined the heat transfer and bubble dynamics related to lateral bubble merger when nucleate boiling transitions from partial to fully developed. The 3-dimensional Navier/Stokes equation, the continuity equation, and the energy equation are all resolved by the SIMPLE approach. The Level-Set Method (LSM) was employed, and the liquid vapor interface was captured. The LSM involves solving a partial differential equation (PDE) that tracks the growth of the interface corresponding to time. Numerous bubble mergers in a line and in a plane were calculated, and the wall heat transfer and bubble dynamics were contrasted to the case study of a single bubble. The outcomes demonstrated that the total wall heat transmission is greatly increased by the merger of several bubbles. This improved wall heat transfer results from colder liquid being drawn toward the wall throughout contraction after bubble merger, which traps a layer of liquid between the bubble bases.

In 2005, a connected Level-Set and VOF technique for the simulation of incompressible two-phase flows with surface tension was presented by Tomar et al. [16]. The paired algorithm saves bulk while accurately capturing the intricate interfaces. For various superheat levels, a planar simulation of bubble development in water at a critical pressure was conducted. Analysis has been done on how superheat affects the frequency of bubble formation. Additionally, simulations of bubble formation and film boiling in the refrigerant R134a are carried out at pressures close to and far from the critical pressure points. The ways in which saturation pressure affects frequently bubble development have been investigated. When water is heated to over 15 K, a divergence from the sporadic bubble release is seen. Additionally, the impact of heat flow on instability was studied. For water in close-to-critical state, it has been discovered that a drop in super-heat between 15 K and 10 K causes fluctuations that affect the length of the ebullition cycle.

In 2010, a computer simulation of a vapor bubble boiling in a sub-cooled pool was achieved by Wu and Dhir [17]. By contrasting the findings of the bubble growth domain, while considering bubble departure diameter, the numerical process combining level-set function with moving mesh technique was proven to be effective. For various sub-cooling and varied gravity levels, the estimates of bubble dynamics and heat transmission are shown. The drop in gravity for a certain sub-cooling reduced the heat transfer around the bubble.

In 2016, the progression dynamics of a bubble population were studied by Marchal et al. [18] in a yield stress fluid matrix (a liquid medium (bitumen). A uniform generation of gas was used to drive the bubble population. The situation involves the gradual expansion of bitumen barrels over an extended period of time, inside of which radioactive salts were distributed. Radioactivity results in the creation of hydrogen in a consistent volume throughout the radiolysis of bitumen chains. The results ascertained that the utilised fluids can deform indefinitely if they are subjected to stress over a specific value. Furthermore, it has been demonstrated that, in the event that the population is stationary, yield stress interrupts the classic paradigm of nucleation, bubble formation, and Ostwald ripening in a Newtonian fluid. The study by Marchal et al. [18] was characterised by the use of two different nucleation models (threshold nucleation, continuous nucleation) to explore the impact of the yield stress on the most important metrics of the settling time to the steady state, the max. swelling, and the stationary swelling, as represented in Figures 1–3, respectively. These figures indicate that the two nucleation models produce the same steady state swelling of up to 6 Pa of the yield stress (Figure 2).

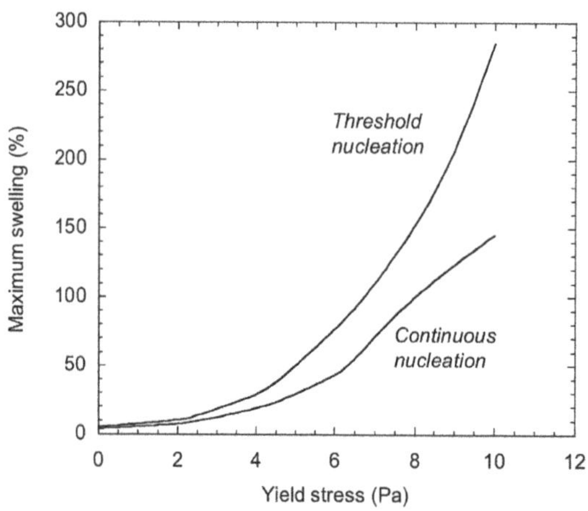

Figure 1. Maximum swelling vs. yield stress [18] "(Reproduced with permission from Journal of Non-Newtonian Fluid Mechanics by Elsevier, 2024; Order Number: 5858171492020)".

Goel et al. [19] used high-speed vision tests with sub-cooled de-mineralized water at 1 atm under nucleate pool boiling conditions to evaluate the bubble departure features such as the bubble departure frequency and diameters. To achieve the parametric examinations of bubble departure behavior, the water pool was designed with dimensions of 300 × 135 × 250 mm and four distinct heating components. The results indicated that the degree of sub-cooling (T_{sub} = 520 K), super-heat (T_{sat} = 110 K), as well as heat flux, parameters of heater configuration, heater inclination, and heater surface roughness were studied.

It was revealed that the heater size, inclination angle, and wall super-heat all increased the departure diameter while liquid sub-cooling and surface roughness dampened it. It was also exposed that the departure rate reduced with an upsurge in heater size but enlarged wall superheat and inclination angle.

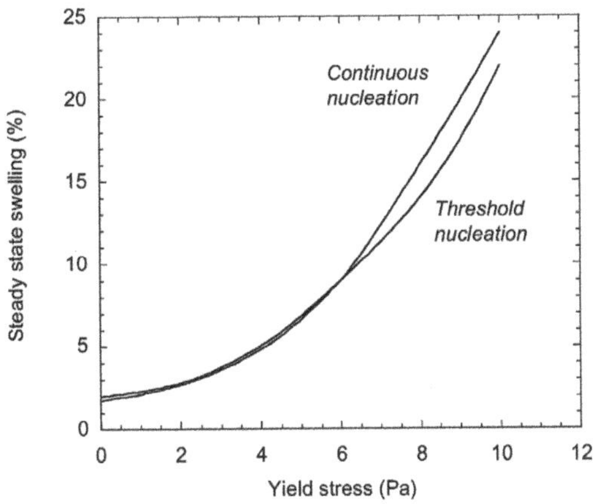

Figure 2. Steady state swelling vs. yield stress [18] "(Reproduced with permission from Journal of Non-Newtonian Fluid Mechanics by Elsevier, 2024; Order Number: 5858171492020)".

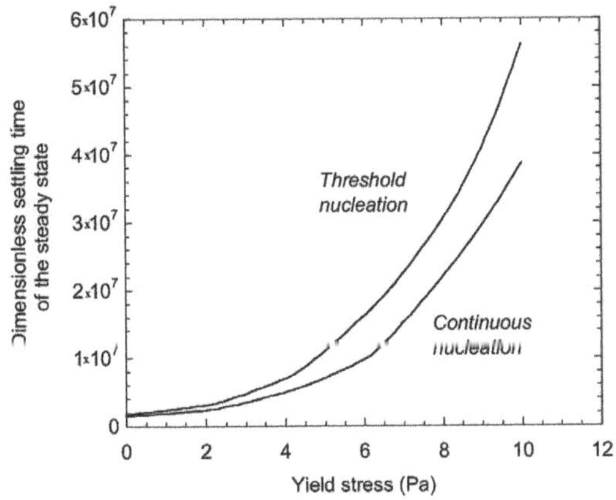

Figure 3. Settling time of the steady state vs. yield stress [18] "(Reproduced with permission from Journal of Non-Newtonian Fluid Mechanics by Elsevier, 2024; Order Number: 5858171492020)".

Thermal cavitating flows in liquid nitrogen surrounding a 2D hydrofoil were inspected by Zhang et al. [20]. The mass transfer homogeneous cavitation model connected to the energy equation and the RNG k-ε turbulence model with an adapted turbulent eddy viscosity were utilised to account for thermal effects. The Clausius-Clapeyron equation was utilised to enhance the saturated vapour pressure in the cavitation model process. Particularly in the thermal area, the enhanced Zwart–Gerber–Belamri model comes to an agreement with the experimental data of Hord et al. of NASA, as established by the expected pressure and temperature within the cavity under cryogenic circumstances. The cavitation dynamics during the phase-change process were significantly affected by the temperature effect, which may cause a delay or suppression in the onset and development of cavitation behavior.

Pahk et al. [21] invented the creation and dynamic performance of a boiling vapor-bubble that happens under boiling histotripsy insolation. Unambiguously, the researchers observed how the bubble expands and contracts over time. Numerical and experimental studies were attained using a high-speed camera to examine the dynamic of bubbles that were created in optically transparent tissue imitating gel phantoms and subjected to 2.0 MHz of the field of a High Intensity Focused Ultrasound (HIFU) transducer. The results indicated that the shockwave's asymmetry and water vapor transport can cause incorrect bubble formation, that might be accountable for HIFU-tempted tissue decellularization.

Using in situ synchrotron radiography, Lu et al. [22] studied bubble development, intermetallic compounds (IMCs) dissolution, and their connections throughout heating of an Al-5 wt.% Mn alloy. As the melt was heated, the pores underwent a slow transition from an irregular shape to a subglobose shape, and then eventually a transformation into spherical bubbles. The radii (r) of bubbles 1 and 3 become smaller as time (t) progresses, and the connection between r and t may be effectively described by a Gaussian function in terms of its shape (Figure 4a). Bubbles 2, 4, and 5 all exhibit a rise in their radii with time, with bubble 5 exhibiting the most rapid expansion rate among the three (Figure 4b). With a rise in temperature, the hydrogen should become more soluble in the aluminum melt, causing the bubble size to diminish throughout the course of the heating process and eventually vanish altogether.

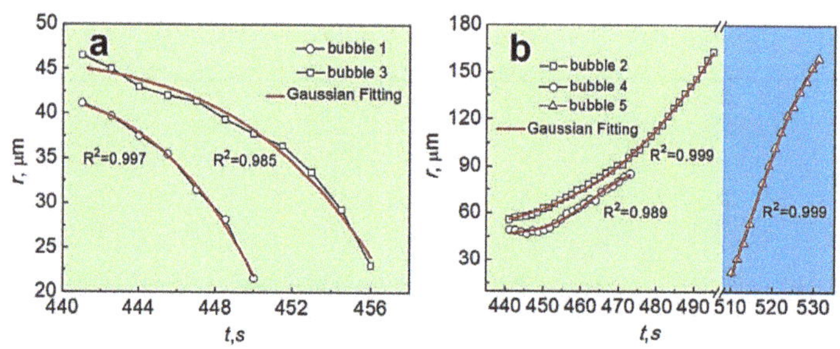

Figure 4. (**a**,**b**) Bubble radii changes against operational time for the five representative bubbles [22]. The correlation of bubble 5 starts after 510 seconds as represented in blue background "(Reproduced with permission from Journal of Alloys and Compounds by Elsevier, 2024; Order Number: 5858181017625)".

In 2020, Zeng et al. [23] exhibited concrete proof for the development of growth styles, shape, mass transfer channels, and lateral growth rate when thickening hydrates on gas (CH_4-C_3H_8) bubbles are suspended in water with inhibitors. These findings were found during the process of thickening growth. During the thickening growth process, an identical and dense-hydrate coating was detected on the bubble as well as at the gas-liquid planar contact. This happened when an inhibitor was not present. The heterogeneous nature of the film instigated a number of circular and cross-shaped craters to perform on its surface. These craters represented as a new conduit for the transfer of mass and sped up the initial thickening and expansion of the hydrate. The hydrate's viscosity was created in the reaction setup and the capacity of the aqueous solution to spread over the reactor wall was instituted to be connected with the enlargement of the morphology of the hydrate.

Allred et al. [24] created a conceptual framework for the wetting and de-wetting processes that occur throughout bubble generation relying on dynamic contact angles. To accurately illustrate the impact of surface wettability on contact angle dynamics and contact lines throughout bubble development and departure, this structure is integrated into VOF simulation. Computers were used to conduct these simulations. It has been established that the early stages of bubble formation are controlled by the declining contact angle. This happens as the contact line becomes farther and farther from a bubble's centre. The primary wetting parameter that controls the greatest departure size and contact diameter is the receding contact angle. Better reduced-order models were then constructed, which resulted in correlations between the greatest departure diameter and contact diameter as functions of the dynamic contact angles, independent of fluid properties.

Zhao et al. [25] employed a phase change lattice Boltzmann to assess the impacts of 2-computational fields with constrained space and a free outlet boundary while observing conjugate heat transfer on the whole nucleate boiling. The model introduced was tested, notably in the enhanced convective outlet boundary for multi-phase flow in detailed information, using Laplace's law and the droplet wetting process. A thorough investigation was carried out into the dynamic behaviors of the whole nucleate boiling process, considering the nucleation, growth, coalescence, departure, ascent, and ultimate rupture. Although growth and departure were significantly assisted when more isolated bubbles coalesced, the dynamic flow behaviors of bubbles' coalescence are essentially similar. Additionally, the researchers demonstrated that the volume of escalating bubbles in the free outlet field increased over time. It was found that the pressure impact between the inside of the bubble and its surroundings was the main cause of this.

Winder et al. [26] used a stainless-steel framework called the SNS target module which is designed to hold mercury. In order to lessen stresses and prevent cavitation, bubbles of helium are inserted into the mercury. This gas makes it more challenging to precisely guesstimate the physical reaction of the target module. It is essential to have the skill to simulate the reaction of the target vessel before endeavoring to estimate its life. A forceful model for mercury with injected gas bubbles has been offered with the intention of delivering a high-accuracy mercury–vessel reply behavior without the obligatory level of processing expenditures to screen individual bubbles.

Li et al. [27] appraised the influence of thermodynamic effects on the cavitation flow in a liquid oxygen turbo-pump. To couple the physical parameters with temperature, the thermodynamic model based on the Zwart–Gerber–Belamri cavitation model and the shear stress transport k–ω turbulence model were utilised to numerically simulate the entire flow passage. Referring to numerical studies, cavitation mostly occurs at the head of the impeller inlet's central blades and at the leading edge of the inducer inlet's suction surface. At low cavitation numbers ($\sigma = 0.07$), thermodynamic impacts meaningfully lessen the vapor-phase volume fraction of the cavitation zone; yet, at high cavitation numbers, they increase it.

In 2022, the pool boiling tests were employed on vertical surfaces by Zhou et al. [28] in sub-cooled and nearly saturated conditions. Deploying a high-speed camera with great time-based and latitudinal resolution, the behavior of bubbles on micro-pin-finned surfaces was considered. Plentiful forms of bubble growth on many surfaces were created as a result of this research, which observed the impacts of the micro-pin-fins on the bubble creation and collapse. When scheming the bubble development rate, the influences of heat transfer from the micro-pin-fins and thermal boundary layer were considered. Further investigation was done to see how surface tension, reaction forces, and inertia affected the movement of bubbles. Following these components, a model for the growth and motion of bubbles was shaped. In this aspect, the micro-pin-fins used three key mechanisms to withstand bubble collapse. These include the inertia deduced by the variation in bubble growth rate, the pin influence on the contact radius, and the response force caused by the local differential pressure.

Lai et al. [29] formed a model for the construction of a single bubble in liquid hydrogen, while bearing in mind how temperature is disseminated throughout the bubble. By simultaneously resolving the Rayleigh Plesset equation, the thermal equilibrium correlation, the thermal diffusion correlation, and the heat conduction correlation in semi-infinite space, one may define and forecast the formation of a single bubble in liquid hydrogen. The model investigates the growth pattern of the bubble radius, the bubble radius growth rate, the vapor pressure, the thickness of the thermal boundary layer, and the temperature differential between the boundary and the centre; it inspects how variables like the liquid's temperature and the surrounding pressure affected the growth of a single bubble in liquid hydrogen. It was possible to examine if a single bubble may transit from experiencing dynamic expansion to facing thermal growth by comparing the important moments of the numerous physical metrics described above.

Cheng et al. [30] built an examination setup with an observable flow channel in order to study bubble generation in vertical rectangular narrow channels. At the Onset of Nucleate Boiling location, bubbles were discovered at one atmosphere in sub-cooled flow boiling water. Two different common bubble development schemes were chosen to be simulated using numerical methods. During the experiment, thermocouples and a high-speed camera were utilised to capture the bubble shapes as well as the wall temperatures. Approximately 30% of the total amount of evaporation is attributed to the microlayer evaporation that takes place close to the base of the bubble.

3.2. Solid–Gas Studies

Hu et al. [31] outlined the Phase Field Model (PFM) of gas bubble growth that comprises micro-structure dependent cluster dynamics to discover the performance of gas bubble swelling in the recrystallization zone of a solid matrix Uranium Monolithic (UMo) fuel. The PFM is an influential numerical approach for simulating multi-phase flows with complex interface dynamics. It can precisely capture the interface between various fluids, handle complex geometries, and allow for easy implementation of boundary conditions. The PFM includes solving the model's equations that recognize the progress of the scalar field representing the interface, along with the motion's equations for the fluid-phases. This model was then utilised to discover gas bubble growth. The phase field coupled model permits the systematic simulation of the effect that the rate of defect creation, the rate of clustering, the rate of interstitial emission, and the rate of sink rates on grain boundaries have on the advance of gas bubbles. The findings specified that interstitial clustering is the prominent key physical process that leads to rapid gas bubble swelling kinetics in the recrystallization zone.

Zhou et al. [32] utilised molecular dynamic simulation to search for the dynamic progress features of a helium (He) bubble throughout spallation in single crystal aluminum (a solid matrix) and its consequences on void nucleation growth and thermomechanical features. Before and after the bubbles merge, the connection between bubble radius and time is nearly linear, and the expansion rates that are parallel to those rates were instituted. The expansion towards adjoining bubbles sources an overlap of plastic deformation zones, which endorses local melting. This happens due to the bubbles' proximity to one another. The bubble avoids the nucleation and development of voids in the surrounding area, and when the expanding bubble comes into contact with developed voids, it absorbs them. This suppression becomes more operative as the bubble size or density is bigger, but it becomes less operative as the shock force grows stronger. When compared to the sample without He bubbles in solid state, this one has a much lower number of nucleated voids and a meaningfully smaller average radius of voids. Following the melting process, there is a destruction of the continued nucleation of voids, which grounds the radius to rise somewhat, while concurrently causing a lesser drop in quantity.

In 2020, the growth process of nano-solid structured cobalt foams was investigated by Arévalo-Cid et al. [33]. These nanostructured cobalt foams were generated via hydrogen bubble template electrodeposition. In order to accomplish this goal, cobalt foams have been created utilising a diversity of electro-deposition periods and inspected using a variability of characterization approaches, namely confocal microscopy in reflection mode and scanning electron microscopy (SEM). Throughout the creation of the foam, agar-agar was mixed into the electrolytic bath to examine the significance of the chemical additives on the growth process and concluding qualities of the foams. Figure 5 introduces the combination of agar occasioned in a successful adjustment of the micro-structure of the final foams. This resulted in a denser, more porous structure that had a lesser pore area and a slower rate of growth, which decoded into improved mechanical features.

In 2022, the time evolution of an inter-granular bubble was reconstructed by Prudil et al. [34] utilising the Included Phase Model (IPM). The IPM involves resolving a number of partial differential equations that define the dynamics of the dispersed phase, including transport, diffusion, and reaction processes. The model incorporated complicated interface morphology while accounting for the creation and associated transit of gas atoms and vacancies. Interfacial energy, internal energy, and elasticity are the driving forces behind these two processes. The four processes that the model predicts ultimately resulted in the generation of linked high-aspect ratio bubbles on the grain face are over-pressure, bubble development, coarsening, and coalescence. The results compared favorably with both the CAGR/UOX/SWELL SEM data and the current SIFGRS model when sets of 250 bubbles were simulated over time using the model's computing efficiency. Analysis of the simulation's results revealed that the two-point autocorrelation function captured the development of an inter-connected network, perhaps leading to improved release thresholds. The findings of this investigation further showed the significance of bubble density as a measure of the bubble structure created by coalescence within a solid matrix. Furthermore, this study ascertained that the IPM is a powerful numerical technique for simulating multi-phase flows. It can accurately capture the dynamics of the dispersed phase, and its impacts on the continuous phase, while handling high volume fractions and complex physical phenomena.

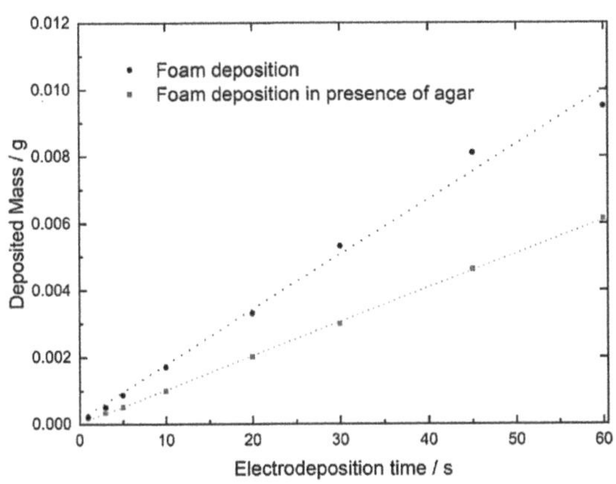

Figure 5. Obtainable electrodeposited foam mass at various times [33] "(Reproduced with permission from Materials Characterization by Elsevier, 2024; Order Number: 5858181471771)".

Table 1 depicts the summarization of the associated research on bubble growth studies while distinguishing the most interesting results.

Table 1. Outline of research on bubble growth indifferent applications.

Authors (Year) [Reference]	Sector/Area of Process Engineering	Configuration	Medium	Type of Study	Parameters Under Investigation	Highlighted Findings
Agarwal et al. (2004) [14]	Thermal conductivity and specific heat of vapor	Film boiling and bubble formation in water.	Water	Numerical	Effect of film thickness.	It is discovered that the scale and position of minimal film thickness are related to the location and magnitude of greatest heat transfer from the wall surface.
Mukherjee and Dhir (2005) [15]	Bubble dynamics and wall heat transfer	Lateral bubble merger throughout transition from partial to fully-developed nucleate boiling.	Water	Numerical	Effect of merger of multiple bubbles.	The overall wall heat transmission is considerably increased when numerous bubbles merge.
Tomar et al. (2005) [16]	Incompressible two phase flows with surface tension	Modelling incompressible two phase flows with surface tension.	Water	Numerical	The impact of heat flux on the instability.	When the superheat is reduced from 15 to 10 K for water that is almost at critical condition, oscillations with sub-harmonics affect the length of the ebullition cycle.
Wu and Dhir (2010) [17]	Bubble boiling in a sub-cooled pool	Vapor bubble in sub-cooled pool boiling.	Water	Numerical	Effect of gravity.	The heat transfer around the bubble reduces while gravity dropping for a given sub-cooling.
Marchal et al. (2016) [18]	Macroscopic swelling of bitumen drums	In a fluid matrix under yield stress, consistent gas generation drives the population of bubbles.	Yield stress fluids	Numerical	Impact of yield stress value.	In a static population (no movement owing to buoyancy), bitumen's yield stress disrupts the conventional state of nucleation, bubble development, and Ostwald ripening in a Newtonian fluid.
Goel et al. (2017) [19]	Water pool	Sub-cooled demineralized water at one atmosphere.	Water	Experimental	Effect of wall super-heat, heater size and the inclination angle.	It was discovered that the heater size, inclination angle, and wall superheat all increased the departure diameter while liquid sub-cooling and surface roughness dampened it.

Table 1. *Cont.*

Authors (Year) [Reference]	Sector/Area of Process Engineering	Configuration	Medium	Type of Study	Parameters Under Investigation	Highlighted Findings
Zhang et al. (2018) [20]	Thermal cavitating flows.	Cavitating flows in liquid nitrogen around a 2D hydrofoil.	Liquid nitrogen.	Numerical	Thermal effects.	The cavitation dynamics during the phase-change process are greatly affected by the temperature effect, which may deduce a delay or suppression in the onset and evolution of cavitation behavior.
Pahk et al. (2019) [21]	Boiling histotripsy exposure	Boiling histotripsy insolation causes a vapor bubble to boil.	Viscoelastic medium	Experimental and Numerical	Shockwave asymmetry and water vapor transportation.	Asymmetric shockwave and water vapor transfer may cause corrected bubble development, which may cause HIFU-induced tissue decellularization.
Lu et al. (2020) [22]	In-situ synchrotron radiography for heating an Al-5 wt.%Mn alloy	In situ synchrotron radiography, bubble development, IMC dissolution, and interactions occur throughout heating of an Al-5 wt.% Mn alloy.	Intermetallic compounds	Experimental	Effects of bubble existence.	A logistic model may explain IMC dissolution, which bubbles can speed up.
Zeng et al. (2020) [23]	Hydrates in different reaction systems	On a gas bubble (CH_4-C_3H_8) floating in water with inhibitors, hydrates are thickening in growth.	Kinetic inhibitors added in the water	Experimental	Hydrate growth mode impacts the inhibitor concentration.	Since hydrate particles do not form films, they will not hinder flow substantially.
Allred et al. (2021) [24]	Boiling surface wetting	During bubble development, wetting and dewetting processes depend on the dynamic contact angles.	Wetting surface	Numerical	Contact angle.	The early stages of bubble development are controlled by the receding contact angle, which also establishes the greatest contact diameter and departure size as the contact line moves away from the bubble centre.
Zhao et al. (2021) [25]	Nucleate boiling	Boiling of the whole nucleate.	Convective and confinement space with solid wall boundary condition	Numerical	The pressure difference.	Coupled pressure from the bubble's interior and surroundings causes the volume of bubbles in the free outlet domain to increase.

Table 1. *Cont.*

Authors (Year) [Reference]	Sector/Area of Process Engineering	Configuration	Medium	Type of Study	Parameters Under Investigation	Highlighted Findings
Winder et al. (2021) [26]	Sudden energy input	Mercury is filled with helium bubbles to lessen stresses and cavitation in the SNS target module.	Structure containing liquid and gas	Numerical	The energy deposit and helium gas injection into the mercury.	The protons' deposited energy has the potential to cause harm to the target. It is possible to lessen the harm by injecting helium gas into the mercury. On the other hand, the addition of this gas makes the bodily reaction more complicated.
Li et al. (2021) [27]	Turbo-pump.	Cavitation flow of a liquid oxygen turbopump.	Liquid oxygen.	Numerical	The influence of thermodynamic effects.	At low cavitation numbers, the thermodynamic effects significantly decrease the vapor-phase volume fraction of the cavitation zone. Hitherto, at large cavitation numbers, they increase it.
Zhou et al. (2022) [28]	Nucleate pool boiling	Experiments on vertical surfaces with pools boiling in subcooled, almost saturated environments.	Vertical surfaces	Experimental and Numerical	Location of the bubble, size of the micro-pin-fins, pin effect, and force of response.	The formation of bubble and motion, as well as its departure time and diameter, are all influenced by the size of the micro-pin-fins. Through the pin effect and response force, micro-pin-fins encourage bubble departure.
Lai et al. (2022) [29]	Liquid hydrogen delivery	Liquid hydrogen single bubble growth model with internal temperature distribution.	Liquid hydrogen	Numerical	The impact of ambient pressure and super-heat on the development of a single bubble in liquid hydrogen.	Comparing the crucial times of the various physical indicators outlined above sheds light on the process by which a single bubble might go from experiencing dynamic expansion to experiencing thermal growth.
Cheng et al. (2022) [30]	Compact type reactors	Bubble development in small, vertical, rectangular channels.	Vertical rectangular narrow flow channels	Experimental and Numerical	Investigate the behaviour of bubble growth in vertical rectangular narrow channels.	Approximately thirty percent of the entire quantity of evaporation may be attributed to the micro-layer evaporation that occurs at the bubble's base.

Table 1. *Cont.*

Authors (Year) [Reference]	Sector/Area of Process Engineering	Configuration	Medium	Type of Study	Parameters Under Investigation	Highlighted Findings
Hu et al. (2020) [31]	Polycrystalline UMo fuels	Gas bubble evolution incorporating cluster dynamics based on microstructure.	Irradiated UMo fuels	Numerical	Rates of defect clustering, interstitial emission, defect production, and sinking.	Gas bubble swelling kinetics rises with defect formation, interstitial emission, and clustering rates, but decreases with grain boundary defect sink rates.
Zhou et al. (2020) [32]	Spallation of aluminium	During spallation, a helium bubble forms in a single Al crystal.	Aluminium	Numerical	Size, density, and shock resistance of the bubbles.	The bubble prevents the formation and growth of near voids, and its expansion absorbs grown voids. Suppression increases with bubble size or density and decreases with shock strength.
Arévalo-Cid et al. (2020) [33]	Hydrogen bubble template electrodeposition	Cobalt foams with nanostructures created via hydrogen bubble template electrodeposition.	Nanostructured cobalt foams	Experimental	The addition of agar.	Agar has effectively affected the micro-structure of the final foams, providing a denser porous structure with smaller pore area and slower growth rate, which improves mechanical qualities.
Prudil et al. (2022) [34]	Populations of intergranular fission gas bubbles	Bubble development, coarsening, and coalescence lead to connected, high-aspect-ratio bubbles on the grain face.	Oxide fuels	Numerical	Populations of intergranular fission gas bubbles that are significant.	Bubble density is a metric of bubble structure shaped by coalescence, and the two point auto-correlation function records the expansion of a connected network, that might decrease release thresholds.

According to the studies that were previously addressed, the creation of corrected bubbles may be the source of HIFU-induced tissue decellularization due to the potential imbalance in shockwaves and the transfer of water vapor. One may have a better understanding of the procedure by which a single bubble converts from dynamic development to thermal growth by linking the critical moments of the aforementioned physical indications. Furthermore, a cluster dynamics and phase field model can be stretched to train the accompanying growth of defect and second phase precipitates often detected in treated materials.

4. Bubble Collapse Studies

Plentiful academics have, over the past few decades, attempted to recognize the fluid dynamical mechanisms in charge of the impulses connected to the cavitation collapse of bubbles. However, it has been proved that at the three distinct processes—including the shockwave, the splashing properties, and the impinging liquid jet—are at work here. This section proposes conducting a critical assessment of the associated studies on bubble collapse in a wide range of industrial applications.

In 2013, Gong and Cheng [35] conducted numerical simulations in two dimensions using the novel phase-change lattice Boltzmann method (LBM) to study the nucleation of water on a micro-heater with constant heat flux and constant wall temperature, respectively. Under conditions of constant wall temperature, the effects of gravity, contact angle, and superheat on bubble departure diameter and release duration were tested. The analysis included the three-phase movement of the vapor bubble's contact line, as well as the flow fields and temperature profiles both within and outside the bubble during the boiling process.

In 2015, using wavelets and Fourier transforms to analyse the piezoelectric waveforms, Navarrete et al. [36] investigated the dissipation of acoustic energy. The main experimental findings show that as compression progresses: (1) the liquid meniscus attains the conical zone in cavitation conditions; (2) the liquid meniscus undertakes a symmetrical alteration from two dimensions to three dimensions, becoming a "bulb with a nozzle", where the instabilities are drawn and controlled; (3) the nozzle is maintained as part of the new meniscus that persists in pushing the gas pocket; and (4) furthermore, as the collapse process proceeds, the bulb eventually changes into a cloud of bubbles, and the bubble structures display their distinct expansion–contraction rate. The authors arrived at the conclusion that the breaking of a conical bubble activates a variety of distinct processes for the emission of light after looking at the dynamics and acoustics.

In 2016, Gong and Klaseboer [37] developed a fluid–solid integrated algorithmic scheme to consider the relationship between a deformable viscoelastic solid and a collapsing bubble. A Zener type viscoelastic model was utilised using the linked Boundary Element (BE) and Finite Element Method (FEM) to achieve this. The FEM was utilised to determine the viscoelastic solid response to the spontaneous pressure brought on by the bubble's collapse, whereas the boundary element method was utilised to mimic bubble growth, contraction, and collapse. The Zener type viscoelastic model was employed to appropriately accommodate for the change in a viscoelastic solid's stiffness over time as a result of pressure. The study indicated how variables, including the solid's viscoelastic properties, the range of max. bubble sizes, and the locations of the bubbles affect the viscoelastic dynamic response of a solid.

In 2016, Tang et al. [38] conducted a study on the impact of an ultrasonic field on the deflation and condensation of vapor bubbles in a sub-cooled pool that was at rest. The effects of the ultrasonic field were examined in this study. When outcomes were introduced at liquid sub-cooling temperatures of 15–30 K, the vapor bubbles progressively disintegrated, but at temperatures over 40 K, they collapsed into a significant number of smaller bubbles. The limit of liquid sub-cooling for bubble break was lowered to 20–26 K while utilising ultrasonic vibration, causing capillary waves to appear on the bubble surface. The enhanced sensitivity of bubbles to the ultrasonic vibration made this possible. Additionally, the region of the bubble that was in touch with the cold bulk liquid was extended due to the capillary wave's existence. The temperature boundary layer close to the vapor–liquid interface was also disturbed by the capillary wave's presence, which augmented the condensation process. Accordingly, at the same amount of liquid sub-cooling, the liquid's inertial shock to the vapor bubble with ultrasonic vibration was a bit more robust than what would have been detected in the absence of ultrasonic vibration, as proved in Figure 6.

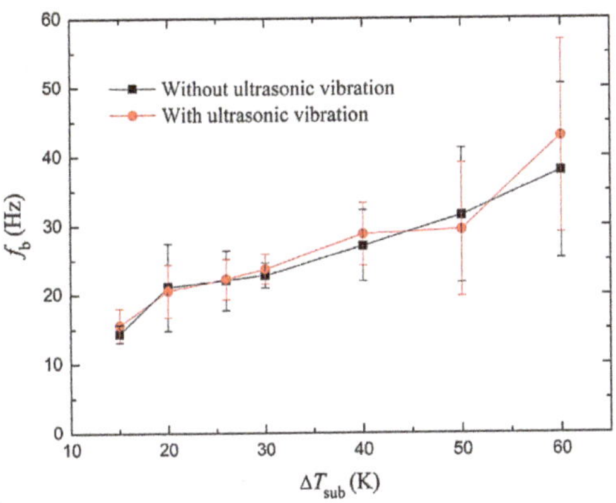

Figure 6. Frequency of bubble collapse at various sub-cooling without and with ultrasonic vibration [38] "(Reproduced with permission from Experimental Thermal and Fluid Science by Elsevier, 2024; Order Number: 5858190317013)".

In an underwater launching test, Wang et al. [39] testified to a distinct internal collapse mechanism that happens from the tail to the head of an axisymmetric projectile. This collapse phenomenon was discrete from other collapse scenarios and is on occasion seen in water tunnel experiments. The cavities swiftly contract upstream, which results in high-impact pressure. Moreover, fast re-entry jets occur, which causes substantial instability. Referring to the findings of the analysis, this phenomenon is associated with sudden changes in the cavitation number and vehicle speed. The quick acceleration of the vehicle causes the creation of thin cavitation bubbles, and the consequent rapid condensation of vapor during the deceleration phase results in the formation of an extra body force at the point where the cavity is closed. Hence, a major amount of velocity that is engaged toward the wall is created, and the layer of liquid water that is moving toward the wall remains to travel toward it and collide with it as an interior collapse. Additionally, numerical calculations were achieved with the effect of slowing down the development of the cavity. Referring to the findings, an enlargement in pressure was a crucial component, which brought about a nonlinear shift in the cavity's overall length.

Qin and Alehossein [40] introduced a model of heat transfer allowing for the use of the Rayleigh/Plesset (RP) correlation and Computational Fluid Dynamics (CFD) with the aim of precisely forecasting the change of temperature and heat transfer that happens throughout a bubble's collapse. CFD comprises solving a number of equations that outlines the dynamics of the fluid phases and their collaborations, while utilising numerical approaches. A two-phase compressible CFD model was formed to correctly mimic the bubble collapse to establish the changes in velocity, temperature, and pressure distribution between the liquid and the bubble. After being revised to take into account the impacts of radiation and conduction, the RP equation's conclusions mostly settled with those of the numerical CFD. Further investigation was achieved on the temperature rise and collapse of bubbles, the rate of heat transmission by radiation and conduction, and the cumulative heat transfer of bubbles of altered sizes. It was detected that the bubble burst more violently as the initial max. bubble size was increased. The max. temperature was extended inside the bubble, which quickly enlarged, as depicted in Figure 7, along with the amount of heat that was accumulated and stored. The authors also confirmed the superiority of CFD as an influential instrument for appreciating and predicting the behaviour of complex multi-phase flows. The simulations are carefully validated against experimental data.

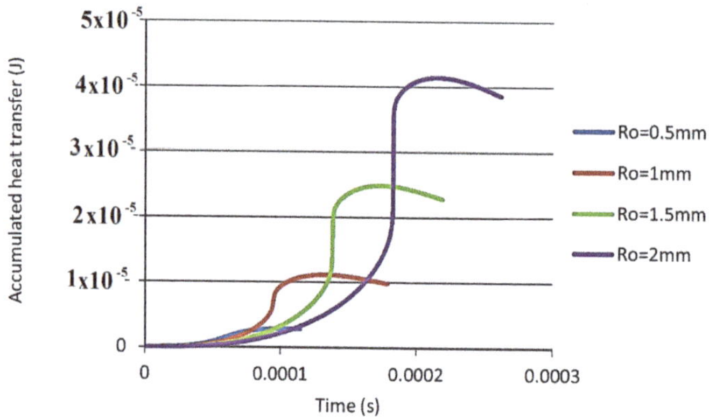

Figure 7. Accumulation of heat transfer throughout different bubble collapse (Ro is the bubble radius) [40] "(Reproduced with permission from Heat transfer during cavitation bubble collapse by Elsevier, 2024; Order Number: 5858191386212)".

Li et al. [41] calculated the dynamic bubble pressure using the Potential Flow Theory (PFT) and BEM. PFT is an analytical technique that describes fluid flow as an irrational and inviscid fluid motion. This technique assumes that the fluid particles move in such a way that the velocity vector at each point is the gradient of a scalar potential function. The pressure was specifically estimated using the Bernoulli equation and the auxiliary function approach. The potential function satisfying Laplace's equation is defined such that the fluid velocity and pressure at any point in the flow can be obtained from the potential function. The BEM involves discretizing the boundary of the domain into a set of components and approximating the solution to the governing equations within each element. Bubble-induced pressure had two sources, and dynamic pressure had two portions. The first component, ρg, monitored the uneven pressure between bubble gas and ambient flow, which contributes to dynamic pressure. The second term of decomposed pressure, pm, is created via bubble motion, which evaluates jet impact. ρg fluctuation mirrored gas pressure. After the jet impact, pm at the wall centre peaks and subsequently drops owing to jet velocity and may increase when the toroidal bubble migrates into the wall.

Rowlatt and Lind [42] prolonged the Spectral Element Marker Particle (SEMP) technique to integrate a third fluid phase to investigate the bubble collapse problems in a fluid–fluid interface. The method is thought to be validated when a two-phase bubble explodes close to a stiff barrier. SEMP involves resolving a number of partial differential correlations that define the dynamics of the fluid and solid phases, including momentum conservation, mass conservation, and energy conservation. A wide set of fluid parameters and geometric arrangements were considered for understanding the impact of bubble collapse in bioengineering applications. In an effort to gain visions into the flow dynamics behind sono-poration and microbubble improved targeted drug administration, the authors considered a straightforward model of collaboration between the bubbles and cells. The bubbles deflated into spherical shapes at higher interface heights away from the rigid wall as a direct outcome of the compact contact that existed between the stiff wall and the bubbles. Indeed, SEMP has allowed for accurate modelling of the dynamics of the fluid–solid phases and can handle complex geometries.

Ye and Zhu [43] suggested the effect of a near wall acoustic bubble collapse micro-jet on an aluminum 1060 sheet, the cavitation threshold formula, and micro-jet velocity formulation. The Johnson Cook rate equation was used to create a 3-dimensional fluid–solid coupling model of the micro-jet effect on a wall. The utilised ultrasonic pressure amplitude was far higher than the liquid cavitation threshold, causing cavitation in the liquid. Micro-jets caused surface micro-pits. Clearly, micro-jet velocity affected the wall pressure. Ultrasonic cavitation examination evaluated tiny pits generated by micro-jet on aluminum 1060. Inversion analysis showed that pit diameter-to-depth ratio affected equivalent stress, equivalent strain, impact strength, and micro-jet velocity.

In 2017, calculations were made by Pavlov [44] to acquire the kinetics of a bubble's collapse under the influence of an abruptly utilised external pressure. The bubble was initially in a state of thermal and mechanical balance with the nearby liquid. The condensation of vapor inside the bubble correlates to the bubble's compression creating a pressure of saturated vapor inside the bubble at the interface temperature. Adiabatic processes are likely to cause a temperature increase in the main body of the bubble. Both the heat fluxes field and the temperature in the instant area of contact were deliberated. The mass flow coming from the bubble was calculated from the thermal balance. The assumption used in the calculation of the liquid's pressure field was that the liquid is both incompressible and non-viscous. Several equations associated with the radius of bubble, compression rate, and internal vapor pressure were developed.

In 2017, a numerical simulation of the collapse of a cavitation bubble enclosed inside a diesel droplet was performed by Ming et al. [45] using the volume of fluid (VOF) method of the open source CFD toolbox OpenFOAM. The outcome exhibited that the collapse process was composed of several phases of rebound and collapse, which is alike to the vibration process of a damping spring oscillator. Figure 8 depicts that the pace of collapse accelerates as the pressure of the environment around it rises, whereas Figure 9 depicts that the collapse's rate accelerates as the pressure of the saturated vapor falls. The impact of the surrounding pressure on the collapse process is far greater than the impact of the saturation vapor pressure.

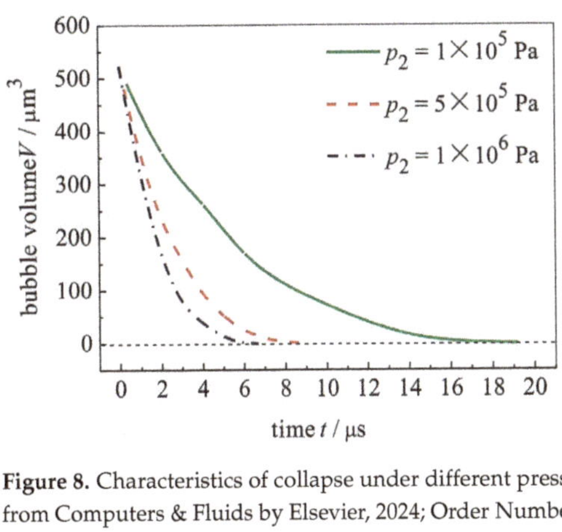

Figure 8. Characteristics of collapse under different pressures [45] "(Reproduced with permission from Computers & Fluids by Elsevier, 2024; Order Number: 5858200400261)".

Rosselló et al. [46] created a water hammer to achieve managed bubble collapses. A laser-generated bubble is enlarged and then squeezed utilising an electro-mechanical piston. It permits high energy concentrations in collapses and autonomous regulation of system parameters. To expand experimental findings, simulations of prototype bubble dynamics were conducted. The expansion ratio, consisting of the increased bubble radius over the equilibrium bubble radius, was the most important metric for collapse strength.

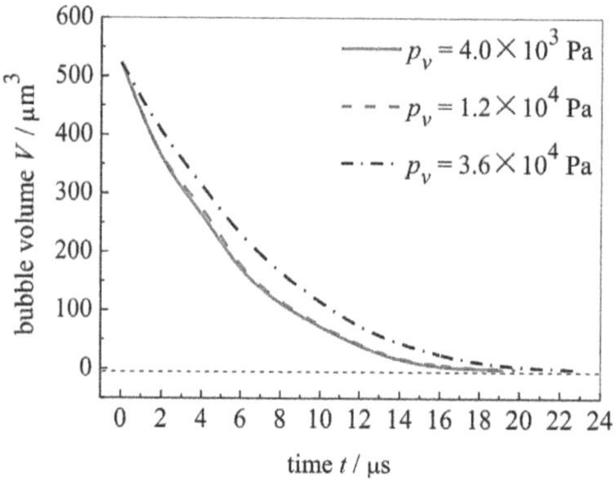

Figure 9. Characteristics of collapse under different saturation vapor pressures [45] "(Reproduced with permission from Computers & Fluids by Elsevier, 2024; Order Number: 5858200400261)".

Phan et al. [47] developed an algorithm for compressible interface flows with and without mass transfers. The mathematical technique was built upon dual time fully compressible multi-phase homogeneous mixture flow modelling. The authors measured the boundaries that showed a substantial change in velocity, density, pressure, and temperature while deploying a compressive high-resolution interface capturing technique. It was possible to observe the formation of the wave-like boundary between the liquid and vapor around the sphere as the film boiled. The average heat fluxes generated by the simulation were contrasted against forecasted ones using analytical correlations and numerical data. Also considered were the impacts of the superheated wall on the heat fluxes. A pattern similar to the one seen in the previously published data was discovered as a result of a bigger heat flux on the sphere caused by a rise in the superheated wall's temperature (Figure 10).

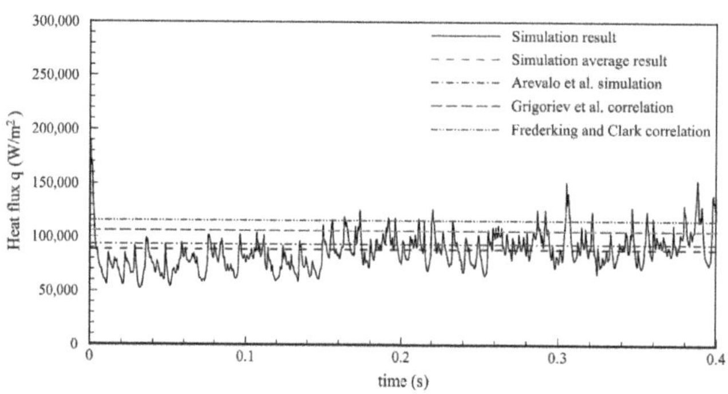

Figure 10. Wall heat fluxes against operational time [47–50] "(Reproduced with permission from International Journal of Heat and Mass Transfer by Elsevier, 2024; Order Number: 5858200884075)".

Kerabchi et al. [51] introduced computational results while considering the impact of depths between 0 and 10 m on acoustic cavitation. The study provided particular attention to the impact of the ultrasonic wave's attenuation in regard to the dramatic conditions that emerged inside the bubbles as they burst. The liquid's surface tension, compressibility, viscosity, depth, and attenuation of the ultrasonic wave are all reflected by the mathematical model that was utilised. The greatest pressure (p_{max}) and bubble temperature (T_{max}) at the bubble collapse were instituted to meaningfully decrease when the depth of the water was increased up to 10 m, with the ultrasonic wave attenuation significantly contributing to the overall reduction event. However, when the bubbles were under the same pressure, this was the case. Lower temperature (10 °C) and higher frequency (1000 kHz) caused the decrease in T_{max} and p_{max} with depth to be more dramatic; in these conditions, losses beyond 72% in T_{max} and up to 94% in p_{max} (as opposed to values at z = 0) were achieved at z = 10 m. When the temperature and frequency were lower (10 °C) and higher (1000 kHz), the decrease in T_{max} and p_{max} with depth was more dramatic.

In 2019, a Smoothed Particle Hydrodynamics (SPH) solver was established by Joshi et al. [52] to model the deflation of a single cavitation bubble near an elastic plastic material and examine the origins of plasticity which resulted in material degradation. The results highlighted the significance of the material deformation caused by the micro-impact jets and the shockwave produced during the collapse. An event dominated by shockwaves has the potential to degrade material at a far faster rate than a micro-jet hit. It was found that, in the case of the strain rate of stainless steel A2205, there is a significant impact on plastic deformation that, if ignored, may cause an over-estimation of the plastic deformation by as much as 60%. Additionally, they demonstrated that the biggest plastic strain happens at a radial offset from the symmetry axis, even though the impact pressure is the strongest below the collapsed bubble. This is due to the effects of inertia, which signify an influence on the position and the size of the plastic domain inside the material.

Brujan et al. [53] examined the dynamic of a laser-prompted cavitation bubble utilising 100,000 frame per second high-speed photography. A bubble near a vertical solid wall collapsed into a flat jet. Both radial and planar jets were created in the transition zone. The planar jet penetration and secondary cavitation construction creation were seen throughout bubble oscillation when the bubble diameter at max. expansion was lower than the distance between parallel solid walls. Planar jet height grew with bubble size relative to stiff wall distance. The jet velocity decreased as the bubble size and distance from the vertical wall increased.

In 2019, a unique experimental setup was reported by Hopfes et al. [54]. This novel experimental setup made use of a shock tube and applied a jellylike mixture as a water-like carrier medium. The effects of an instantaneous pressure rise on air bubbles of mm size that were injected in gelatin were investigated from two distinct angles. First, they demonstrated that individual bubbles in the gelatin behaved correspondingly to bubbles in water throughout the collapse of the structure, and that the concentration of the gelatin did not greatly impact the behavior of the bubbles. They also investigated four primary forms of bubble interactions that may also be defined by non-dimensional factors. It was observed that if there was a big size ratio between two bubbles, the interaction between them would either be controlled by the larger bubble if the bubbles were separated or would result in a prominent liquid jet if the bubbles were near to each other.

Bardia and Trujillo [55] offered a mathematical study of the intermediary bubble collapse. In contrast to thermally generated or inertia controlled collapse, intermediary bubble collapse was influenced by both liquid/vapor interfacial heat transfer and surrounding liquid advection. The bubble pressure and interfacial temperature were constantly shifting all the way through the process of this particular kind of collapse, which was one of the most identifying characteristics of this form of collapse. To be more specific, the work demonstrated in a quantitative manner that the collapse behavior shifted, when the process was happening, by a regular rate of pressurization in the adjacent liquid phase, rather than the commonly deployed approximated method of a step change. This was in contrast to the previous approximation, which assumed that the change would occur all at once.

Gan et al. [56] investigated a hull girder constructed to mimic a real warship and subsequently exposed to underwater explosion loads to see how it deformed. Experimental and computational techniques were both used in this study. The study also considered how a non-spherical underwater explosion bubble might affect things. The numerical computation was carried out by Coupled Eulerian–Lagrangian (CEL) method of CFD, and the experiment was carried out in an underwater environment with explosions that occurred in a swimming pool. Analysis of the associated results led to the conclusion that when detonation distances increase, the hull girder's deformation form shifts from rigid-body motion to whipping, leading to the creation of a plastic hinge. The dynamic reactions of the hull girder will be the same for a range of charge weights when measured at the same dimensionless distance if the charge weights are within a given range.

Chudnovskii et al. [57] directed an examination of the boiling of a substantially sub-cooled liquid in which they explained how the liquid's local heating by concentrated laser radiation via a thin optical fibre caused the boiling to begin. They investigated the growth of a vapor–gas bubble established at the surface of an optical fibre tip extensively. They presented estimations of these jets' velocities, which appear to be in perfect agreement with the experiment.

Using a tank of water inside a cold room, Yuan et al. [58] studied the dynamics of a collapsed bubble close to a floating ice cake and its potential to be harmed by a high-speed jet and shockwave. A cryostat produced an ice cake from clean water, which was subsequently managed and positioned in a water tank inside a frigid room. The bubble was simultaneously created at a distance underneath the floating ice-cake by an electric spark that was produced at the same time. The two of them were interacting while being observed and captured by a fast camera. After seizing the bubble's dynamics, which comprise the bubble jet and the shockwave, the authors calculated the negative impacts of bubble loads on the ice, counting fractures, breakage, and movement. Relying on the level of damage and the propagation of fractures, it was detected that crevasse patterns, radial and circumferential crack patterns, and radial crack patterns were produced on the ice. However, the most distinctive type of damage pattern was crevasse patterns.

Han et al. [59] deliberated the association between two bubbles of diverse sizes as well as the consequence of a wall on the process of bubble collapse using the Volumetric Optical Force (VOFE) microscopy method. VOFE microscopy is a prevailing experimental procedure used to train complex multi-phase flows in three dimensions. VOF microscopy contains and presents particles into the fluid and illuminates them with a laser beam. The laser light exerts optical forces on the particles, which can be utilised to operate their motion and track their trajectories. Thus, VOFE permits the quantification of the velocity, acceleration, and trajectory of individual particles or droplets within the flow. The authors observed connections between two bubbles of diverse sizes in free water without a wall in between. The authors guaranteed that these bubbles would move close to the wall and close to each other as they collapsed when there was a wall nearby. This has specifically occurred concurrently. They revealed that, in both a single bubble system and a double bubble system, the presence of a wall would increase the time necessary for the bubble to rupture. However, the wall's existence can result in a more powerful process of the bubble collapsing and then rebounding. By inspecting the bubble's centre of displacement, they revealed that the double bubble system may upsurge the dislocation of both bubbles. The VOFE technique is restricted to studying small-scale flows and requires careful calibration and validation against other experimental procedures to guarantee accurateness and reliability of the results.

Fursenko et al. [60] delivered a simulation of vapor bubble collapse close to a micro/fiber submerged in a sub-cooled liquid to suggest the process of high velocity liquid jet creation. Numerical simulations showed that jet generation is driven by bubble surface dynamics, not bubble shrinkage. The jet's ideal fibre thickness was discovered.

Tan and Yeo [61] introduced a particle velocity estimate model relying on mass loss. Micro-abrasive particles measuring 5–50 μm were subjected to 20 kHz ultrasonic irradiation in deionized water for 10 min. In a well-controlled experiment, targets positioned 0.5 mm from the ultrasonic horn surface captured pushed particles. A reverse solid particle erosion model may estimate particle impact velocity from specimen mass loss. The results revealed particle velocity was 8–40 m/s and depended on the particle size and ultrasonic amplitude.

Orthaber et al. [62] conducted a number of tests in which they inspected the behavior of a single cavitation bubble on two sides of an oil–water interface. They looked at the specifics of the interaction between bubbles and surfaces (deformation, penetration). The anisotropy (the property of being directionally dependent) parameter makes the correct prediction that the bubble would permanently shoot to the interface if it was formed in low-density liquid. On the other hand, the bubble will continually shoot away from the interface if it was formed within high-density liquid. They expanded the scope of the study to include the correlations between a variety of bubble features and the anisotropy parameter.

Zhang et al. [63] performed an experiment to investigate the dynamics of a laser-prompted cavitation bubble that was near to the edge of a solid wall. The authors used high-speed photography equipment. Three typical scenarios were identified for the purpose of categorizing the phenomenon, relying on the deformation of the bubble interface that happened throughout the process of the bubble collapsing. The movements of the bubble centroid and interface in various directions were identified using a numerical analysis of the captured high-speed photographs. There was visible neck development in the middle of the bubble when it was extremely near to the edge. The drive of the bubble interface at the edge was hampered by this clear neck. In this case, the edge increased the time needed for the bubble to burst by up to 22% over Rayleigh's entire time requirement (Rayleigh number is large, somewhere around 10^6 to 10^8).

Ezzatneshan and Vaseghnia [64] appraised the collapse dynamics of cavitation bubbles of various liquids in the vicinity of the solid surface with variable wettability conditions using a computational methodology based on the pseudo-potential multiphase lattice Boltzmann method (LBM). The simulation of the Redlich–Kwong–Soave equation of state (EoS) with an acentric factor was used to address the physical features of water (H2O), liquid nitrogen (LN2), and liquid hydrogen (LH2). The H_2O, LN_2, and LH_2 fluids were considered while evaluating the cavitation bubble collapse close to the solid wall. Also, the effect of the surface's wettability on the dynamics of the collapse was tested. The findings showed that the water's cavitation bubble collapses faster when it comes across a strong liquid jet, but the LN2 bubble collapses more gently because of its bigger bubble radius at the rebound.

Huang et al. [65] explored the acoustic waves caused by the collapse and oscillation of a single bubble. The Schlieren Method was employed throughout the trials to document both the temporal growth of the bubble formations and the equivalent acoustic waves. A three-dimensional (3D) model was employed throughout the computational simulation to train the dynamics of a single bubble. The acoustic waves were the main topics of this study. The results showed that compression waves were produced as a result of the bubble's rapid growth. These waves progressively steepened into a single shockwave that propagated outward through the liquid, and during the final phase of the collapse of the bubble, another strong shockwave was produced. The liquid field's pressure and velocity increased following the shockwave. The liquid's pressure dropped as a result of the rarefaction wave, and its spatial distribution became dispersive. As the distance over which the acoustic waves travel increased, the pressure that they exert and the effect they have on the liquid's speed were observed to increase.

In 2021, to better understand the method of air entrainment used in the hydraulic aspect to lessen cavitation, Xu et al. [66] examined the influence of a single air bubble on the volume of noise produced by the collapse of a cavitation bubble. When a rising air bubble interacted with an underwater low-voltage electric discharge unit, it resulted in the formation of a cavitation bubble. For this analysis, only circumstances where the hydrophone probe, cavitation bubble centroid, and air bubble centroid were all almost on the identical horizontal line were taken into account. Referring to the findings, an air bubble's size, roundness, and distance all considerably impacted the cavitation bubble's direction of collapse and the sound generated pressure. The results also revealed a strong correlation between the direction of collapse and the associated pressure that the collapse produced.

Peng et al. [67] outlined the concept of the collapse of a single and a dual cavitation bubble while utilising the Thermal Lattice Boltzmann (TLBM) and double distribution function (DDF) approach. This approach uses a lattice-based model to simulate the movement of fluid particles and their interactions with solid particles and interfaces. Despite LBM including a thermal lattice model that defines the transfer of heat between fluid and solid phases, LBM affords an exceptional density-to-viscosity ratio. The approach also involves solving a number of partial differential correlations that define the dynamics of the fluid–solid phases, including mass conservation, momentum conservation, energy conservation, and temperature evolution. The correlations were solved using mathematical approaches such as the lattice Boltzmann method, and the simulation findings were authorized against experimental data. The outcomes of the simulation were in decent consistency with those of the theoretical analysis and the experiments. Peng et al. [67] dignified the max. collapse pressure, the lifetime of the cavitation bubble, the maximum collapse temperature, and the max. collapse velocity as corresponding to the distance between the cavitation bubble and the rigid barrier. Moreover, investigations were achieved to determine how the viscosity of the fluid and the starting pressure differential affect the max. temperature at which a system can collapse. The outcomes of the LBM simulation completely reproduced the slowing influence of a hard boundary on the cavitation bubble, the distance between the cavitation bubble, and the hard barrier growth; this resulted in the growth of the max. collapse pressure, max. collapse temperature, and max. collapse velocity.

In 2021, the behavior of cavitation bubbles when they collapse close to walls under conditions of high ambient pressure was highlighted by Trummler et al. [68]. Generic setups with various stand-off lengths were analysed using mathematical simulation with a solver that incorporates phase change. The findings showed that the stand-off distance and the collapse dynamics, the production of micro-jets, the rebound, and the maximum wall pressure were significantly related. The pressure impact data gathered at the wall may be used to establish a connection between cavitation-related material damage and the corresponding collapse processes. The findings on six different grid resolutions were contrasted, and it was exposed that, while the peak pressures were meaningfully dependent on the resolution, the key collapse characteristics were already documented on the coarsest resolution.

Long et al. [69] observed and investigated the collapse and formation of cavitation bubbles created throughout the pulsed-laser ablation of submerged targets in three different liquid environments: water, ethanol, and n-butanol, utilising a stroboscopic shadowgraph system. These bubbles were produced when submerged targets were being ablated by a pulsed laser. The images of collapsed bubble transitory phases were obtained. The bubble volume was at its lowest at this phase. Both ethanol and n-butanol might undergo cavitation-induced reactions, which generated gaseous molecules that prevented the bubbles from collapsing explosively. In contrast, the penultimate step of collapse was when cavitation bubbles produced by a laser in water acquired their lowest radii, leading to a higher upsurge in localized pressure and temperature that might cause a second round of etching.

Wang et al. [70] investigated the association between the various physical factors and the max. dimensionless radius of the bubbles by employing single factor analysis in addition to multiple regression analysis. Only a very weak linear relationship exists between the following quantities, for example, between density and viscosity, and between sound velocity and surface tension. Additionally, the regression models developed based on the random forest and neural network types of machine learning algorithms were utilised to forecast how the combined physical factors affect the severity of the bubble collapse. The outcomes confirmed that each of these models was workable and efficient. Additionally, the strength of the collapse was clearly influenced by the viscosity and surface tension, whereas density and sound velocity had the least effect.

Gai et al. [71] adapted the Conventional Lattice Boltzmann Method (CLBM) to allow for the statistical analysis of ice nucleation resulting from the collapse of a cavitation bubble. To examine cavitation bubble dynamics, including growth and collapse, the authors examined two distinct real-world ice nucleation application scenarios: scenario I used a solid barrier, while scenario II involved a pressured environment. Some of the significant system variables whose impacts on the max. collapse pressure (p_{max}) were inspected included the stand-off distance (λ), initial bubble size (R_0), and differential pressure (Δp). The findings showed that scenario II yields a substantially higher p_{max} than scenario I, leading to more direct initiation of ice nucleation.

In 2021, a three-dimensional computational tool based on the multiphase lattice Boltzmann method (LBM) with multiple relaxation time (MRT) was used by Ezzatneshan and Vaseghnia [72] to study the dynamics of cavitation bubble clusters produced by sound near a solid wall under various wetting circumstances. To correctly impose the physical features of real fluids, the Peng–Robinson Stryjek–Vera equation of state with an acentric component was integrated into the LBM. The obtained findings for the cavitation bubble cluster dynamics driven by sound showed that the pressure pulse cannot reach the lower bubbles due to the shielding effect of top bubbles. As such, the destruction of the top layer bubbles has a greater impact on the cluster core and the bubbles close to the solid surface than it does on the acoustic field.

In 2022, to better perceive the association between the cavitation bubble's proximity to the wall and the collapse features, Xu et al. [73] created models with various distances from the benzamide wall to the cavitation bubble. The bubble placed exactly in the middle of two parallel walls experienced a major rate of volume change at the beginning of the collapse, according to the research. The bubble's rate of volume change during the acceleration phase is proportional to its proximity to the wall and, thus, the pressure at its peak is higher. The initial spherical bubble regularly changed into an ellipsoid shape, and the surface of the bubble closest to the wall had a rather smooth texture. The wall penetrated the area with the strongest pressure field created by the collapse of the bubble when d was smaller than 1.59 R_0.

In 2022, the dynamic behavior of bubble collapses, pressure loads, and water jets throughout the collapse of the bubble between walls and a free surface was mathematically inspected by Nguyen et al. [74] using a VOF-based simulation approach. This was done by mathematically analysing the pressure loads, water jets, and bubble collapses that occurred when the bubble collapsed close to the walls. The mathematical approach was relying on the compressible Navier–Stokes correlations, which were represented in a conservative formula. These equations explain how compressible viscous fluids flow. Investigations of spherical and non-spherical bubbles, as well as grid dependence studies of a spark-generated bubble were conducted. This study modelled the burst of a bubble. The outcomes indicated that the dynamics of the bubble during its collapse were highly consistent between the experiments and simulation's results. The inspection of a single bubble that was situated near to a wall was done utilising a range of standoff distances. The pressure loads caused by the jets interacting with the walls were evaluated and analysed. A more difficult situation of bubble collapse along an inclined wall and free surface was also developed and experienced.

Yang et al. [75] evaluated the influences of wall temperature on the collapse of a bubble and formed a robust model for a cavitation bubble close to a cooled or heated wall. The process of the micro-jet and the cavitation bubble itself affecting the solid wall heat transfer, as well as the thermodynamic behavior traits of the cavitation bubble collapse close to the wall, were resolved by the researchers. This makes it possible to better realize how the micro-jet affects the cavitation bubble. By modifying the wall's temperature, T_w, the cavitation's thermal effects could also be influenced. To further explore the heat transfer intensity, a dimensionless temperature parameter was added. Research and analysis were conducted on the impact that p, and R_0 have on the heat transfer. The outcomes showed that the ideal values of p, R_0, and T_w could be utilised to recover heat conduction and device cooling or heating treatments on several surfaces.

He et al. [76] looked into the thermodynamics of bubble interactions using the double distribution function thermal lattice Boltzmann technique. Strong and weak interaction regimes of double cavitation bubbles were methodically explored, and the whole development and collapse processes during an interaction were effectively replicated. Also, the essential lengths between two equal-sized bubbles that allow for the distinction of various interaction types were proposed. When there are two unequal-sized bubbles in a liquid film, imbalanced forces operate on them, making the film harder to rupture than when the bubbles are equal in size. This means that the coalescence regime requires a smaller starting distance between the bubbles.

Ezzatneshan et al. [77] used a multiphase lattice Boltzmann method (LBM) to investigate the beginning of the boiling phenomena and the bubble dynamics that resulted. The researchers used two equations of state (EoS), the Peng–Robinson (P–R) and the Redlich–Kwong–Soave (R–K–S). Their predictive power for the nucleation, growth, and departure of bubbles was measured in comparison to the existing literature data. The influence of the wettability and form of the micro-heater surface, namely the grooved and cavity areas, was appraised in relation to the detached bubble's separation time and velocity under various flow conditions. The region with larger temperature gradients—the groove and cavity's sharply angled edges—were where the boiling phenomena originated, according to the results. The bubble continues to develop on the hydrophilic heated surface's sharp edges where there are not many dry spots.

The implication of a dual distribution function multiphase lattice Boltzmann technique (DDF-MLBM) to examine the collapse of cavitation bubbles in cryogenic liquids was addressed by Ezzatneshan et al. [78]. The exact difference method (EDM) was used in the scheme to apply fluid–solid adhesion forces and interparticle interactions, while counting the energy equation. The Peng–Robinson (PR) equation of state was utilised in conjunction with an acentric factor to properly simulate phase shifts and the molecular complexity of cryogenic fluids such as liquid nitrogen (LN_2) and hydrogen (LH_2). The research observed the impacts of pressure, temperature, collapse duration, and the separation between a cavitation bubble and a nearby solid wall. The results indicated how vital the distances between the cavitation bubbles in the cluster and between the bubbles and the surrounding solid surface are in defining micro-jet velocity.

Table 2 shows a summary of the accompanying research of bubble collapse while demonstrating the most important conclusions.

Table 2. Summary of studies on bubble collapse in different applications.

Authors (Year) [Reference]	Sector/Area of Process Engineering	Configuration	Medium	Type of Study	Parameters Under Investigation	Highlighted Findings
Navarrete et al. (2015) [36]	Biomedicine	Collapse of a conical bubble.	The liquid meniscus	Experimental	Effect of collapse on various light emission.	When a conical bubble bursts, a number of light-emitting mechanisms—thermal, chemical, and electrical in nature—are engaged.
Gong and Klaseboer (2016) [37]	Biomedicine or marine industry	Interaction between a deforming viscoelastic solid and a bubble that is in the process of bursting.	Viscoelastic solid	Numerical	The standoff distance, bubble size, and form, as well as the viscosity of the plate.	The outcomes showed that standoff distance and bubble size have a substantial impact on viscoelastic solid deformation and bubble collapse.
Tang et al. (2016) [38]	Bubble condensation with and without ultrasonic	The process of vapor bubbles condensing and bursting in an otherwise still and cold pool with the application of ultrasonic field.	Vapor bubble surface	Experimental	Impact of capillary wave.	The existence of the capillary wave enlarged the area of the bubble's contact with the cold bulk and disturbed the thermal boundary layer near the vapor liquid interface, which aided in the condensation process.
Wang et al. (2016) [39]	Underwater vertical launching	The axisymmetric projectile experiences a phenomenon known as internal collapse that moves forward from the tail to the head.	Water	Experimental and Numerical	Effect of increasing pressure.	A crucial element that causes a non-linear variation in the cavity length is the pressure rise.
Qin and Alehossein (2016) [40]	Water jet drilling and rock cutting	The fluctuation in temperature and the transfer of heat that occurs when a bubble bursts.	Water	Experimental and Numerical	Bubble size and temperature.	The bubble crumbled more quickly as the original maximum bubble size grew. The inside bubble's maximum temperature rose sharply, and the total transmitted heat also rose greatly.

Table 2. *Cont.*

Authors (Year) [Reference]	Sector/Area of Process Engineering	Configuration	Medium	Type of Study	Parameters Under Investigation	Highlighted Findings
Li et al. (2016) [41]	Ship propellers and hydroturbine	The fluctuating pressure that is caused as a bubble deflates.	Water	Experimental and Numerical	The impacts of the stand-off parameter, which is denoted by γ, the strength parameter, which is denoted by, and the ratio of specific heats, which is denoted by κ.	The maximum pm progressively reduces as γ rises from 1 to 1.4. A crucial standoff parameter, γc, exists. The pressure peak of *pg* rises as κ rises, whereas the pressure peak of pm falls.
Rowlatt and Lind (2017) [42]	Bioengineering	Near a fluid–fluid contact, bubbles collapse.	Stiff wall	Numerical	Interface heights.	Due to the lessened contact between the bubble and stiff wall, the bubble collapsed spherically at increasing interface heights from the rigid wall.
Ye and Zhu (2017) [43]	Propeller blades	Acoustic bubble collapse micro-jet near the wall.	Al 1060	Numerical	Micro-jet velocity.	The wall pressure rises with increasing micro-jet velocity.
Pavlov (2017) [44]	Implosion phenomenon	The bursting of a vapor bubble caused by the immediate application of an external pressure.	Water	Numerical	Effect of vapor pressure.	Vapor offers shock absorption to the compression pressure and lengthens the period until the empty bubble bursts. In this situation, the temperature increases rapidly and the vapor pressure peaks.
Ming et al. (2017) [45]	Fuel jet when supercavitation phenomenon	The process of the collapse of the cavitation bubble that was occurring within the diesel droplet.	Water droplet	Numerical	Environmental pressure and saturation vapor pressure.	As ambient pressure rises, the rate of collapse accelerates, while saturation vapor pressure falls, the rate of collapse accelerates.
Rosselló et al. (2018) [46]	Water hammering	A device that uses water to create well-ordered collapses of a single cavitation bubble is called a water hammer.	Water	Experimental	The expansion ratio.	The expansion ratio and the equilibrium bubble radius, is the most important metric in relation to collapse strength.
Phan et al. (2018) [47]	Free surface flow	Compressible interface flows with and without mass transfers.	Water	Numerical	Superheated wall temperature.	The super-heated wall temperature increases heat flow on the sphere, comparable to previously reported findings.

Table 2. *Cont.*

Authors (Year) [Reference]	Sector/Area of Process Engineering	Configuration	Medium	Type of Study	Parameters Under Investigation	Highlighted Findings
Kerabchi et al. (2018) [51]	Lab-scale sonoreactors	Concentrated acoustic cavitation.	Liquids subjected to ultra-sound	Numerical	The influence of depth, ranging from 0 to 10 m, on acoustic cavitation, with a particular emphasis on this factor.	The drop in T_{max} and p_{max} with depth was more severe at 1000 kHz frequency and 10 °C, with 72% losses in T_{max} and 94% in p_{max} at z = 10 m.
Joshi et al. (2019) [52]	Cavitation erosion	Cavitation bubble collapsing in close proximity to an elastic-plastic substance.	Elasto-visco-plastic material	Numerical	Shockwave and Strain rate.	Shockwave impacts erode material more than micro-jet impacts. Strain rate overestimates plastic deformation by 60%.
Brujan et al. (2019) [53]	Medicine	The dynamics of a cavitation bubble created by a laser in the centre of two hard, horizontal parallel walls.	Microfluidics	Experimental	Size of the bubble, size of the bubble in relation to other bubbles, and the starting position of the bubble.	Planar jet height grows with bubble size relative to stiff wall distance. The jet velocity declines as the bubble size and distance from the vertical wall increase.
Hopfes et al. (2019) [54]	Material science to medical applications	Gelatine was filled with air bubbles that were mm-sized and subjected to an immediate pressure rise.	Gelatinous fluids	Experimental	During the collapse, there was motion of single bubbles of gelatine, each of which had a distinct concentration.	Single gelatine bubbles collapse similarly to water bubbles, and gelatine concentration does not influence their behaviour.
Bardia and Trujillo (2019) [55]	Heat exchanger and petroleum industries	Intermediate bubble collapse.	Homogeneous surround-ing liquid phase	Numerical and Theoretical	Effects of the slow but steady increase in pressure.	Instead of a step shift, a progressive rate of liquid phase pressurization affects the collapse behaviour.
Gan et al. (2019) [56]	Underwater explosion	A hull girder that replicates the construction of a genuine warship and is subjected to the stresses of an underwater explosion.	Water	Experimental and Numerical	Examined how the charge weight and the distance from the detonation effect the interactions between loads from an underwater explosion and a hull girder.	Deformation characteristics are the same over a particular charge weight range.
Chudnovskii et al. (2020) [57]	Liquid aerosol above the sea surface	Liquid that has been significantly subcooled begins to boil as a result of local heating.	Subcooled liquid	Experimental	Radiation power and duration.	The relationship between maximum bubble size and absorbed radiation power (and time, if pulsed) is nonlinear.

Table 2. *Cont.*

Authors (Year) [Reference]	Sector/Area of Process Engineering	Configuration	Medium	Type of Study	Parameters Under Investigation	Highlighted Findings
Yuan et al. (2020) [58]	Floating ice cake	A bubble bursts close to an ice cake that's drifting.	Cake	Experimental	Effect of bubble collapse on ice.	Referring to damage degree and crack growth, three damage patterns may be created on ice: crevasse, radial and circumferential, and radial.
Han et al. (2020) [59]	Chemical industry and medicine	Two different-sized bubbles interact and the wall affects bubble collapse.	Water	Numerical	The presence of the wall as well as the quantity of the bubbles.	Walls delay bubble collapse. Walls might enhance the collapse and rebound of the bubble. The number of bubbles does not directly affect collapse times.
Fursenko et al. (2020) [60]	Biological tissues and surgery	Near the microfibre submerged in a subcooled liquid, a vapor bubble collapses.	Micro-fibre immersed in a liquid	Experimental	The jet velocity and intensity.	Jet velocity is maximum during formation and declines with time. Jet speed depends on fibre radius. Maximum jet intensity be contingent on initial vapor bubble radius and fibre thickness. Water jet speed directly affects gas bubble volume decrease.
Tan and Yeo (2020) [61]	Ultrasonic device	Microabrasive particles with an average diameter of 5 µm to 50 µm were subjected to 20 kHz of strong ultrasonic irradiation in a deionized water medium.	Deionized water medium	Experimental	The influence on particle velocity of both the particle size and the ultrasonic amplitude.	Particle velocity is 8–40 m/s, depending on particle size and ultrasonic amplitude.
Orthaber et al. (2020) [62]	Biology, chemistry, and medicine	Dynamics of a single cavitation bubble at the oil–water contact.	Oil-water	Experimental and Numerical	Effect of liquid type (light or dense).	In the bubble develops in the lighter liquid, it jets toward the interface; if the denser liquid, it jets away.
Zhang et al. (2020) [63]	Biomedical treatment	Using a high-speed photography equipment, the dynamics of a laser-induced cavitation bubble collapsing close to a stiff wall.	Near the edge of a rigid wall	Experimental	Bubble location.	When a bubble is near to the edge, the interface's mobility is constrained by a distinct neck in the centre. The edge might delay bubble collapse by 22% of the Rayleigh time.

Table 2. *Cont.*

Authors (Year) [Reference]	Sector/Area of Process Engineering	Configuration	Medium	Type of Study	Parameters Under Investigation	Highlighted Findings
Ezzatneshan and Vaseghnia (2020) [64]	Different wettability conditions	Collapse dynamics of cavitation bubbles of various liquids.	Water, liquid nitrogen, and liquid hydrogen.	Numerical	Wettability effect of the surface.	The water's cavitation bubble collapses faster due to a powerful liquid jet, whereas the LN2 bubble takes a longer time to collapse due to its bigger bubble radius at the rebound.
Huang et al. (2021) [65]	Noise and erosion	The collapse and oscillation of a single bubble produce acoustic waves.	Water	Experimental and Numerical	Rarefaction wave pressure of acoustic waves.	After a shockwave, liquid velocity and pressure rise. Rarefaction waves disperse and lower liquid pressure. Acoustic waves' pressure and influence on liquid velocity decrease with distance.
Xu et al. (2021) [66]	Ships	Air entrainment cavitation bubble collapse to reduce cavitation.	Water	Experimental	The roundness, size, and distance of an air bubble.	The collapse pressure determines the collapse direction.
Peng et al. (2021) [67]	Medicine	Single and dual bubbles of cavitation seen in close proximity to a stiff border that has a significant density-to-viscosity ratio.	Near a rigid boundary	Numerical	The impacts of the gap in space between the bubble of cavitation and the solid border.	As the bubble's distance from the hard boundary grows, so will its maximum collapse pressure, velocity, and temperature. Cavitation bubble lifespan decreases.
Trummler et al. (2021) [68]	Ships	The way cavitation bubbles collapse in close proximity to walls when there is a lot of atmospheric pressure.	Vertical wall	Numerical	Effect of stand-off distance.	Stand-off distance affects collapse dynamics, micro-jet generation, rebound, and wall pressure.
Long et al. (2021) [69]	Biomedical	Throughout the pulsed laser ablation of sub-merged targets in three liquids, cavitation bubbles were created.	Three organic liquids	Numerical	Effect of liquid media on cavitation.	In ethanol and n-butanol, cavitation-induced reactions produce gaseous gases that prevent bubble collapse. In water, cavitation bubbles may be crushed into a microscopic area, increasing localized temperature and pressure and triggering a second etching.

Table 2. Cont.

Authors (Year) [Reference]	Sector/Area of Process Engineering	Configuration	Medium	Type of Study	Parameters Under Investigation	Highlighted Findings
Wang et al. (2021) [70]	Ships	Bubble collapse strength.	Water	Numerical	Viscosity, surface tension, density and sound velocity.	Viscosity and surface tension affect collapse strength more than density and sound speed.
Gai et al. (2022) [71]	Ice formation	Ice formation brought on by the cavitation bubble collapsing.	Water	Numerical	The stand-off distance, differential pressure, and initial bubble size on the maximum collapse pressure.	Scenario II's greater p_{max} initiates ice nucleation more easily than scenario I.
Xu et al. (2022) [73]	Biomedical	Various separations (d) between the cavitation bubble and the benzamide wall.	Benzamide wall	Numerical	Influence of time on maximum pressure.	When one stays in the field of maximum pressure for a longer period of time, the pressure at its peak will be higher.
Nguyen et al. (2022) [74]	Ships	Bubble collapse near walls and a free surface: dynamic behaviour of water jets, pressure loads, and bubble collapses.	Water	Experimental and Numerical	Water jets and pressure loads.	The pressure stresses that were produced as a result of the jets colliding with the walls.
Yang et al. (2022) [75]	Heating or cooling system	Near the heated/cooled wall, a cavitation bubble develops.	Water	Numerical	The mechanism of influence that the micro-jet and the cavitation bubble itself have on one another.	The values of λ, Δp, R_0 and T_w that are ideal may be employed to improve heat transmission and bring about heating or cooling treatments on a variety of various surfaces.
Ezzatneshan et al. (2024) [78]	Cryogenic fluids.	Collapse of cavitation bubbles in cryogenic liquids.	Cryogenic liquids	Numerical	The influence of the distance between a cavitation bubble with an adjacent solid wall.	The collapsing bubble cluster's micro-jet velocity and collapse intensity are both enhanced by raisng the solid surface's contact angle.

Despite the fact that it has a very tiny influence on the contour of the bubble, the analysis of a large number of previous studies revealed that the viscosity of the fluid may lower the pressure that is created by the bubble right before it bursts. The bubble deflated into a spherical shape as it moved further from the rigid wall's interface heights as a further effect of the reduced contact between the bubble and the hard wall. Although nonlinear, the relationship between the max. bubble size and power is connected to the radiation power absorbed. In addition, the max. bubble size is proportionate to the absorbed radiation power.

5. Bubble Growth and Collapse Studies

This section develops the evaluation of the bubble growth and collapse studies that were encountered to enhance the heat transmission of specific set of industries.

In 2017, a Sharp-Interface Level-Set (SILS) approach was addressed by Lee and Son [79] for modelling the development and contraction of a compressible vapor bubble. This method utilises a level set function to characterize the interface and a number of partial differential correlations to express the dynamics of the fluid and solid phases. Referring to the findings, it was revealed that the temperature of the surrounding environment was an extensive factor in describing the pattern of bubble development and collapse. Throughout the phase of bubble expansion, it was also detected that the bubble's pressure does not fall to a level lesser than the saturation pressure corresponding to the phase. The impacts of ambient temperature and phase change on the temporal discrepancies of the mathematical outcomes for the bubble growth and burst close to a wall, considering the radius of bubble and pressure of wall against time, are exhibited in Figure 11. Figure 11 shows that the bubble growth and collapse in close vicinity to a wall established that a re-entrant liquid jet was designed as a result of irregular liquid inertia and was influenced by the phase change and ambient liquid temperature. This phenomenon was observed when the bubbles formed near the wall. Furthermore, it was determined that the SILS Approach is a prevailing numerical technique for simulating complex multi-phase flows that involve fluid–solid interactions. It permits accurate modelling of the sharp interface between the fluid and solid phases and can handle complex geometries.

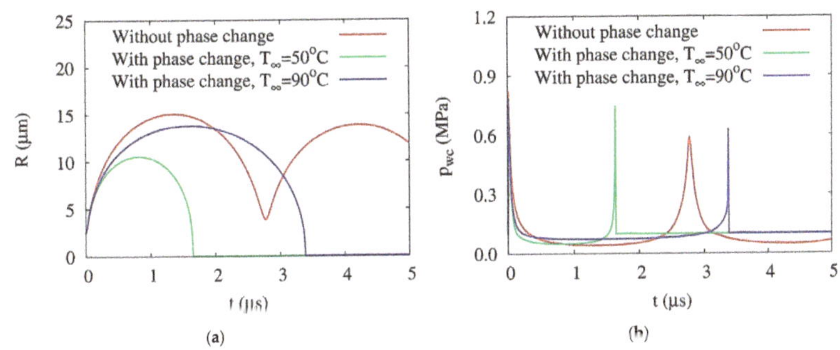

Figure 11. The consequences of ambient temperature and phase change on the temporal variations of numerical results for the bubble growth and shrinkage near the wall: (**a**) radius of bubble vs. time and (**b**) pressure of wall vs. time [79] "(Reproduced with permission from International Communications in Heat and Mass Transfer by Elsevier, 2024; Order Number: 5858201250142)".

Wang et al. [80] utilised a comprehensive set of ordinary differential equations to observe bubble production and collapse in micro-gravity situations. A dimensionless fitting constantly considers both the area changes of the moving vapor/liquid interface and the diffusive nature of the interface layer was added. By comparing predicted temporal fluctuations of bubble radii with actual data, one may define the fitting coefficients for bubble production and collapse in water and ethanol. The original conditions of a sharp lessening in temperature caused some "fluctuations" in the simulated bubble radii during the initial phase of bubble collapse. The "re-bound influence" of pressure equilibrium in the bubble, which occurred due to the original state, was linked to these "fluctuations".

In 2019, to power micro-synthetic jets, Sourtiji and Peles [81] calculated the viability of using bubble growth and collapse over a wide set of operational frequencies between 0.1 and 2.5 Hz and a heating rate between 3 and 4.5 W. A primary micro-channel was linked to a micro heater that was placed inside a chamber with dimensions of 3.5 mm in radius and 220 μm in height. The nozzle's aperture was 300 μm. The artificial jet was formed by switching the micro-heater on and off at regular intervals, which led to the implosion and explosion of bubbles inside the container. The authors captured successive images of bubble nucleation, growth, and eventual collapse using high speed DSLR photography and a microscope. The characterization of the synthetic jet comprised the use of a momentum coefficient. It was indicated that the synthetic jet's average value was more than unity over a wide set of operational frequencies, which recommends that it could progress the performance of a number of micro-systems, including the micro-mixing that takes place in microfluidic devices.

In 2019, continued visual observations were prepared by Sun et al. [82] of the morphological development of a hydrate shell on suspended methane bubbles in super-saturated water. Referring to recently discovered information and results, the hydrate shell's dynamic growth progressively went through three phases: rapid development, ongoing collapse, and growing stasis. These stages can be further separated into their individual sub-stages. A novel mechanistic model for the development of hydrate shells on bubble surfaces was established. This model measured the mechanical features of hydrate collapse as well as the inter-coupling of several hydrate-forming mechanisms and multi-phase mass transfer processes. This model was created based on the surface properties of hydrated bubbles at various phases. The model not only described the growth of hydrates on each side of the hydrate shell, it also provided the first quantifiable account of the capillary holes within the hydration shell.

Zimnyakov et al. [83] employed image analysis to pinpoint when foam samples started to form at various temperatures. The transition divided the collaborative of surface film cells into sub-ensembles that are developing and depopulating. With the depopulation of unstable sub-ensembles, a specific rate of cell rearrangements exponentially dropped and got closer to a fixed value during the growth. The association between the temperature dependent "macroscopic" growth collapse rate for average cell size, a "microscopic" rate parameter for individual cell growth–collapse, and the probability density function for cell size distribution was established for growth.

In 2020, a cavitation bubble produced by the nanosecond pulsed laser ablation of a solid at shallow liquid depths was explored by Nguyen et al. [84]. When the bubble radius was between 0.4 and 9 times the max. the liquid was presented. Photo-elasticity laser stroboscopic videography was utilised to track bubble dynamics. The bubble breaks through liquid when the liquid depth is slighter than the bubble radius. The duration of the bubbles is shortened but no further shock was felt. The bubbles expanded during contraction and flattened during expansion at liquid depths that are half or less of the max. bubble radius. The bubbles became toroidal at the lowest contraction, emitting more shockwaves.

Puncochar et al. [85] proposed a simple and straightforward physical model for the description of the bubble formation process. This concept covered bubble expansion, bubble ascent, bubble separation, and bubble deformation. Buoyancy, inertia, and elasticity are the three effects that were considered. The relationship between the final form of the bubble and its Weber number was assessed during the detachment procedure using the energy conservation concept. For this investigation, the bubble shapes, including oblate spheroid, prolate spheroid, spherical cap, and lens shape were taken into account. The data that were gathered are examined and contrasted with earlier studies. Table 3 shows different applications of the accompanying studies that delineated the bubble growth and collapse.

In the publications that were analysed above, it was found that ambient temperature was a key aspect in defining the pattern of bubble development and collapse. A re-entrant liquid jet arises as a result to the irregular liquid inertia and is substantially influenced by the phase shift and ambient liquid temperature, as seen by the bubble expansion and collapse close to the wall. The fact that the bubbles became larger before collapsing close to the wall served as evidence for this. Additionally, the secondary shockwaves become enlarged as the liquid depth increases, and a durable impulse is generated into the solid target.

Table 3. Outline of studies on bubble growth and collapse in different applications.

Authors (Year) [Reference]	Sector/Area of Process Engineering	Configuration	Medium	Type of Study	Parameters Under Investigation	Highlighted Results/Findings
Lee and Son (2017) [79]	Ultrasonic cleaning and medical treatment	The expansion and subsequent deflation of a compressible vapor bubble.	Liquid-vapor phase change	Numerical	The influence of the ambient temperature, the phase shift, and the wall on the development and collapse of the compressible bubble.	It was demonstrated that the ambient temperature had a substantial influence on the bubble formation and collapse patterns. Additionally, it was noted that throughout the bubble growth phase, the bubble pressure did not descent lower the comparable saturation pressure.
Wang et al. (2018) [80]	Subcooled boiling	The expansion and contraction of bubbles when subjected to microgravity.	Water	Numerical	Behaviour of bubble radii.	The bubble radii exhibited some "variations" throughout the early stages of bubble collapse, which may be ascribed to the "re-bound effect" of pressure balance in the bubble due to the starting condition of a sharp decrease in temperature.
Sourtiji and Peles (2019) [81]	Microbiology and micro-sensors	Micro synthetic jets that function by using bubble development and collapse throughout a spectrum of operational frequencies (0.1–2.5 Hz) and heating powers.	Microfluidics	Experimental	Operating frequencies.	The average value is greater than unity over a wide range of operational frequencies, proposing that this jet may advance the functionality of a number of micro-systems, including micro-mixing in microfluidic devices.
Sun et al. (2019) [82]	System sub-cooling	Variation in the morphology of the hydrate shell that forms on methane bubbles floating in supersaturated water.	Water	Experimental	Effect of continuous collapse and hydrate shell.	The opposition between the hydrate shell's critical collapse pressure and the bubbles' differential pressure has a significant impact on how frequently continuous collapse occurs.

Table 3. *Cont.*

Authors (Year) [Reference]	Sector/Area of Process Engineering	Configuration	Medium	Type of Study	Parameters Under Investigation	Highlighted Results/Findings
Zimnyakov et al. (2019) [83]	Fresh constrained foams	The surface film cells that are located at the interface between the new liquid foam and the confining wall.	Liquid foam	Numerical and Analytical	Effect of splitting process.	The averaged specific rate of local re-arrangements in the surface film cells and clusters of bulk bubbles decays exponentially during the splitting process until it reaches a constant value. This happens as the rate of re-arrangements in the clusters of surface film cells and bulk bubbles slows down.
Nguyen et al. (2020) [84]	Nanoparticles synthesis	A nanosecond pulsed laser is used to ablate a solid at varying depths of a finite liquid, which causes a bubble of cavitation to form.	Finite liquid	Experimental	liquid depth.	The secondary shockwaves grow more concentric as the liquid depth rises, and a larger impulse is produced into the solid target as a result of this change.
Puncochar et al. (2022) [85]	Water desalination	Process of bubble production and subsequent deformation.	Water	Numerical	Buoyancy, inertia, elasticity.	The principle of energy conservation is used throughout the detachment process in order to define the association that exists between the final form of the bubble and its Weber number.

6. Evaluation of the Numerical Methods Used in Bubble Growth and Collapse Studies

Many industrial and environmental activities involve binary and multi-phase flow phenomena. These procedures entail the simultaneous flow of multiple phases (gas, liquid, and solid) in varied arrangements. To accurately represent the complicated physical interactions between the phases in the numerical simulation of these processes, specialized models and techniques are needed. The current review recognizes the utilization of different numerical methods to mimic multi-phase flow phenomena. Each method has its advantages and shortcomings, depending on the specific presentation and the level of accurateness needed. Some of the commonly used numerical techniques for multi-phase flow simulation are discussed below:

The Finite Volume Method (FVM) is a generally utilised mathematical method for resolving complex multi-phase flow problems. It is a conventional approach that can handle complex geometries and irregular boundaries. FVM can also handle non-uniform grids and is computationally effective. However, FVM needs careful selection of numerical schemes to avoid numerical instability [86]. Specifically, FVM is an adaptable method that can be used in both liquid–gas and solid–gas systems as a result of its capacity to handle complicated geometries, non-uniform grids, and multi-phase interactions. However, the careful selection of numerical patterns is important in both cases to guarantee stability and precision in the simulations.

The Level-Set Method (LSM) is a general numerical method for pursuing the development of a free surface or interface between two fluids in a multi-phase flow. It indicates the interface as a level set of a higher dimension corresponding to the computational domain. The LSM delivers a number of benefits, including the ability to handle complex interface topologies and geometries, and making it simple to apply boundary conditions. The dynamics of the interface, such as merging and splitting events, can also be precisely captured [87].

The numerical method known as the Phase-Field Model (PFM) is utilised to mimic multiphase flows that entail the progress of a diffuse interface between various fluids. The interface between the two fluids is characterised by a non-conserved scalar field in the model's mathematical foundation. The PEM delivers a number of benefits, including the ability to handle complex interface topologies and geometries and make boundary conditions simple to implement. Furthermore, it may faithfully depict the dynamics of the interface, such as coalescence and break-up processes [88].

A numerical technique known as the Included Phase Model (IPM) is utilized to simulate multiphase flows involving the presence of bubbles in a fluid. Instead of representing the dispersed phase as discrete particles, it signifies it as a continuum and uses a series of transport equations to chart its behavior. The IPM offers many benefits, including the ability to handle high volume fractions of the dispersed phase and the ability to consider the impacts of the dispersed phase on the continuous phase. Moreover, it can record intricate physical processes like turbulence modulation [89].

A powerful tool for modelling complicated multi-phase flows in which the collaboration between distinct phases is important is CFD. Multi-phase flows can be simulated using a variety of CFD techniques, counting the Eulerian–Eulerian, Eulerian–Lagrangian, and Eulerian–Granular approaches. The specific parameters of the multi-phase flow being simulated determine the CFD that should be used [90].

The complex multi-phase flow phenomenon can be accomplished using two analytical and numerical methods called the Potential Flow Theory (PFT) and the Boundary Element Method (BEM). Understanding the underlying notions regulating fluid dynamics is made easier by PFT. The BEM is a mathematical approach to solve partial differential equations that contain discretization within the domain of interest's borders. When dealing with complex geometries, where more standard numerical tactics like the Finite Element or Finite Difference approaches can become computationally expensive, BEM is particularly helpful. It is crucial to remember that both PFT and BEM methods include restrictions and presumptions that might not always hold true in practical settings. Thus, suspicious validation of the results against experimental data is necessary to safeguard the accurateness and reliability of the simulation results [91].

Spectral Element Marker Particle (SEMP) is a mathematical method used in general to simulate complex multi-phase flows that include the interaction between fluid and solid phases to accurately model fluid and solid dynamics. SEMP discretizes the fluid and solid domains, which permits a high order of accuracy and efficient computation of the solution. SEMP has several benefits over other numerical techniques, such as its ability to accurately capture the dynamics of the fluid–solid interface and the interaction between the particles [92].

The Finite Element Method (FEM) is a mathematical approach that can solve complicated multi-phase flow problems by dividing the domain into small subdomains. The

FEM can handle complex geometries and asymmetrical boundaries and can provide accurate results. However, the FEM is computationally expensive, particularly for large-scale problems [93].

A numerical method called the Thermal Lattice Boltzmann Method (TLBM) is utilised to model intricate multiphase flows involving heat transfer. This technique is based on a mesoscopic approach that simulates the behavior of individual particles within a lattice to address fluid flow problems. In comparison to other numerical methods, LBM provides a number of benefits, including the ability to precisely model the dynamics of multi-phase flows and heat transfer events. Also, it is computationally active and enables the simulation of intricate geometries. To settle the correctness of the results, it is essential to compare the simulation results to experimental data [94].

Smoothed Particle Hydrodynamics (SPH) is a meshless numerical method that can simulate multi-phase flow phenomena with complicated geometries and boundaries. SPH is computationally efficient and can handle large deformations and complex free-surface flows. However, SPH requires a large number of particles to achieve accurate results, which can raise computational cost [95].

A prevalent numerical approach to simulate free-surface flows is Volume of Fluid (VOF). It is capable of handling complicated geometries and boundaries as well as properly capturing the interface between several phases. To accurately capture the contact, VOF may need a fine grid resolution, which can be computationally expensive, especially for large-scale problems [96].

A numerical method known as Sharp-Interface Level-Set (SILS) is also used to model complex multi-phase flows that incorporate the collaboration of the fluid and solid phases. The approach relies on the Level-Set Technique, which numerically tracks the boundary between two fluids in a flow. In comparison to other numerical methods, the SILS Approach has several benefits, such as the ability to accurately represent the sharp boundary between fluid and solid phases. Moreover, complex multi-phase flows incorporating surface tension and other interfacial phenomena can be simulated using this method. To ensure the correctness of the simulation outcomes, the outcomes should be contrasted with experimental data [97].

The various discussed numerical methods, including the Finite Volume Method (FVM), Level-Set Method (LSM), Phase-Field Model (PFM), Included Phase Model (IPM), Computational Fluid Dynamics (CFD), Potential Flow Theory (PFT), Boundary Element Method (BEM), Spectral Element Marker Particle (SEMP), Thermal Lattice Boltzmann Method (TLBM), Smoothed Particle Hydrodynamics (SPH), Volume of Fluid (VOF), and Sharp-Interface Level-Set (SILS) are all useful tools to simulate both liquid–gas and solid–gas systems. Each method affords exceptional strengths, particularly when capturing the complexities of multiphase flows. These include an accurate modeling of fluid–solid interactions, managing complex geometries, and efficiently tracking phase interfaces. Despite some numerical methods being prominent in specific cases—like SPH for large deformations or VOF for free-surface flows—others can provide detailed contexts for perceiving the dynamics in both types of systems, such as CFD and IPM. This flexibility can highlight the prosperity of these numerical methods to be used in a wide sector of applications in fluid dynamics and materials science.

7. Recommendations for Further Research

The following are some suggestions that may be derived from the thermal performance analysis of each research study conducted, as well as from a careful examination of the information shown in Tables 1–3. In other words, the review came up with the limitations of past research on bubble growth and collapse that should be resolved following these suggestions:

There is a necessity to utilise the Bayesian approach to modelling bubble growth and collapse because it is still extremely pertinent today.

As the use of high-speed underwater acoustic communications shifts from military to commercial applications, it has absorbed a lot of attention. The implementation of these vehicles necessitates a speed variation in the cavitation flow, which introduces a significant challenge. Furthermore, deceleration may contribute to the worsening of cavitation instability, making it an important factor that has to be researched further. Thus, more exact experimental and numerical methodologies are necessary in the future to quantify the necessities of the maximum heat transfer.

The creation of bubbles with variable amounts of gas within them and the subsequent investigation of how this affects the behavior of the bubbles, as well as the investigation of how a bursting bubble interacts with a flexible substance, are both interesting possibilities for future research.

8. Conclusions

These are the most important conclusions that can be made:

The interaction between the ultimate form of the bubble and its Weber number is assessed throughout the detachment process using the energy conservation concept.

Throughout the splitting process, the average specific rate of local rearrangements in surface film cells and bulk bubble clusters decreases rapidly until it achieves a fixed value. This occurs as the intensity of rearrangements in the bulk bubbles and clusters of surface film cells declines.

A key element in deciding whether continuous collapse takes place is the degree of competitiveness between the critical collapse pressure of the hydrate shell and the bubble differential pressure.

It was found that the pattern of bubble formation and collapse was significantly influenced by the ambient temperature.

Collapse strength is most strongly influenced by viscosity, followed by surface tension, and then by density and sound velocity, in that order of diminishing importance.

Inside the bubble of ethanol and n-butanol, cavitation-induced reactions result in the production of gaseous by-products that prevent the bubble from collapsing.

The dynamics of the collapse, the growth of the micro-jet, the rebound, and the maximum wall pressure are all significantly influenced by the stand-off distance.

The shockwave sources a growth in the velocity and pressure of the liquid field. The rarefaction waves trigger a reduction in the liquid's pressure, which causes a dispersive change in the liquid's spatial distribution.

A visible neck will form in the middle of the bubble when the bubble is near the edge. This apparent neck will restrict the bubble interface's mobility close to the edge.

Bubble density, a metric of bubble structure created by coalescence, and the two/point autocorrelation function show the expansion of a linked network, that may result in lower release thresholds.

The micro-layer that forms at the bubble bottom is responsible for around thirty percent of the entire quantity of evaporation.

Numerical methods can accurately simulate complex physical interactions between different phases while providing detailed information on the flow field and phase distribution.

Careful selection of numerical schemes is important to avoid numerical instability.

Author Contributions: F.L.R.: Conceptualization, Formal analysis, Writing—original draft, Project administration. M.A.A.-O.: Formal analysis, Investigation, Writing—review & editing, Visualization, Project administration. A.K.H.: Formal analysis, Investigation, Visualization. S.A.: Formal analysis, Visualization. M.S.A.: Formal analysis. I.M.M.: Investigation, Writing—review & editing, Visualization, Project administration. All authors have read and agreed to the published version of the manuscript.

Funding: This research received no external funding.

Conflicts of Interest: The authors declare that they have no known competing financial interests or personal relationships that could have appeared to influence the work reported in this paper.

Nomenclature

ANOVA	analysis of variance
BE	boundary-element
BEM	boundary-element method
CEL	Coupled Eulerian–Lagrangian
CFD	Computational Fluid Dynamics
CLBM	Conventional Lattice Boltzmann Method
DDF	double distribution function
FEM	Finite Element Method
HIFU	High Intensity Focused Ultrasound
IMCs	intermetallic compounds
IPM	Included Phase Model
LBM	Lattice Boltzmann method
LSM	Level-Set Method
MRT-LBM	multi-relaxation-time lattice Boltzmann method
ONB	Onset of Nucleate Boiling
PDE	partial differential equation
PFM	Phase-Field Model
PFT	Potential Flow Theory
RP	Rayleigh-Plesset
SEM	scanning electron microscopy
SEMP	Spectral Element Marker Particle
SILS	Sharp-Interface Level-Set
SPH	Smoothed Particle Hydrodynamics
TLBM	Thermal Lattice Boltzmann method
UMo	Uranium Monolithic
VOF	Volume of Fluid
VOFE	Volumetric Optical Force

References

1. Wang, H.; Zhang, C.; Xiong, H. Growth and collapse dynamics of a vapor bubble near or at a wall. *Water* **2020**, *13*, 12. [CrossRef]
2. Mehralizadeh, A.; Shabanian, S.R.; Bakeri, G. Effect of modified surfaces on bubble dynamics and pool boiling heat transfer enhancement: A review. *Therm. Sci. Eng. Prog.* **2020**, *15*, 100451. [CrossRef]
3. Hussein, E.Q.; Rashid, F.L.; Hussein, A.K.; Younis, O. Hydrodynamics of single bubble rising through water column using volume of fluid (VOF) method. *J. Therm. Eng.* **2021**, *7*, 2107–2114. [CrossRef]
4. Mohanty, R.L.; Das, M.K. A critical review on bubble dynamics parameters influencing boiling heat transfer. *Renew. Sustain. Energy. Rev.* **2017**, *78*, 466–494. [CrossRef]
5. Hong, Y.; Ashgriz, N.; Andrews, J.; Parizi, H. Numerical simulation of growth and collapse of a bubble induced by a pulsed microheater. *J. Microelectromech. Syst.* **2004**, *13*, 857–869. [CrossRef]

6. Al-Obaidi, M.A.; Rasn, K.H.; Aladwani, S.H.; Kadhom, M.; Mujtaba, I.M. Flexible design and operation of multi-stage reverse osmosis desalination process for producing different grades of water with maintenance and cleaning opportunity. *Chem. Eng. Res. Des.* **2022**, *182*, 525–543. [CrossRef]
7. Al-Obaidi, M.A.; Alsarayreh, A.A.; Mujtaba, I.M. Reduction of energy consumption of brackish water reverse osmosis desalination system via model based optimisation. *J. Tech.* **2023**, *5*, 1–7. [CrossRef]
8. Avvaru, B.; Venkateswaran, N.; Uppara, P.; Iyengar, S.B.; Katti, S.S. Current knowledge potential applications of cavitation technologies for the petroleum industry. *Ultrason. Sonochem.* **2018**, *42*, 493–507. [CrossRef]
9. Rajoriya, S.; Carpenter, J.; Saharan, V.K.; Pandit, A.B. Hydrodynamic cavitation: An advanced oxidation process for the degradation of bio-refractory pollutants. *Rev. Chem. Eng.* **2016**, *32*, 379–411. [CrossRef]
10. Edwards, D.H.; Jones, I.R. An experimental study of the forces generated by the collapse of transient cavities in water. *J. Fluid Mech.* **1960**, *7*, 596–609.
11. Fujikawa, S.; Akamatsu, T. Experimental investigations of cavitation bubble collapse by a water shock tube. *Bull. JSME* **1978**, *21*, 223–230. [CrossRef]
12. Karri, B.; Pillai, K.S.; Klaseboer, E.; Ohl, S.W.; Khoo, B.C. Collapsing bubble induced pumping in a viscous fluid. *Sens. Actuators A Phys.* **2011**, *169*, 151–163. [CrossRef]
13. Koynov, A.; Khinast, J.G.; Tryggvason, G. Mass transfer and chemical reactions in bubble swarms with dynamic interfaces. *AIChE J.* **2005**, *51*, 2786–2800. [CrossRef]
14. Agarwal, D.K.; Welch, S.W.J.; Biswas, G.; Durst, F. Planar Simulation of Bubble Growth in Film Boiling in Near-Critical Water Using a Variant of the VOF Method. *J. Heat Transf.* **2004**, *126*, 329–338. [CrossRef]
15. Mukherjee, A.; Dhir, V.K. Study of Lateral Merger of Vapor Bubbles During Nucleate Pool Boiling. *J. Heat Transf.* **2005**, *126*, 1023–1039. [CrossRef]
16. Tomar, G.; Biswas, G.; Sharma, A.; Agrawal, A. Numerical simulation of bubble growth in film boiling using a coupled level-set and volume-of-fluid method. *Phys. Fluids* **2005**, *17*, 112103. [CrossRef]
17. Wu, J.; Dhir, V.K. Numerical simulations of the dynamics and heat transfer associated with a single bubble in subcooled pool boiling. *J. Heat Transf.* **2010**, *132*, 1–15. [CrossRef]
18. Marchal, A.; Vergnes, B.; Poulesquen, A.; Valette, R. Competitive growth and rising of bubbles in a yield stress fluid. Consequences on the macroscopic swelling of bitumen drums. *J. Nonnewton Fluid Mech.* **2016**, *234*, 162–169. [CrossRef]
19. Goel, P.; Nayak, A.K.; Kulkarni, P.P.; Joshi, J.B. Experimental study on bubble departure characteristics in subcooled nucleate pool boiling. *Int. J. Multiph. Flow* **2017**, *89*, 163–176. [CrossRef]
20. Zhang, S.; Li, X.; Zhu, Z. Numerical simulation of cryogenic cavitating flow by an extended transport-based cavitation model with thermal effects. *J. Cryog.* **2018**, *92*, 98–104. [CrossRef]
21. Pahk, K.J.; de Andrade, M.O.; Gélat, P.; Kim, H.; Saffari, N. Mechanical damage induced by the appearance of rectified bubble growth in a viscoelastic medium during boiling histotripsy exposure. *Ultrason. Sonochem.* **2019**, *53*, 164–177. [CrossRef] [PubMed]
22. Lu, W.; Zhang, N.; Ding, Z.; Yang, F.; Hu, Q.; Zeng, L.; Li, J. Bubble growth, intermetallic compounds dissolution and their interactions during heating of an Al-5wt.%Mn alloy by in-situ synchrotron radiography. *J. Alloys Compd.* **2020**, *822*, 153554. [CrossRef]
23. Zeng, X.; Wu, G.; Wang, J.; Yang, C.; Meng, Q.; Chen, G.; Chen, D. Effects of inhibitors on the morphology and kinetics of hydrate growth on surface of bubble. *J. Nat. Gas. Sci. Eng.* **2020**, *74*, 103096. [CrossRef]
24. Allred, T.P.; Weibel, J.A.; Garimella, S.V. The Role of Dynamic Wetting Behavior during Bubble Growth and Departure from a Solid Surface. *Int. J. Heat Mass Transf.* **2021**, *172*, 121167. [CrossRef]
25. Zhao, W.; Liang, J.; Sun, M.; Wang, Z. Investigation on the effect of convective outflow boundary condition on the bubbles growth, rising and breakup dynamics of nucleate boiling. *Int. J. Therm. Sci.* **2021**, *167*, 106877. [CrossRef]
26. Winder, D.; Lin, L.; Mach, J. Incorporating bubble growth volume feedback to improve simulation of the response of a structure containing liquid and gas to sudden energy input. *Nucl. Instrum. Methods. Phys. Res. A* **2021**, *1005*, 165371. [CrossRef]
27. Li, D.; Ren, Z.; Li, Y.; Gong, R.; Wang, H. Thermodynamic effects on the cavitation flow of a liquid oxygen turbopump. *Cryogenics* **2021**, *116*, 103302. [CrossRef]
28. Zhou, J.; Xu, P.; Qi, B.; Zhang, Y.; Wei, J. Effects of micro-pin-fins on the bubble growth and movement of nucleate pool boiling on vertical surfaces. *Int. J. Therm. Sci.* **2022**, *171*, 107186. [CrossRef]
29. Lai, T.; Yan, S.; Zhao, Q.; Zhang, Z.; Ding, W.; Liu, Y.; Hou, Y. Numerical study on thermodynamic growth of single hydrogen bubble in an infinite space. *Int. J. Hydrogen Energy* **2022**, *47*, 24406–24420. [CrossRef]
30. Cheng, N.; Yu, S.; Xiao, J.; Peng, C. Numerical simulation of single bubble growth in vertical rectangular narrow flow channels. *Nucl. Eng. Des.* **2022**, *391*, 111749. [CrossRef]

31. Hu, S.; Setyawan, W.; Beeler, B.W.; Gan, J.; Burkes, D.E. Defect cluster and nonequilibrium gas bubble associated growth in irradiated UMo fuels-A cluster dynamics and phase field model. *J. Nucl. Mater.* **2020**, *542*, 152441. [CrossRef]
32. Zhou, T.T.; He, A.M.; Wang, P. Dynamic evolution of He bubble and its effects on void nucleation-growth and thermomechanical properties in the spallation of aluminum. *J. Nucl. Mater.* **2020**, *542*, 152496. [CrossRef]
33. Arévalo-Cid, P.; Adán-Más, A.; Silva, T.M.; Rodrigues, J.A.; Maçôas, E.; Vaz, M.F.; Montemor, M.F. On the growth mechanical properties of nanostructured cobalt foams by dynamic hydrogen bubble template electrodeposition. *Mater. Charact.* **2020**, *169*, 110598. [CrossRef]
34. Prudil, A.A.; Welland, M.J.; Ofori-Opoku, N. Modelling the growth and evolution of statistically significant populations of intergranular fission gas bubbles by IPM. *J. Nucl. Mater.* **2022**, *566*, 153777. [CrossRef]
35. Gong, S.; Cheng, P. Lattice Boltzmann simulation of periodic bubble nucleation, growth and departure from a heated surface in pool boiling. *Int. J. Heat Mass Transf.* **2013**, *64*, 122–132. [CrossRef]
36. Navarrete, M.; Godínez, F.A.; Castellanos, F.; Mejía-Uriarte, E.V.; Naude, J.L.; Méndez, F. Dynamics and acoustic energy dissipation in conical bubble collapse. *Int. J. Multiph. Flow* **2015**, *76*, 86–100. [CrossRef]
37. Gong, S.W.; Klaseboer, E. Interaction between collapsing bubble and viscoelastic solid: Numerical modelling and simulation. *Appl. Math. Model.* **2016**, *40*, 122–132. [CrossRef]
38. Tang, J.; Yan, C.; Sun, L. Enhanced vapor bubble condensation and collapse with ultrasonic vibration. *Exp. Therm. Fluid Sci.* **2016**, *70*, 115–124. [CrossRef]
39. Wang, Y.; Huang, C.; Fang, X.; Wu, X.; Du, T. On the internal collapse phenomenon at the closure of cavitation bubbles in a deceleration process of underwater vertical launching. *Appl. Ocean. Res.* **2016**, *56*, 157–165. [CrossRef]
40. Qin, Z.; Alehossein, H. Heat transfer during cavitation bubble collapse. *Appl. Therm. Eng.* **2016**, *105*, 1067–1075. [CrossRef]
41. Li, S.; Han, R.; Zhang, A.M.; Wang, Q.X. Analysis of pressure field generated by a collapsing bubble. *Ocean. Eng.* **2016**, *117*, 22–38. [CrossRef]
42. Rowlatt, C.F.; Lind, S.J. Bubble collapse near a fluid-fluid interface using the spectral element marker particle method with applications in bioengineering. *Int. J. Multiph. Flow* **2017**, *90*, 118–143. [CrossRef]
43. Ye, L.; Zhu, X. Analysis of the effect of impact of near-wall acoustic bubble collapse micro-jet on Al 1060. *Ultrason. Sonochem.* **2017**, *36*, 507–516. [CrossRef] [PubMed]
44. Pavlov, P.A. Fast collapse of a vapor bubble. *Int. J. Heat Mass Transf.* **2017**, *104*, 627–633. [CrossRef]
45. Ming, L.; Zhi, N.; Chunhua, S. Numerical simulation of cavitation bubble collapse within a droplet. *Comput. Fluids* **2017**, *152*, 157–163. [CrossRef]
46. Rosselló, J.M.; Urteaga, R.; Bonetto, F.J. A novel water hammer device designed to produce controlled bubble collapses. *Exp. Therm. Fluid Sci.* **2018**, *92*, 46–55. [CrossRef]
47. Phan, T.H.; Ha, C.T.; Park, W.G. Numerical simulation of bubble collapse between two parallel walls and saturated film boiling on a sphere. *Int. J. Heat Mass Transf.* **2018**, *127*, 116–125. [CrossRef]
48. Arévalo, R.; Antúnez, D.; Rebollo, L.; Abánades, A. Estimation of radiation coupling factors in film boiling around spheres by mean of Computational Fluid Dynamics (CFD) tools. *Int. J. Heat Mass Transf.* **2014**, *78*, 84–89. [CrossRef]
49. Grigoriev, V.A.; Klimenko, V.V.; Shelepen, A.G. Pool film toiling from submerged spheres. In Proceedings of the International Heat Transfer Conference Digital Library, Munich, Germany, 6–10 September 1982; Begel House Inc.: Danbury, CT, USA, 1982.
50. Frederking, T.H.; Clark, J.A. Natural convection film boiling on a sphere. In *Advances in Cryogenic Engineering: Proceedings of the 1962 Cryogenic Engineering Conference University of California Los Angeles, CA, USA, 14–16 August 1962*; Springer: New York, NY, USA, 1963; pp. 501–506.
51. Kerabchi, N.; Merouani, S.; Hamdaoui, O. Depth effect on the inertial collapse of cavitation bubble under ultrasound: Special emphasis on the role of the wave attenuation. *Ultrason. Sonochem.* **2018**, *48*, 136–150. [CrossRef]
52. Joshi, S.; Franc, J.P.; Ghigliotti, G.; Fivel, M. SPH modelling of a cavitation bubble collapse near an elasto-visco-plastic material. *J. Mech. Phys. Solids* **2019**, *125*, 420–439. [CrossRef]
53. Brujan, E.A.; Takahira, H.; Ogasawara, T. Planar jets in collapsing cavitation bubbles. *Exp. Therm. Fluid Sci.* **2019**, *101*, 48–61. [CrossRef]
54. Hopfes, T.; Wang, Z.; Giglmaier, M.; Adams, N.A. Collapse dynamics of bubble pairs in gelatinous fluids. *Exp. Therm. Fluid Sci.* **2019**, *108*, 104–114. [CrossRef]
55. Bardia, R.; Trujillo, M.F. An improved categorization of vapor bubble collapse: Explaining the coupled nature of hydrodynamic thermal mechanisms. *Int. J. Heat Mass Transf.* **2019**, *145*, 118754. [CrossRef]
56. Gan, N.; Yao, X.L.; Liu, L.T.; Xiao, W.; Wang, X.L. Research on overall damage characteristics of a hull girder under explosion bubble collapse. *Ocean. Eng.* **2019**, *188*, 106315. [CrossRef]

57. Chudnovskii, V.M.; Levin, A.A.; Yusupov, V.I.; Guzev, M.A.; Chernov, A.A. The formation of a cumulative jet during the collapse of a vapor bubble in a subcooled liquid formed as a result of laser heating. *Int. J. Heat Mass Transf.* **2020**, *150*, 119286. [CrossRef]
58. Yuan, G.Y.; Ni, B.Y.; Wu, Q.G.; Xue, Y.Z.; Zhang, A.M. An experimental study on the dynamics and damage capabilities of a bubble collapsing in the neighborhood of a floating ice cake. *J. Fluids Struct.* **2020**, *92*, 102833. [CrossRef]
59. Han, S.; Li, Y.; Zhu, H. The similarities and differences between the bubble collapse near a solid wall and in free water. *Eur. J. Mech.-B/Fluids* **2020**, *84*, 553–561. [CrossRef]
60. Fursenko, R.V.; Chudnovskii, V.M.; Minaev, S.S.; Okajima, J. Mechanism of high velocity jet formation after a gas bubble collapse near the micro fiber immersed in a liquid. *Int. J. Heat Mass Transf.* **2020**, *163*, 120420. [CrossRef]
61. Tan, K.L.; Yeo, S.H. Velocity estimation of micro-particles driven by cavitation bubble collapses through controlled erosion experiments. *Int. J. Multiph. Flow* **2020**, *127*, 103271. [CrossRef]
62. Orthaber, U.; Zevnik, J.; Petkovšek, R.; Dular, M. Cavitation bubble collapse in a vicinity of a liquid-liquid interface—Basic research into emulsification process. *Ultrason. Sonochem.* **2020**, *68*, 105224. [CrossRef]
63. Zhang, Y.; Qiu, X.; Zhang, X.; Tang, N. Collapsing dynamics of a laser-induced cavitation bubble near the edge of a rigid wall. *Ultrason. Sonochem.* **2020**, *67*, 105157. [CrossRef] [PubMed]
64. Ezzatneshan, E.; Vaseghnia, H. Simulation of collapsing cavitation bubbles in various liquids by lattice Boltzmann model coupled with the Redlich-Kwong-Soave equation of state. *Phys. Rev. E* **2020**, *102*, 053309. [CrossRef] [PubMed]
65. Huang, G.; Zhang, M.; Han, L.; Ma, X.; Huang, B. Physical investigation of acoustic waves induced by the oscillation and collapse of the single bubble. *Ultrason. Sonochem.* **2021**, *72*, 105440. [CrossRef] [PubMed]
66. Xu, W.; Zhu, R.; Wang, J.; Fu, Q.; Wang, X.; Zhao, Y.; Zhao, G. Molecular dynamics simulations of the distance between the cavitation bubble and benzamide wall impacting collapse characteristics. *J. Clean. Prod.* **2022**, *352*, 131633. [CrossRef]
67. Peng, H.; Zhang, J.; He, X.; Wang, Y. Thermal pseudo-potential lattice Boltzmann method for simulating cavitation bubbles collapse near a rigid boundary. *Comput. Fluids* **2021**, *217*, 104817. [CrossRef]
68. Trummler, T.; Schmidt, S.J.; Adams, N.A. Effect of stand-off distance and spatial resolution on the pressure impact of near-wall vapor bubble collapses. *Int. J. Multiph. Flow* **2021**, *141*, 103618. [CrossRef]
69. Long, J.; Eliceiri, M.H.; Wang, L.; Vangelatos, Z.; Ouyang, Y.; Xie, X.; Zhang, Y.; Grigoropoulos, C.P. Capturing the final stage of the collapse of cavitation bubbles generated during nanosecond laser ablation of submerged targets. *Opt. Laser Technol.* **2021**, *134*, 106647. [CrossRef]
70. Wang, X.; Ning, Z.; Lv, M.; Sun, C. Machine learning for predicting the bubble-collapse strength as affected by physical conditions. *Results Phys.* **2021**, *25*, 104226. [CrossRef]
71. Gai, S.; Peng, Z.; Moghtaderi, B.; Yu, J.; Doroodchi, E. LBM study of ice nucleation induced by the collapse of cavitation bubbles. *Comput. Fluids* **2022**, *246*, 105616. [CrossRef]
72. Ezzatneshan, E.; Vaseghnia, H. Dynamics of an acoustically driven cavitation bubble cluster in the vicinity of a solid surface. *Phys. Fluids* **2021**, *33*, 1075290. [CrossRef]
73. Xu, W.L.; Li, J.B.; Luo, J.; Zhai, Y.W. Effect of a single air bubble on the collapse direction and collapse noise of a cavitation bubble. *Exp. Therm. Fluid Sci.* **2021**, *120*, 110218. [CrossRef]
74. Nguyen, T.T.P.; Tanabe-Yamagishi, R.; Ito, Y. Effects of liquid depth on the expansion and collapse of a hemispherical cavitation bubble induced in nanosecond pulsed laser ablation of a solid in liquid. *Opt. Lasers Eng.* **2020**, *126*, 105937. [CrossRef]
75. Yang, Y.; Shan, M.; Su, N.; Kan, X.; Shangguan, Y.; Han, Q. Role of wall temperature on cavitation bubble collapse near a wall investigated using thermal lattice Boltzmann method. *Int. Commun. Heat Mass Transf.* **2022**, *134*, 105988. [CrossRef]
76. He, X.; Peng, H.; Zhang, J.; Yuan, H. Multiple vapor cavitation bubble interactions with a thermal lattice Boltzmann method. *Ocean. Eng.* **2022**, *266*, 113058. [CrossRef]
77. Ezzatneshan, E.; Salehi, A.; Vaseghnia, H. Study of micro-heater shape and wettability effects on inception of boiling phenomenon using a multiphase lattice Boltzmann method. *Int. J. Therm. Sci.* **2023**, *184*, 107913. [CrossRef]
78. Ezzatneshan, E.; Salehi, A.; Vaseghnia, H. Numerical study on collapsing cavitation bubble dynamics in cryogenic fluids. *Cryogenics* **2024**, *141*, 103879. [CrossRef]
79. Lee, J.; Son, G. A sharp-interface level-set method for compressible bubble growth with phase change. *Int. Commun. Heat Mass Transf.* **2017**, *86*, 1–11. [CrossRef]
80. Wang, Q.; Yao, W.; Quan, X.; Cheng, P. Validation of a dynamic model for vapor bubble growth and collapse under microgravity conditions. *Int. Commun. Heat Mass Transf.* **2018**, *95*, 63–73. [CrossRef]
81. Sourtiji, E.; Peles, Y. A micro-synthetic jet in a microchannel using bubble growth and collapse. *Appl. Therm. Eng.* **2019**, *160*, 114084. [CrossRef]
82. Sun, B.; Liu, Z.; Wang, Z.; Chen, L.; Li, H.; Duan, W. Experimental and modeling investigations into hydrate shell growth on suspended bubbles considering pore updating and surface collapse. *Chem. Eng. Sci.* **2019**, *207*, 1–16. [CrossRef]

83. Zimnyakov, D.A.; Yuvchenko, S.A.; Isaeva, A.A.; Isaeva, E.A.; Tsypin, D.V. Growth/collapse kinetics of the surface bubbles in fresh constrained foams: Transition to self-similar evolution. *Colloids Surf. A Physicochem. Eng. Asp.* **2019**, *579*, 123693. [CrossRef]
84. Nguyen, V.T.; Phan, T.H.; Duy, T.N.; Kim, D.H.; Park, W.G. Modeling of the bubble collapse with water jets and pressure loads using a geometrical volume of fluid based simulation method. *Int. J. Multiph. Flow* **2022**, *152*, 104103. [CrossRef]
85. Puncochar, M.; Ruzicka, M.C.; Simcik, M. Bubble formation and deformation. *Chem. Eng. Sci.* **2022**, *260*, 117729. [CrossRef]
86. Castro, J.M. *Modelling and Numerical Simulation of Combustion and Multi-Phase Flows Using Finite Volume Methods on Unstructured Meshes*; Universitat Politècnica de Catalunya: Barcelona, Spain, 2018.
87. Van der Pijl, S.P.; Segal, A.; Vuik, C.; Wesseling, P. A mass-conserving Level-Set method for modelling of multi-phase flows. *Int. J. Numer. Methods Fluids* **2005**, *47*, 339–361. [CrossRef]
88. Alpak, F.O.; Riviere, B.; Frank, F. A phase-field method for the direct simulation of two-phase flows in pore-scale media using a non-equilibrium wetting boundary condition. *Comput. Geosci.* **2016**, *20*, 881–908. [CrossRef]
89. Kono, K.; Ishizuka, T.; Tsuda, H.; Kurosawa, A. Application of lattice Boltzmann model to multiphase flows with phase transition. *Comput. Phys. Commun.* **2000**, *129*, 110–120. [CrossRef]
90. Afzali, S.; Rezaei, N.; Zendehboudi, S.; Chatzis, I. Computational fluid dynamic simulation of multi-phase flow in fractured porous media during water-alternating-gas injection process. *J. Hydrol.* **2022**, *610*, 127852. [CrossRef]
91. Sichani, A.B.; Mehring, C. Generalized pressure-correction for viscous potential flow simulations of multiphase flows using boundary element method. *Int. J. Multiph. Flow* **2021**, *141*, 103646. [CrossRef]
92. Pei, C.; Vahab, M.; Sussman, M.; Hussaini, M.Y. A Hierarchical Space-Time Spectral Element and Moment-of-Fluid Method for Improved Capturing of Vortical Structures in Incompressible Multi-phase/Multi-material Flows. *J. Sci. Comput.* **2019**, *81*, 1527–1566. [CrossRef]
93. Efendiev, Y.; Ginting, V.; Hou, T.; Ewing, R. Accurate multiscale finite element methods for two-phase flow simulations. *J. Comput. Phys.* **2006**, *220*, 155–174. [CrossRef]
94. Sahu, A.; Bhowmick, S. Applications of Lattice Boltzmann in method in multi-component and multi-phase flow: A review. In Proceedings of the AIP, Woodbury, Long Island, NY, USA, 6–7 December 2019.
95. Szewc, K.; Tanière, A.; Pozorski, J.; Minier, J.-P.A. Study on Application of Smoothed Particle Hydrodynamics to Multi-Phase Flows. *Int. J. Nonlinear. Sci. Numer. Simul.* **2012**, *13*, 383–395. [CrossRef]
96. Kim, M.S.; Lee, W.I.L. A new VOF-based numerical scheme for the simulation of fluid flow with free surface. Part I: New free surface-tracking algorithm and its verification. *Int. J. Numer. Methods Fluids* **2003**, *42*, 765–790. [CrossRef]
97. Han, L.H.; Hu, X.Y.; Adams, N.A. Scale separation for multi-scale modeling of free-surface and two-phase flows with the conservative sharp interface method. *J. Comput. Phys.* **2015**, *280*, 387–403. [CrossRef]

processes

MDPI

Review

A Review of Key Technologies for Green and Low-Carbon Future Buildings in China

Lei Xu [1,†], Dong Yu [2,*,†], Jinyu Zhou [1] and Chaowu Jin [3]

[1] College of Mechanical and Electrical Engineering, Jinling Institute of Technology, Nanjing 211169, China
[2] School of Electrical Engineering, Southeast University, Sipailou No. 2, Nanjing 210096, China
[3] College of Mechanical and Electrical Engineering, Nanjing University of Aeronautics and Astronautics, Nanjing 210016, China
* Correspondence: yudong@seu.edu.cn
† These authors contributed equally to this work.

Abstract: Green and low-carbon transformation has become a global development theme. This paper provides a systematic survey of the literature related to key technologies for green and low-carbon future buildings, including the green design method system of low-carbon future buildings; the low-carbon future buildings construction system and green building materials; the low-carbon future buildings energy system; the low-carbon intelligent operation of buildings and carbon emission control in the operation stage; and the low-carbon future buildings assessment system. This paper analyzes and summarizes the whole chain of "carbon monitoring-carbon accounting-carbon assessment-carbon control" of the whole life cycle of buildings. We systematically surveyed the development overview and the latest development trend of scientific and technological innovation; the existing starting level of science and technology; and the existing intellectual property, competition, and industrialization prospects in each stage of the whole life cycle. We conducted an overall analysis of the characteristics, thematic context, and knowledge evolution of low-carbon future building-related research, both domestically and internationally; referring to the relevant standards and specifications for new types of buildings, such as green buildings, a preliminary theoretical framework for low-carbon future buildings is constructed from the perspective of the entire life cycle.

Keywords: green and low-carbon future buildings; green design method system; construction system and green building materials; energy system; intelligent operation and carbon emission control; evaluation system

1. Introduction

Currently, the green and low-carbon transition has become a global development theme [1]. China has announced that carbon emissions are striving to reach a carbon peak by 2030 and striving to achieve carbon neutrality by 2060. China is the world's top carbon emission country, with a total carbon emission of nearly 10 Gt, accounting for nearly 30% of the world. According to statistics, carbon emissions from the whole process of construction account for 50.6% of China's carbon emissions and are the main source of China's carbon emissions. Therefore, the promotion of green and low carbon in the building sector is the key link to achieving the goal of "carbon peak and carbon neutrality" in China [2]. Low-carbon buildings and green buildings have the same value orientation but different connotations. With the double carbon target reduction schedule of each country, the transformation from green to low-carbon buildings is inevitable in the next phase of construction [3]. At present, several studies have been conducted in the field related to low-carbon future buildings in academia; this paper has conducted a systematic research survey of the literature related to key technologies of green and low-carbon future buildings. We have reviewed the depth of current academic research in the planning and design of low-carbon future buildings, green building materials and intelligent construction, building energy systems, intelligent operation and maintenance, and evaluation systems.

At present, academics have already conducted a considerable depth of research and obtained rich results in low-carbon future building planning and design; green building materials and intelligent construction; building energy systems; intelligent operation and maintenance; and evaluation systems [4]. In terms of building planning and design, the use of artificial intelligence and big data for building form generation; the early intervention of energy-saving and emission-reducing elements in the design process; and assisting architects in positive design have become the direction of research on low-carbon future building design [5]. In the direction of green building materials and intelligent construction, new building materials with high performance and low emissions will continue to be developed to meet the needs of the construction industry, and the use of digital technology to enhance assembly-type buildings and intelligent construction is the main direction of future research and development [6]. In the direction of building energy, improving electrification, renewable energy building integration, green power Direct Current (DC) drive equipment, etc., are the main directions of future research and development [7]. In the building operation stage, energy consumption and carbon emission limit control require systematic and in-depth research in the areas of building energy diagnosis; intelligent and optimized operation and carbon emission monitoring; and carbon emission limit. Breakthroughs are made in relevant key technologies to provide a reliable guarantee for low-carbon building operations [8]. In terms of building evaluation systems, European and American countries have taken the lead in carrying out low-carbon building practices, establishing a foundation for the clarification of the concept of low-carbon buildings and the establishment of evaluation systems.

However, there are no accepted methods and mature conclusions for research on low-carbon design, low-carbon construction, low-carbon energy systems, low-carbon operation strategies, and low-carbon evaluation methods for buildings [9]. Most studies have applied the concepts of green building, energy-efficient building, and low-energy building to study low-carbon future buildings, which has brought some trouble to the current development of low-carbon future buildings [10]. Existing green and energy-efficient building construction technologies mainly focus on safety and clean construction, which cannot be directly transposed to low-carbon future buildings, and there is a lack of cutting-edge construction technology research for low-carbon goals [11]. The development requirements of low-carbon future buildings bring serious challenges to the traditional building materials industry, which mostly presents a rough growth mode, and the problem of overcapacity is becoming more and more prominent and lacks perfect guidelines for the transformation of the low-carbon building materials development mode. Renewable energy utilization is the key engine of low-carbon city and building development, but there is still a need to make efforts in source, network, load, and storage, especially with the lack of a multi-energy complementary supply system and active–passive low-carbon technology integration methods for low-carbon cities, building design, and operation and maintenance [12]. The high urbanization rate and large industrial output value in some regions still dominate "high energy consumption and high emissions", and carbon emissions from building operation and maintenance continue to grow, lacking effective carbon emission control tools and low-carbon- and zero-carbon-oriented intelligent carbon control technologies. The evaluation purpose of low-carbon future buildings is not clear enough; the evaluation index is relatively single; a unified list of basic data, such as energy consumption and carbon emission coefficients in each life stage of buildings and a systematic evaluation system, have not been formed; and an operable and relatively accurate evaluation method has not been formed.

Several studies have been conducted in the field of low-carbon future buildings, resulting in several scientific and technological achievements. This review has conducted a systematic research survey of the literature related to key technologies of green and low-carbon future buildings. The main contributions of this review can be summarized as follows:

1. The research survey of this review will take each stage of the whole life cycle of the building (planning and design, building materials construction, energy system, operation and maintenance management, evaluation system, etc., as shown in Figure 1) as the main focus, and systematically analyze the development overview and latest development trend of science and technology innovation in each stage of the whole life cycle of low-carbon future buildings, the existing starting point of the science and technology level, and the existing intellectual property rights, as well as the global competition and industry prospects.
2. We point out the current troubles of low-carbon future buildings, judge the future development trend of low-carbon future buildings, determine the development goal and implementation path of low-carbon future buildings, and provide support and reference for the development of low-carbon future buildings-related fields.
3. We conduct an overall analysis of the characteristics, thematic context, and knowledge evolution of low-carbon future building-related research, both domestically and internationally; referring to the relevant standards and specifications for new types of buildings, such as green buildings, a preliminary theoretical framework for low-carbon future buildings is constructed from the perspective of the entire life cycle.

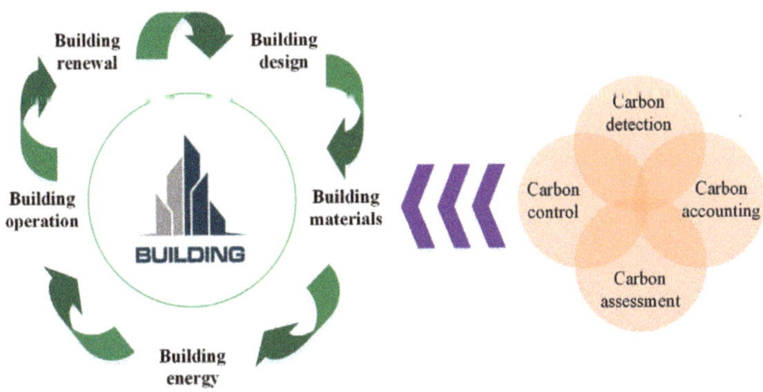

Figure 1. Carbon emission control mechanism in the whole life cycle of buildings.

This paper reviews the characteristics of low-carbon future buildings design; reviews construction, energy, operation, and maintenance systems; points out the current troubles of low-carbon future buildings; examines the future development trend of low-carbon future buildings; determines the development goal and implementation path of low-carbon future buildings; and provides support and reference for the development of low-carbon future buildings-related fields.

The research objectives of this paper are as follows: to comprehensively examine and explore the key technologies and development trends in green and low-carbon buildings and provide a theoretical basis and practical guidance for the sustainable development of the construction industry. Firstly, this article clarifies the basic concept of green and low-carbon buildings, including the relevant definitions and connotations of low-carbon buildings, green buildings, and energy-saving buildings, as well as their relationships and differences. This helps readers have a clear understanding and comprehension of green and low-carbon buildings. By reviewing the development history of green and low-carbon buildings, the entire process of this field from its inception to development can be demonstrated, including key time nodes, important milestone events, and the main factors driving its development. We also explore the future development trends and challenges in green and low-carbon buildings. This includes the direction of technological innovation, the impact of policies and regulations, the reference to international experience, and opportunities for cooperation.

Through these discussions, we can help with understanding the future development direction and potential challenges in green and low-carbon buildings, providing a reference for future research and practice. By comprehensively and systematically introducing and analyzing the key technologies and development trends of green and low-carbon buildings, valuable reference information can be provided for researchers, designers, engineers, and policymakers to better understand and apply green and low-carbon building technologies and promote the sustainable development of the construction industry. The purpose of this article is to comprehensively examine and explore the key technologies and development trends in this field, providing a theoretical basis and practical guidance for the sustainable development of the construction industry. This paper can provide a deeper understanding of the core concepts, development history, key technologies, development trends, and challenges of green and low-carbon buildings, providing reference and inspiration for future research and practice.

2. Materials and Methods

This paper contains an analysis of data from open sources, including publications and data from internet portals, Web of Science, Scopus scientometric databases, websites of patent offices, libraries, and reading rooms.

In the process of analyzing the scientific, technical, and patent-licensed literature, the following were considered: the key technologies for green and low-carbon future buildings, including the green design method system of low-carbon future buildings; the low-carbon future buildings construction system and green building materials; the low-carbon future buildings energy system; the low-carbon intelligent operation of buildings and carbon emission control in the operation stage; and the low-carbon future buildings assessment system. The above key technologies are analyzed from three perspectives: the general situation and the latest development trend of scientific and technological innovation; the existing starting point of the science and technology level and the existing intellectual property rights; and the global competition and industrialization prospects.

In writing this paper, a total of 150 papers (in the past ten years) were analyzed on the development of the key technologies for green and low-carbon future buildings. In this paper, 86 papers were selected for in-depth analysis and summary, including the green design method system of low-carbon future buildings; the low-carbon future buildings construction system; green building materials; the low-carbon future buildings energy system; the low-carbon intelligent operation of buildings and carbon emission control in the operation stage; and the low-carbon future buildings assessment system. The keywords used in the search for scientific and technical information were the following: low-carbon buildings; building design; buildings construction system; green building materials; buildings energy system; intelligent operation of buildings and carbon emission control; and buildings assessment system.

3. Results and Discussion

3.1. Green Design Method System of Low-Carbon Future Buildings

4. General Situation and the Latest Development Trend of Scientific and Technological Innovation

In terms of low-carbon future buildings design, the 2017 International Energy Agency (IEA) report raised an urgent need to reduce the proportion of carbon emissions from urban construction. In recent years, some scholars have proposed the relationship between the impact of the construction environment and building planning on building carbon emission and urban block comfort [13]. Many Chinese scholars have explored the correlation between urban form, microclimate, and building energy performance and its form generation and optimization strategy [14]. In the direction of building a single design, as early as the beginning of this century, professional institutions in developed countries in Europe and the United States developed a large number of tools for building carbon emission calculations based on ISO 14040 Life Cycle Assessment (LCA) [15], such as BEAT 2002 in Denmark, Eco-Quantum in the Netherlands, GaBi in Germany, etc. However, these tools are all developed based on the carbon emission database of industrial products, and they have some problems in the applicability of buildings, so they have not become mainstream in the construction market. At the same time, the green building evaluation tools based on the scoring and evaluation system represented by Leadership in Energy and Environmental Design (LEED) in the United States have dominated the market. Although such tools are very successful in the commercial market, their scientificity is constantly questioned by professionals. In recent years, countries have developed their own carbon emission calculation standards based on the whole life cycle of the carbon emission calculation method, which has become the basis of [16–18] for building carbon emission evaluation. In 2002, the Building Information Model (BIM) technology proposed by the Autodesk company tried to integrate information about building design, construction, and operation. After more than ten years of development, this technology has been gradually improved, providing a technical basis for computing carbon emissions. BIM's unified format standards provide the basis for secondary development, and commercial tools based on BIM have been developed. In the future, this software will be further integrated with existing mainstream design tools to provide strong support for low-carbon future building design. In addition, the forward design supported by intelligence has also become the latest design development trend.

5. Existing Starting Point of Science and Technology Level and Existing Intellectual Property Rights

The field of low-carbon future building planning and design is still in its infancy. Researchers have made some achievements in low-carbon urban planning [19–21]. In the 2017 International Energy Ownership Report, Thibaut Abergel et al. proposed an urgent need to reduce the proportion of carbon emissions from urban building construction. Abdo Abdullah Ahmed Gassar studied the relationship between the impact of the built environment and building planning on building carbon emissions and urban block comfort [22,23]. Gauch, Hannes L. et al. aimed to evaluate the influence of building design variables on the efficiency of building designs and devised a model that combines parametric optimized design of structural frames with thermal building modeling. In terms of architectural design, it is in the stage of transformation from energy-saving buildings and green buildings to low-carbon future buildings. The research on green buildings is on the rise. China has set up a special project of the "13th Five-Year Plan" on the topic of "Green building and building industrialization" and carried out dozens of special studies related to green building, with fruitful results [24–26]. At the same time, many scholars have also begun to actively explore intelligent generation design methods emphasizing the whole chain workflow from data collection to generation design and even related to Computerized Numerical Control (CNC) construction. The current artificial intelligence is from the information integration, mapping modeling, prediction, and decision level of architectural design and generates design methods to make intelligent algorithms involved in the designer's design thinking and space operation process. It also combines environmental elements, physical elements, and other conditions based on being environmentally friendly as the goal of positive design exploration. In general, these studies are in the early stages, and the research on zero-carbon block planning and design is not comprehensive. Future architectural design methods that are low carbon and based on multi-objective optimization and complex adaptive systems are needed. In addition, there is still a large distance between the industrialization of theory and methodological tools.

6. Competition and Industrialization Prospects

The research on sustainable building design methods is undergoing the process of conversion from energy-saving buildings and green buildings to low-carbon future buildings. At present, the research on low-carbon future building design methods is still in its infancy, but it is developing rapidly [27–29].

The International Energy Agency-Energy in Buildings and Communities (IEA-EBC) has created the Future Buildings Forum (FBF) to explore low-carbon design methods, which has published several collections since its inception and is the authoritative theoretical integration of exploring the methods of future buildings design technology. The Future Cities Laboratory (FCL) at the Singapore-Eidgenössische Technische Hochschule (ETH) Research Center has been continuously researching and developing zero-carbon building models since 2015. Google's X company is developing computer-aided generation of green building design methods. The International Institute for Sustainable Built Environment (IISBE) has been continuously developing and promoting sustainable building methods and sustainable building tools, such as the Sustainable Building (SB) Method and SB Tool. In addition to scientific research institutions, a large number of commercial companies are also continuing to invest in this field. Autodesk has developed the Green Building Studio based on the Revit platform to analyze the projects' monthly and annual carbon emissions. Xu Weiguo from Tsinghua University, Yuan Feng from Tongji University, and Li Biao from Southeast University have made important explorations in promoting digital design direction [30,31]. Under the leadership of Academician Cui Kai, the China Academy of Architectural Design has conducted research on green construction design methods. Based on PKPM, the Institute of China Academy of Architectural Sciences continues to develop design software tools with carbon emission computing functions.

In addition to the traditional method of design before simulation, forward design methods based on computing technology are becoming a major trend in low-carbon future architectural design [32–34]. The intelligent generation design method emphasizes the whole chain workflow from data collection to generation design and is even related to CNC construction. In recent years, Professor Hovestadt of ETH has developed a digital chain control system for the design of an apartment block with 300 household units, which is an early attempt at the generation design method based on the digital chain system. Southeast University took the lead in China to explore the forward-generation design mode of green public buildings by a programming method and obtained independent intellectual property rights.

In general, exploration has made certain achievements but has mostly stayed at the theoretical level; to be used in practical engineering projects and obtain large-scale application conditions, there are still multiple problems to be solved: first, the existing design method tools use a high threshold—for grass-roots architectural designers, learning these is too difficult. Secondly, the connection with the existing construction engineering system is not smooth enough. As a mature industry, the construction industry has a relatively complete set of workflows and corresponding tools and software. The position of the new tools in the existing system is awkward, and they still have not been widely recognized by relevant practitioners. Finally, the use cost of advanced tools is too high, so a large number of small- and medium-sized design institutions with hundreds of thousands of software use costs cannot afford them, and a large number of construction projects are completed by these small- and medium-sized design agencies, which makes it difficult to apply these research results in engineering practice.

3.2. Low-Carbon Future Buildings Construction System and Green Buildings Materials

1. General Situation and the Latest Development Trend of Scientific and Technological Innovation

In terms of low-carbon future building construction technology, the combined carbon emissions of building materials and building construction carbon emissions exceed 50% of the whole process of building carbon emissions. In the Pre-2030 Carbon Peak Action Plan released by China, it is clearly pointed out that green and low-carbon building materials and green construction methods should be promoted, the industrialization of new buildings should be accelerated, prefabricated buildings should be vigorously developed, steel structure housing should be promoted, and the recycling of building materials should be promoted. In terms of building materials production, the high energy consumption and high carbon emission problems in the construction department are largely caused by the extensive production of traditional building materials [35]. Held in 1999, the first green building seminar in China put forward the concept of green building materials, internationally also known as ecological building materials, healthy building materials, or environmental protection building materials. Corresponding to traditional building materials, green building materials use clean production technology and include industrial and agricultural or urban solid waste production, achieving this effect after the recycling cycle, and they can effectively reduce the whole life cycle of building carbon emissions. In the past 20 years, Europe, the United States, Japan, and other developed countries have attached great importance to the development of green building materials and have carried out comprehensive and systematic basic research work, which has also greatly promoted the development of green building materials in China.

At present, green building materials are used in a large number of green building materials production technologies, such as high-quality cement, high-performance concrete, energy-saving doors and windows, and new walls. Green building materials production technology has become a key research field in building carbon emission reduction in the world, which can effectively promote the sustainable, green, and low-carbon development of the whole life cycle of the construction sector [36]. In terms of green construction, prefabricated building technology can reduce on-site operations through standardized design and factory production to effectively reduce energy consumption in the construction stage and achieve carbon emission reduction. The development of Western prefabricated buildings has been relatively mature, and China is also vigorously promoting the development of prefabricated buildings [37]. The development of prefabricated buildings plays a positive role in reducing pollutants and waste emissions and improving labor productivity, and it is also an effective implementation of national policies. It is an important technical path for the construction industry to achieve a "carbon peak" and "carbon neutrality" [38]. A study by Sriskanthan Srisangeerthanan et al. considers modular buildings as those built using prefabricated, fully completed volumetric units called modules, which could be an apartment unit, staircase, structural core component, etc. Such modules are factory-manufactured and fit with mechanical connections for assembly on-site, where they would be stacked vertically and scaled horizontally to form complete buildings. The key structural materials for modules vary among timber, concrete, or steel; however, steel elements are more commonly used.

Intelligent construction, namely digital technology and engineering construction systems, is deeply integrated to form the innovative development mode of engineering construction. Its technology system is mainly built up by artificial intelligence, cloud computing, big data, BIM, and other technologies. Intelligent construction technology can be fully applied to all stages of the whole life cycle of the building, and the use of new information and communication technology can enhance the accuracy of geological analysis in the survey and design stage, strengthen the safety cost progress management in the construction stage, and assist the monitoring and early warning in the building operation and maintenance stage. With the support of new communication technology, intelligent construction technology has effectively improved building production efficiency, land saving, materials saving, and energy saving, and it is also an important means of building carbon emission reduction in various countries.

2. Existing Starting Point of Science and Technology Level and Existing Intellectual Property Rights

In the field of low-carbon construction, after a series of research and engineering practices, there are now more than 30 regional or national building standards formed. Among them, the prefabricated concrete structure is one of the most important systems of industrial buildings. To follow the existing design method of the cast-in-place concrete structure system, the current prefabricated concrete structure specifications and engineering practice are based on the design concept of "equivalent cast-in-place", that is, grouting sleeve and other methods are used to connect steel bars at the splice of the prefabricated components and post-cast concrete. In China's newly released "Prefabricated Concrete Building Technical Standard" (GB/T51231-2016) [39], the "equivalent cast-in-place" design method of prefabricated concrete structure is given, which rapidly promotes the industrialization process of construction industrialization. Transforming the traditional site pouring into factory manufacturing is an important step forward in the low carbonization of buildings. However, due to the small gap between the grouting sleeve and the connecting steel bar, it is difficult to install the prefabricated components. Grout sleeve connection must be carried out in very detailed grouting operations, but because it is buried deep in the prefabricated components, the actual construction quality is difficult to test, resulting in blind spots in quality supervision and causing concern for the government and the owners. Therefore, a large number of vertical components in the current prefabricated building practice do not use prefabricated components. At the same time, there are still a large number of structures that have to be built in cast-in-place mode due to their structural particularity, such as large span, high durability, complex function, and shape [40]. These factors lead to the current low proportion of industrialization, and the actual assembly rate is not high, which affects the low-carbon process of the building construction process.

In addition, in the field of green building materials, cement production involves silicate decomposition and kiln calcination processes, and the carbon dioxide emitted accounts for about 7% of the total global carbon emissions. Seeking new green building materials has become the primary priority of low-carbon emission reduction in the current construction industry [41]. As the largest city source of waste in China, construction waste has an annual output of more than 1.5 Gt, accounting for about one-third of the total municipal waste. At present, the utilization rate of construction waste resources in China is less than 10%. It is particularly important to realize the sustainable development of the construction industry.

Based on China's "double carbon" strategic goal, current construction waste recycling is mainly for the following: (1) Industrial by-products (such as fly coal, steel slag, etc.) used as cement concrete mineral admixture and fine aggregate to reduce the use of cement; the construction industry also issued a series of technical standards, such as the mineral blending division application technical specification (GB/T51003-2014) [42]. However, in this process, it is still impossible to completely avoid the use of high energy consumption and high carbon emission product cement; thus, real green building material production is not realized. (2) To produce recycled aggregate to replace natural aggregate and reduce the exploitation of natural resources. The conversion rate between construction waste and recycled aggregate is extremely high, and 1 Mt of construction waste can generate 0.88 Mt of recycled aggregate. However, in the production of recycled aggregate, high energy consumption calcination, such as high-temperature crushing and grinding and high-energy drying, cannot achieve low-carbon emission processing.

Therefore, in the development of cement-free green building materials, construction waste should be used to produce low-carbon emission recycled aggregate free from high-temperature calcination to realize the resource utilization of construction waste and to promote the green development of the construction industry. The low-carbon and green development of the construction industry depends not only on the research and development of green building materials but also on the use of green construction technology. Compared with the traditional concrete cast-in-place process, 3D printing technology can achieve formwork-free support in the molding process, saving materials and greatly improving construction efficiency, which has a sustainable development prospect. Using 3D printing and construction technology can reduce the waste of labor and resources in the traditional construction process and promote the transformation of the construction industry to industrialization.

3. Competition and Industrialization Prospects

The new low-carbon construction mode takes "green" as the goal, "intelligent" as the technical means, "industrialization" as the production mode, project general contracting as the implementation carrier, and green building materials as the material basis to realize the goal of "energy conservation, environmental protection, efficiency, quality, and safety" in the construction process [43].

In recent years, under the guidance of the state and the housing and construction departments policy, China's construction industrialization process is in full swing. In the current engineering practice, the main industrial building systems that have been adopted all follow the concept of "equivalent cast-in-place" through the bonding of the reinforcement at the end of the component and the post-pouring area or the concrete of the adjacent components. Under the principle of "equivalent cast-in-place", design and construction technicians can apply standards and design theory to the whole series of cast-in-place structures, effectively promoting the application of prefabricated buildings. This has formed a relatively mature prefabricated concrete shear wall structure and prefabricated frame structure system, which is reflected in the engineering practice of good economic benefits and social and environmental benefits.

While protecting the indoor space and reducing energy loss, the building skin receives renewable energy, such as solar energy and wind energy, recovers rainwater resources, and carries the ecological greening to create the carbon sink [44]. Among the receiving devices for renewable energy for building epidermis structures, solar energy is the most widely used [45]. Professor FelixTrombe, director of the French Solar Energy Laboratory, and his collaborators first proposed and tested the successful Trumber wall, which relies on the interaction of a bilateral epidermis, no mechanical power, no traditional energy consumption, and only the passive collection of solar energy to heat buildings. In addition, by solar collectors, pipes, fans or pumps, heat storage devices, indoor heat dissipation ends, and other active solar energy utilization systems, many countries have developed a lot of new solar thermal utilization products using photovoltaic tiles, photovoltaic wall panels, etc. and the development of buildings combined with photovoltaic technology. In 2002, the United States of America Department of Energy launched and hosted the international Solar Decathlon (SD) competition, combining solar energy, energy saving, and architectural design with a new way of integrating, designing, building, and running a functional, perfect, comfortable, livable, sustainable solar living space and showcasing the world's latest energy technology achievements.

In terms of technology, many countries have been trying since the 1990s and have relatively mature technology and experience. The "world capital of fashion and design", Milan, Italy, has built two Ferris "tree towers", becoming the world's first pair of green apartments—green from the top and the ground to the vertical wall extension [46]. In recent years, China has also begun to consciously use green epidermis, and relevant research and practice are still in the initial stages, with a lack of standard and technical crossover research. When using it in specific projects, inconsistent design concepts and technologies and even difficult integration occur, most of which still stay in the roof garden mode.

3.3. Low-Carbon Future Building Energy System

1. General Situation and the Latest Development Trend of Scientific and Technological Innovation

In terms of the low-carbon future building energy system, the building energy system can provide support for the future low-carbon/zero-carbon building in the three major directions of energy efficiency improvement, building production capacity, and building electrification [47]. Figure 2 shows a block diagram of the energy system of a building that uses a composite energy system to provide cooling and heat for the building. In terms of energy efficiency, there are two main methods [48]: first, more use of low-grade renewable energy; countries have successively used ground, water, and air sources in the renewable energy category. Solar photovoltaic photothermal energy, low-grade waste heat-driven refrigeration heating, and power generation technology are increasingly mature; the main bottleneck is how to improve source uncertainty or discontinuity (such as photovoltaic, thermal) and building energy system characteristics and how to overcome the source, water, sewage source, and heat resources constrained by the spatial and temporal distribution problem [49]. At present, new energy storage media and systems are being developed, and the construction method of low-carbon future building energy systems according to local conditions and multi-energy complementary methods is being studied to solve the problem of continuity and the spatial and temporal distribution of renewable low-grade energy [50].

In addition, the air source heat pump, as the main building cooling and heating mode in the Yangtze River Basin in China and the main method of clean heating in northern China, is also an important solution. We need to study its winter frost suppression and frost-free and summer efficient operation technology and to develop cold and heat source technology not limited by geographical and geological conditions. Second, with more intelligent and efficient optimization control methods and platforms, more and more buildings have begun to access the energy monitoring platform, mainly to monitor the energy system node parameters and parameter visualization; some have feedback regulation ability according to the outdoor parameters and load, set key parameters to achieve the purpose of energy saving, and in frontier research, have been used for model-based optimization regulation. With big data, artificial intelligence aims to solve the three problems of building load accurate prediction, building energy system modeling, and global optimal control [51]. In terms of load prediction, from the calculation method based on building a thermal model to the actual operation process of the data-driven method, data characteristic engineering, gradient algorithm, and memory network have the building energy consumption prediction level at about 70%. The future higher proportion of renewable energy scenarios, days, and schedules puts forward better requirements for energy consumption prediction.

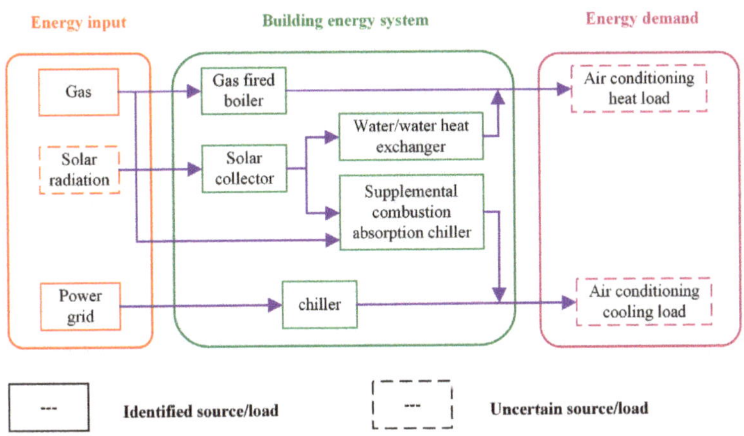

Figure 2. Block diagram of a building energy system.

Building energy system modeling mainly adopts the black box model (data-driven) and white box model (physical model), which are suitable for existing buildings with a large amount of operating data and new buildings with only physical parameters, respectively [52]. The development of a model suitable for the whole cycle of the building is the direction of future development. In terms of optimization control, heuristic algorithms, such as genetic algorithms and particle swarm algorithms, have received more and more attention and application [53]. Under the situation of a large increase in cooling, heat storage, and power storage systems, dynamic planning and other algorithms that can solve multi-stage decision problems will get more and more attention.

Construction production capacity mainly relies on a photovoltaic roof, photovoltaic glass, photovoltaic wall panels, and other realizations. Its research directions are mainly two: (1) Through advanced materials and manufacturing technology, the photoelectric conversion efficiency is constantly improved. At present, the photovoltaic conversion efficiency of photovoltaic panel products has exceeded 20%, and some laboratory efficiency is more than 25%. But for the photovoltaic products integrated with the buildings, such as photovoltaic roofing and photovoltaic wall panels, the efficiency of their commercial products has not exceeded 16% and 12% because of the need to consider the structure and other factors. (2) Through structural design and building design, there can be better implementation of routine building replacements, installation efficiency can be improved, the cost can be reduced, and the payback period can be shortened. The current investment payback period for the photovoltaic roof is 8–10 years; photovoltaic wall panels are still more than 10 years old. In terms of building electrification, comprehensively promoting building electrification is an important way to absorb building production capacity and green power in the power grid under the increasing proportion of renewable energy generation in the future. The consumption of energy in building hot and cold environment construction accounts for about 40% of the total energy consumption; all kinds of electric drive heat pumps combined with low-grade energy can realize the efficient supply of building heating and cooling and are completely consistent with the direction of building electrification. In addition, the development of DC-driven heat pump equipment further reduces loss in the process of power conversion and is an important development direction.

2. Existing Starting Point of Science and Technology Level and Existing Intellectual Property Rights

In terms of the comprehensive utilization and system construction of low-grade renewable energy in buildings, the model library of TRNSYS 18, MODELICA 4.0, MATLAB 2023b, and other platforms are mainly used for the coupling and simulation of multiple different cold and heat source systems [54]. Combining meteorological parameters and building load, we can evaluate the annual energy consumption and carbon emissions or select the load rate and the system performance parameters under specific working conditions [55]. Most of the above methods can only be used as a reference for scheme design and system construction, and there is still a gap with the actual operation effect, mainly because the load prediction, the dynamic characteristics simulation of the system, and the characteristics extraction of low-grade renewable resources are still not accurate enough [56]. In terms of a universal double high-efficiency frost-free air source heat pump, the FUJIA of Merchantmen University in Japan was the first scholar to propose the use of the solution to achieve the full thermal utilization of air in winter. Research teams from Southeast University, Hunan University, and Tsinghua University combined it with heat pump technology and developed double high-efficiency frost-free air source heat pump technology. Qingdao Haier, Jiangsu Simpson, Jiangsu Yuanze, and other enterprises have also carried out the research and development and production of the technical equipment. At present, it has been applied in Tokyo, Nagoya, Jiangsu, Shandong, Guizhou, Hubei, and other regions. The main bottleneck of this technology is the new cyclic solution with low corrosion and a high-performance, high-efficiency solution regeneration method and system control strategy.

In building capacity and renewable energy building integration, through advanced materials and manufacturing processes, the photoelectric conversion efficiency is constantly being improved. In addition, flexible buildings based on photovoltaic energy storage and DC are also a hot topic of research at present; there have been a small number of demonstrations, such as the future Building R3 module in Shenzhen, but how to configure the building capacity and the energy storage capacity and how to realize DC drive and flexible regulation are still common problems [57].

This is especially true in the scenario of a large proportion of renewable energy access; it can be scheduled between days and even across seasons through load forecasting. During the design or evaluation phase, load forecasting mostly relies heavily on physical model-based platforms, such as DOE2; ENERGY PLUS, funded by the U.S. Department of Energy; DEST, developed by Tsinghua University; and TANGENT, developed by Tianzheng Company. Such software needs to input more building physical parameters, personnel parameters, building schedules, etc. But in practice, these parameters are often more difficult to obtain, and there is large variability; thus, in the operational phase, more data-driven load forecasting methods are needed.

The Building Energy Consumption Forecast Competition hosted by ASHRAE is currently the highest level of competition in the world. Through data feature engineering, gradient algorithms, and memory networks, the level of building energy consumption prediction can be increased to about 70%. Additionally, for new buildings that still lack operational data, Xu Peng's team at Tongji University began to use the transfer learning method to make predictions. Building energy system modeling mainly uses a black box model (data driven) and white box model (physical model), with the support of experimental data or running data. The above modeling method has reached high accuracy in equipment performance simulation prediction; the main bottleneck is in the absence of the above experiment or test or the absence of a large amount of accurate modeling of operating data [58]. Based on load prediction and equipment modeling, gradient algorithms, genetic algorithms, particle group algorithms, and heuristic algorithms get more and more attention and application. In this direction, the Wang Shengwei team from the Hong Kong University of Technology and the Chen Huanxin team from the Hua Zhong University of Science and Technology have deep knowledge, as the future storage of their heat storage system has increased dramatically. Dynamic planning, which can solve the problem of multi-stage decision-making algorithms, will get more and more attention [59].

3. Competition and Industrialization Prospects

Low-grade renewable energy relies on heat pump equipment to improve the quality and utilization to achieve the supply of cold and hot water [60]. More efficient compression forms and heat pump units are the main ways to further improve energy efficiency. Currently, the high efficiency centrifugal unit is mainly used for single cooling. YORK and TRANE were the first to start the development and application of this technology, and Haier and Gree can also manufacture the equipment. However, due to the heating conditions, the pressure ratio of the centrifuge increased substantially, the technical difficulty increased significantly, and the matching of hot and cold abilities between winter and summer is also a new problem, there are very few applications yet. However, for hot summer and cold winter areas, the demand for hot and cold in winter and summer is strong. The development of a new and efficient centrifugal cold water heat pump unit has broad application prospects. It can provide the core equipment for improving the quality and efficiency of low-grade renewable energy sources such as water sources, ground sources, and sewage sources [61].

Air source belongs to the category of low-grade renewable energy. The air source heat pump is in line with the development of building electrification and has become the main/important building cold and heat supply mode in the Yangtze River Basin of China, North China, the United States, Japan, Europe, and other similar climate areas. In terms of air source heat pump defrosting and frost suppression, China Gree, Midea, Haier and Japan DAIKIN, as well as other enterprises, are at the forefront of the world's enterprises, mainly using electric defrosting, reverse defrosting technology, or a lateral heat exchanger structure and control strategy to achieve it. Wing surface hydrophobic and hydrophilic modification is the current research hotspot. The Cooling Technology Research Center of Purdue University and the Beijing Institute of China are the main research teams in this direction. In recent years, frost-free air source heat pump technology has received more and more attention, and for the actual industrialization application, Southeast University, Hunan University, Tsinghua University, and the Tokyo Merchant University of Japan are the main research teams.

In terms of operation and maintenance regulation and management platforms, the air conditioning enterprise Johnson Controls has launched a building automatic control system, and China Nanrui Group, Nanjing Tianjia, and Keyuan Group have also successively launched building energy conservation management platforms for different segments, such as terminals, rail transit, and industrial construction. For regional-level building energy systems, different operating units have independently developed the corresponding monitoring and management platforms, such as China Energy Conservation, CLP Energy Saving, and other companies. However, the current energy-saving management platform gives priority to monitoring and feedback control. The identification of various equipment characteristics and an intelligent global optimization level remains to be improved, especially the proportion of renewable energy access under increasing conditions, the accurate prediction of building loads, and optimization based on time series regulation demand, which are more urgent [62–64].

3.4. Low-Carbon Intelligent Operation and Carbon Emission Control in Operation Stage

1. General Situation and the Latest Development Trend of Scientific and Technological Innovation

In terms of the efficient operation of low-carbon buildings in the future, carbon emissions in the operation phase of buildings account for the largest proportion in the whole life cycle of buildings. They are mainly generated through building energy use and can be controlled by further improving energy conservation standards, increasing the promotion of low-carbon future buildings, and promoting the low-carbon operation of buildings [65,66]. There are often various problems that lead to high energy consumption and carbon emissions in building operations, which can be solved through rapid and effective diagnosis, intelligent control, operation optimization, carbon emission monitoring, control, and other measures. In recent years, domestic and foreign scholars have carried out relevant research [67,68]. Huilong Wang and others developed a prototype integration toolbox based on energy consumption data analysis, which can realize the energy consumption diagnosis of a single building floor, system floor, and equipment floor in different periods [69]. Li Zhisheng summarized the fault detection and diagnosis methods for Heating, Ventilation, and Air Conditioning (HVAC) systems [70]. Jiang Yi, Xue Zhifeng proposed a method for energy-saving diagnosis. Jiang Yi, Jiang Ziyan, Wei Qingpeng, etc. proposed a set of energy-saving diagnostic index systems and diagnostic methods for the energy consumption of each energy consumption system and its subsystems [71,72].

The above diagnosis methods need more data to be tested or analyzed and are highly dependent on the professional experience of personnel, so they are difficult to use widely in practical projects. In terms of the low-carbon operation of buildings, with the rise of the Internet of Things, artificial intelligence, big data, and other technologies, the intelligent operation and maintenance of buildings is of wide concern. As building energy use and carbon emissions are affected by climate, building envelope, equipment efficiency, indoor environment, human behavior, and other factors, low-carbon intelligent operation must establish a working platform that enables all systems to connect, interact, and collaborate, and in-depth research also needs to continue to improve the relevant technologies. In terms of building carbon emission monitoring and control, most provinces and cities in China have established building energy consumption monitoring platforms to monitor the energy consumption of a certain number of buildings. The control of building carbon emissions in the operation stage can be achieved through carbon emission monitoring, carbon emission quota formulation, and relevant management means.

2. Existing Starting Point of Science and Technology Level and Existing Intellectual Property Rights

At present, the research on the low-carbon operation of buildings and carbon emission control in the operation stage has achieved certain results [73–76].

In terms of building energy consumption diagnosis, the OTI method of energy-saving diagnosis proposed by Jiang Yi and Xue Zhifeng can realize the energy-saving diagnosis of various energy-use systems and subsystems of buildings, and the building energy monitoring platform widely used in China can be used to analyze the energy consumption of various energy consumption systems [77]. Other countries mostly use CAD, Plant3D, Revit, 3dsmax, SketchUp, and other software to realize 3D visualization and BIM model creation. Relevant plug-ins are introduced to assist in the management of building space and equipment systems. Combined with necessary on-site tests, equipment failures are detected, potential risks are prevented, and energy consumption problems are analyzed and diagnosed. At present, China has introduced a large number of relevant software and developed commercial software, such as PKPM and Sell, and research software, such as DEST 2.0.

However, most of this software is used for drawing, modeling, energy consumption simulation, and other calculations and analyses in the design and construction stages. There are few research reports on building energy use dynamic model construction in the operation stage and the rapid diagnosis of building energy use problems using dynamic simulation and testing. In terms of the intelligent low-carbon operation of buildings, the operation adjustment and intelligent management of electromechanical systems is the main development direction. Jiang Yi of Tsinghua University has carried out a more in-depth study on the adjustment of the whole building process. For example, Wei Qingpeng summarized and proposed the operation adjustment method of an air-conditioning system in public buildings based on operation data from the aspects of a cold water system, a two-stage pump multi-stage plate-type hot water exchange system, a chiller, a cooling water system, an all-air variable volume system, etc.; and the Jiangsu Academy of Building Sciences proposed and applied an adjustment method for the electromechanical system in existing public buildings that lays a foundation for building operation optimization. As there are many factors affecting the energy use and carbon emissions of buildings, the factors that have a greater impact include weather, envelope, energy-using equipment, indoor environmental requirements, user behavior, etc. Low-carbon smart operation needs to break down the professional barriers between intelligent subsystems, clarify the logical relationship between control and feedback that reflects the main influencing factors, and establish a smart operation platform to achieve the low-carbon goal. At present, research and practice in this area are still in the initial stage, and there are few reports on relevant standardized technical systems and low-carbon intelligent operation platforms with significant application effects.

3. Competition and Industrialization Prospects

There has been a lot of research on the low-carbon intelligent operation of buildings and carbon emission control during operation, but the development and application of applicable technologies are still in the initial stage [78]. For example, in the aspect of building energy consumption diagnosis, the diagnosis methods proposed by relevant scholars often require a lot of on-site tests and data analysis. In terms of the research and development of low-carbon intelligent operation technology, some scholars proposed to use 3D laser scanning, BIM, and Internet of Things technology to adjust the building electromechanical system. China has also carried out research and practice on architectural adaptation. Intelligent equipment and products related to intelligent building management are relatively mature, but there are too many subdivisions among intelligent systems, and the information between isolated islands is not interconnected. For this reason, some enterprises have developed relevant platforms, such as Wanda's "Huiyun" intelligent management platform, which integrates 16 subsystems of five management systems, namely fire protection, security, equipment, operation, and energy conservation, into one operation platform. Tencent, Alibaba, and other Internet enterprises have also gradually launched various smart cities and building platforms, but the application effect remains to be observed.

In terms of building carbon emission monitoring and control, most cities have established online monitoring platforms for public building energy consumption. Some cities and projects have also established carbon emission smart monitoring platforms. For example, Shanghai has proposed a Shanghai building carbon emission smart monitoring platform, and Tsinghua University has applied energy consumption monitoring and carbon emission management platforms to environmental energy-saving buildings that are currently being improved. The relevant research has strong practical needs, solid integration, and strong pertinence and is expected to make a breakthrough. The relevant products will have broad market application prospects.

3.5. Low-Carbon Future Building Evaluation System

1. General Situation and the Latest Development Trend of Scientific and Technological Innovation

In terms of the evaluation system of low-carbon future buildings, developed countries (such as Nordic countries and North American countries) lead in the practice of green low-carbon buildings [79–81]. The practical experience of developed countries provides a reference for China to explore the development of low-carbon cities and buildings. Most developed countries and regions in the world have successively issued their green and low-carbon building evaluation standards, such as LEED in the United States, BREEAM in the United Kingdom, CASBEEB in Japan, etc. These standards reflect the requirements for sustainable development of buildings and low-carbon energy conservation orientation and run through all aspects of the whole life cycle of building planning and design, construction, and operation management. Countries have standardized the construction and management of green and low-carbon future buildings and promoted the construction of high-level low-carbon and zero-carbon buildings by establishing a green and low-carbon building evaluation system. The development of green and low-carbon buildings in China is progressing in an orderly manner.

The development goal and path of green buildings have been defined, low-carbon building technologies and evaluation systems are being explored, and high-quality development support systems are gradually being improved [82,83]. At present, a standard system and corresponding indicator system focusing on green buildings and near-zero energy consumption buildings have been formed, which, to some extent, reflects the guidance of the low-carbon development of buildings. However, these standard systems generally focus on the conservation of building resources and energy, and there is still much room for improvement in the comprehensive and systematic aspects of carbon emission control [84].

In combination with the research results of the classification method of carbon emissions in the life cycle of buildings, we have determined the data correlation for each classification method to determine the evaluation factors of the design, and we have determined the factors that are relevant to the design as the basis for evaluating and optimizing the building design. The carbon emission types and data correlation determined by the three classification methods are shown in Figure 3.

2. Existing Starting Point of Science and Technology Level and Existing Intellectual Property Rights

In terms of building carbon emission calculation, there are many corresponding carbon emission calculation methods for different stages of carbon emissions. These methods have their advantages and disadvantages, but there is no general method applicable to various situations [85–87]. Most studies collect and summarize the energy consumption at each stage of the building life cycle through measurement, estimation, simulation, empirical formula, or standard specifications. They then convert it into carbon emissions according to the carbon emission factor database of energy and consumables. The calculation results of different methods and the carbon emissions of the same building in different regions are quite different.

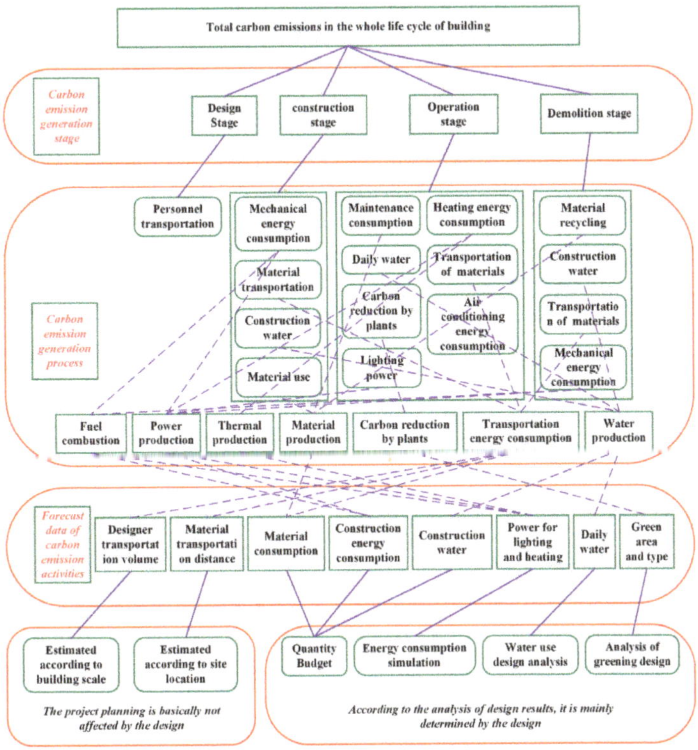

Figure 3. Data correlation in three carbon emission classification methods.

In terms of low-carbon future building indicator systems, there are relatively mature standard systems and corresponding indicator systems mainly oriented to green buildings and near-zero energy consumption buildings, which, to some extent, reflect the requirements for building energy conservation and emission reduction [88–90]. Some countries have formulated technical standards, including indicator systems for zero-carbon buildings, while China has not yet formed building standards and corresponding indicator systems with low carbon as the core. The existing indicator system related to low-carbon future buildings generally focuses on the conservation of building resources and energy. There is still much room for improvement in the comprehensive and systematic aspects of carbon emission control. In addition, the relevant indicator system has not yet conducted targeted research on different building types and different use scenarios.

In terms of low-carbon future building assessment systems, the LCA method is mainly used for research on building carbon emissions [91]. There have been post-evaluation studies and systems for green buildings, but there are no evaluation systems and methods for low-carbon future buildings [92]. Most studies on building carbon emission assessment methods in China and abroad take "carbon emissions per unit area" as an indicator. The indicators are relatively simple. An assessment system covering basic data lists and systems such as building life-cycle energy consumption and carbon emission factors has not been formed, and a highly operable and relatively accurate assessment method has not been formed [93]. At present, there are few building types and a number of cases to be evaluated, and the reliability of the evaluation method needs to be further verified.

In terms of integrated demonstration research, from the methods and research on demonstration project actual measurement research, the focus has been on demonstration project testing research in the fields of building energy efficiency, renewable energy, and green buildings, and the pathway for testing research on this type of demonstration project has been clear. However, there is still a lack of systematic and large-scale demonstration project research in the field of low-carbon future buildings, and there is a lack of engineering guidance on how to comprehensively and systematically test the comprehensive carbon reduction capacity of demonstration projects and the carbon reduction benefits of individual technologies. In addition, Chinese R&D projects often do not combine the findings of technical research with practical engineering applications; the technologies, tools, and products derived from research projects are not well applied in demonstration projects; and there is a lack of relevant verification and analysis of the actual effects of application. Further research and analysis are needed on how to organically combine technical research with engineering applications. Table 1 shows the low-carbon future building demonstration project.

Table 1. Low-carbon future building demonstration project.

Project Name	Function	Project Location	A/m^2	Climate Zones	Major Low-Carbon Technologies
Eco-tech island senior secondary school	School	Nanjing	18,415.06	Hot in summer, cold in winter	1. Adaptation of the site to the climate 2. Optimization of facade shading 3. Solar photovoltaic\photothermal 4. Monitoring of the humid and thermal environment
Peixian wenjing primary school	School	Xuzhou	52,587.38	Cold	1. Control of body coefficient 2. Roof skylights for light 3. Integration of photovoltaic buildings 4. Coupled air source heat pump + solar hot water system
Dongpo middle school, dingshu, yixing	School	Wuxi	16,331	Hot in summer, cold in winter	1. The use of low-carbon materials with local characteristics and recyclability, such as clay pots and tiles 2. Shallow ground heat exchange and fresh air systems 3. Water-saving sanitary ware and water distribution fittings

Table 1. *Cont.*

Project Name	Function	Project Location	A/m^2	Climate Zones	Major Low-Carbon Technologies
Low-carbon demonstration office building at Jiangsu Academy of Building Sciences	Office	Nanjing	14,081	Hot in summer, cold in winter	1. High-efficiency HVAC systems 2. Integration of solar photovoltaic buildings 3. Intelligent micro-grid control system with "light storage and direct flexibility" 4. Energy consumption + carbon emission + indoor environment monitoring system 5. BIM-based intelligent operation and maintenance platform
Xuzhou high-quality zero-carbon research centre	Office	Xuzhou	1245.7	Cold	1. Dynamic shading systems 2. Instant solar hot water 3. Renewable material utilization 4. CO_2 intelligent monitoring
Suzhou high-tech green low-carbon technology industry development co.	Office	Suzhou	10,852.22	Hot in summer, cold in winter	1. Thermal performance enhancement of the envelope 2. Efficient lighting energy saving 3. BIPV photovoltaic roof 4. BIPV facade photovoltaics 5. Vertical greening 6. Rainwater recycling
Suzhou chengyi green construction technology co.	Office	Suzhou	9063.02	Hot in summer, cold in winter	1. Highly efficient thermal insulation of the outer envelope 2. Adjustable external shading 3. Efficiency air conditioning and heat recovery fresh air system 4. Air quality monitoring
Langshi green centre building 1	Office	Nanjing	14,900	Hot in summer, cold in winter	1. Building performance simulation optimization 2. Ground source heat pump + magnetic levitation chiller 3. LED energy-saving light + illumination zoning control 4. Efficient lift + energy recovery
South new town international municipal complex	Synthesis	Nanjing	95,246	Hot in summer, cold in winter	1. Sewage source heat pump energy stations 2. Building skins adapted to wind conditions 3. Modular skylights 4. Waste heat recovery from energy stations
Nanjing jiangbei new area talent apartment (lot 1) service centre	Synthesis	Nanjing	2376	Hot in summer, cold in winter	1. New wood structure 2. Wood grille shading 3. DC micro grid 4. Cooling–heating source DC VRF system for air conditioning

3. Competition and Industrialization Prospects

Scholars have conducted a lot of research on the assessment and evaluation methods of building carbon emissions [94]. The most mainstream is the research on building life cycle assessment, which originated in the United States in the late 1960s, was initially applied to the assessment of industrial products, and formed international standards such as ISO 14040. In recent years, LCA has attracted much attention in the field of construction, especially in the field of building carbon emission research. Europe and the United States have long taken the post-construction use evaluation as the project evaluation procedure. In addition to the evaluation of technology and economy, it also pays attention to the evaluation of user satisfaction, such as the evaluation of user satisfaction through Key Performance Indicators.

Mature green building evaluation systems around the world, such as LEED in the United States, BREEAM in the United Kingdom, DGNB in Germany, CASBEE in Japan, and GB/T 50378 [95] in China, are being promoted and applied. In terms of the evaluation of low-carbon future buildings, the United Nations Environment Programme proposed the "Common Carbon Emission Indicator System" at the beginning of this century. USGBC released the LEED Zero Energy, LEED Zero Carbon, and LEED Zero Waste evaluation systems in 2020. The International Living Future Institute has developed the Net Zero Energy Building certification system. These two systems are also competing for user recognition worldwide, especially in the United States. To carry out the quantitative assessment of the whole life cycle of buildings, North America, the European Union, Japan, and other countries have developed relevant software and databases for the quantitative assessment of the environmental load of the whole life cycle of buildings, including LCA tools for building materials, LCA tools for buildings, LCA tools combined with CAD, etc. Typical evaluation software includes BEES, INVEST, and AIJ-LCA.

3.6. Construction of Theoretical Framework for Low-Carbon Future Building

We conducted an overall analysis of the characteristics, thematic context, and knowledge evolution of low-carbon future building-related research both domestically and internationally; referring to the relevant standards and specifications for new types of buildings such as green buildings, a preliminary theoretical framework for low-carbon future buildings was constructed from the perspective of the entire life cycle (as shown in Figure 4). Based on typical low-carbon future building cases at home and abroad, this study verifies and analyzes the theoretical framework of low-carbon future buildings through field investigations, policy text interpretation, in-depth interviews, and the mechanism of building carbon emissions and further improves and optimizes the theoretical system.

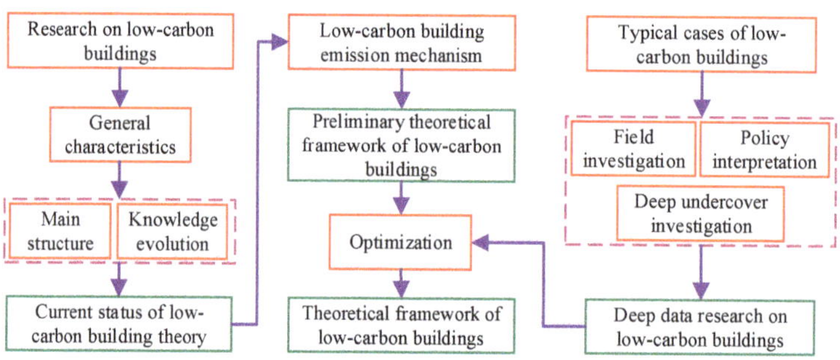

Figure 4. Construction of theoretical framework for low-carbon future building.

4. Conclusions

This paper analyzes and summarizes the whole chain of "carbon monitoring-carbon accounting-carbon assessment-carbon control" of the whole life cycle of buildings. It systematically surveys the development overview and the latest development trend of scientific and technological innovation; the existing starting level of science and technology; and the existing intellectual property, competition, and industrialization prospects in each stage of the whole life cycle and provides a reference for the development of low-carbon future buildings. The main conclusions drawn from this paper are as follows:

1. In the aspect of building planning and design, using artificial intelligence and big data to generate building forms, involving elements of energy conservation and emission reduction in the design process in advance, and assisting architects in positive design have become the research directions in low-carbon future building design.
2. In terms of green building materials and intelligent construction, we will continue to develop new building materials with high performance and low emissions, adapt to the needs of the construction industry, and use digital technology to improve the level of prefabricated buildings; intelligent construction is the main direction of future research and development.
3. In the direction of building energy, improving the level of electrification, "light storage, direct and flexible" buildings, building integration of renewable energy, green power DC-driven equipment, etc., are the main research and development directions of the future.
4. In terms of building operation, the control of energy consumption and a carbon emission quota in the building operation phase are needed. Systematic and in-depth research needs to be conducted on building energy use diagnosis, intelligent optimization operation, carbon emission monitoring, carbon emission quotas, and other aspects to make breakthroughs in relevant key technologies and provide a reliable guarantee for the low-carbon operation of buildings.
5. In terms of building evaluation systems, European and American countries took the lead in carrying out low-carbon building practices, which laid a foundation for the clarification of the concept of low-carbon buildings and the establishment of the evaluation system. Based on the evaluation system, countries are actively developing the construction and evaluation system of green and low-carbon future buildings.

At present, there is a general lack of analysis and even shelving of massive building energy consumption monitoring data in China. With the proposal of the dual carbon target, it is of great significance to explore the data value of building energy consumption monitoring platforms for the establishment of energy consumption and carbon emission quotas, dynamic supervision, and total control. At present, most relevant governments and research institutions are exploring the implementation path of low-carbon transformation in the construction industry from the perspective of top-level design and at a macro level. There is relatively little research on combining the dynamic monitoring of building carbon emissions with regional total control at the macro level. Future research can be conducted in the following aspects:

1. It is important to build a method and technical system for building design based on carbon emission measurement and to provide accurate guidance for building design through convenient and rapid carbon emission calculations. Meanwhile, due to the high urbanization rate in some regions of China and the large existing building stock, there is an urgent need to develop low-carbon design methods and evaluation tools suitable for existing buildings.

2. The use of renewable energy is a key engine for the development of low-carbon cities and buildings in the province, but there is still a need to make efforts in source, network, load, and storage, especially since there is a lack of a multi-dimensional energy complementary supply system and active and passive low-carbon technology integration methods for low-carbon cities, building design, and operation and maintenance.
3. In some regions of China, the urbanization rate is high, the industrial output value is large, and "high energy consumption and high emissions" still dominate. The carbon emissions from building operation and maintenance continue to grow. They lack effective carbon emission control tools and low-carbon, zero-carbon-oriented intelligent carbon control technologies. There is an urgent need to build a building carbon footprint monitoring carbon emission operation and maintenance platform, and there are multiple energy-complementary technologies with core technologies that can be independently grasped.
4. The purpose of the assessment and evaluation of low-carbon future buildings is not clear enough; the assessment index is relatively single; a unified list of basic data, such as energy consumption and carbon emission coefficients for each life stage of the building and a systematic assessment system, has not yet been formed; and a highly operable and relatively accurate assessment method has not yet been formed.

Author Contributions: Conceptualization, D.Y. and L.X.; methodology, D.Y. and L.X.; formal analysis, L.X. and J.Z.; investigation, J.Z.; data curation, C.J.; writing—original draft preparation, L.X. and J.Z.; writing—review and editing, D.Y.; visualization, D.Y. and J.Z.; supervision, D.Y.; project administration, J.Z. All authors have read and agreed to the published version of the manuscript.

Funding: This research were funded by the National Natural Science Foundation of China, grant number 52075232, the Natural Science Research Project of Higher Education Institutions in Jiangsu Province, grant number 18KJD440001, and the Open Project of Key Laboratory of Aircraft Environmental Control and Life Support, Ministry of Industry and Information Technology, grant number KLAECLS-E-201903.

Data Availability Statement: Data sharing not applicable.

Conflicts of Interest: The authors declare no conflicts of interest.

References

1. Zhang, J.; Liu, Z. Low carbon economic scheduling model for a park integrated energy system considering integrated demand response, ladder-type carbon trading and fine utilization of hydrogen. *Energy* **2024**, *290*, 130311. [CrossRef]
2. Hosseinian, S.M.; Faghani, M. Assessing the effect of structural parameters and building site in achieving low carbon building materialization using a life-cycle assessment approach. *J. Build. Eng.* **2021**, *44*, 103318. [CrossRef]
3. Bazzocchi, F.; Ciacci, C.; Di Naso, V. Evaluation of Environmental and Economic Sustainability for the Building Envelope of Low-Carbon Schools. *Sustainability* **2021**, *13*, 1702. [CrossRef]
4. Mei, Q.; Liu, L.; Abu Mansor, M.R. Investigation on spray combustion modeling for performance analysis of future low- and zero-carbon DI engine. *Energy* **2024**, *302*, 131906. [CrossRef]
5. Olkkonen, V.; Hirvonen, J.; Heljo, J.; Syri, S. Effectiveness of building stock sustainability measures in a low-carbon energy system: A scenario analysis for Finland until 2050. *Energy* **2021**, *235*, 121399. [CrossRef]
6. Zhou, W.; Moncaster, A.; Reiner, D.M.; Guthrie, P. Developing a generic System Dynamics model for building stock transformation towards energy efficiency and low-carbon development. *Energy Build.* **2020**, *224*, 110246. [CrossRef]
7. Arsenkin, A.M.; Grigorovich, K.V.; Odesskii, P.D.; Tabakov, Y.I.; Zheleznyi, M.V.; Shibaeva, T.V. Structural nonuniformity: Impact on the cold resistance of low carbon microalloyed steel for building constructions. *Results Mater.* **2020**, *8*, 100141. [CrossRef]
8. Watari, T.; Yamashita, N.; Serrenho, A.C. Net-Zero Embodied Carbon in Buildings with Today's Available Technologies. *Env. Sci. Technol.* **2024**, *58*, 1793–1801. [CrossRef]
9. Usman, M.; Khalid, K.; Mehdi, M.A. What determines environmental deficit in Asia? Embossing the role of renewable and non-renewable energy utilization. *Renew. Energy* **2021**, *168*, 1165–1176. [CrossRef]

10. Guo, Z.; Wei, W.; Chen, L.; Dong, Z.Y.; Mei, S. Impact of Energy Storage on Renewable Energy Utilization: A Geometric Description. *IEEE Trans. Sustain. Energy* **2021**, *12*, 874–885. [CrossRef]
11. Palomba, V.; Borri, E.; Charalampidis, A.; Frazzica, A.; Cabeza, L.F.; Karellas, S. Implementation of a solar-biomass system for multi-family houses: Towards 100% renewable energy utilization. *Renew. Energy* **2020**, *166*, 190–209. [CrossRef]
12. O'Neill, R.; Window, A.; Kenway, S.; Dargusch, P. Integrated operational and life-cycle modelling of energy, carbon and cost for building façades. *J. Clean. Prod.* **2021**, *286*, 125370. [CrossRef]
13. Liu, Y.; Luo, W.; Chen, X. Occupants' Satisfaction of Indoor Environment Quality in Non-Linear Minimum Buildings in Winter Based on the Role of Different Kinds of Perceptions. *Sustainability* **2022**, *14*, 4983. [CrossRef]
14. Chen, C.; Zhao, Z.; Xiao, J.; Tiong, R. A Conceptual Framework for Estimating Building Embodied Carbon Based on Digital Twin Technology and Life Cycle Assessment. *Sustainability* **2021**, *13*, 13875. [CrossRef]
15. *ISO 14040*; Environmental Management—Life Cycle Assessment—Principles and Framework, CEN (European Committee for Standardisation). International Standards Organization ISO: Geneva, Switzerland, 2006.
16. Ang, Y.Q.; Berzolla, Z.M.; Reinhart, C.F. From concept to application: A review of use cases in urban building energy modeling. *Appl. Energy* **2020**, *279*, 115738. [CrossRef]
17. Brooks, M.; Abdellatif, M.; Alkhaddar, R. Application of life cycle carbon assessment for a sustainable building design: A case study in the UK. *Int. J. Green. Energy* **2021**, *18*, 351–362. [CrossRef]
18. Wang, J.; Teng, Y.; Chen, Z.; Bai, J.; Niu, Y.; Duan, H. Assessment of carbon emissions of building interior decoration and renovation waste disposal in the fast-growing Greater Bay Area, China. *Sci. Total Environ.* **2021**, *798*, 149158. [CrossRef]
19. Xiaoxiang, Q.; Junjia, Y.; Haron, N.A.; Alias, A.H.; Law, T.H.; Abu Bakar, N. Status, Challenges and Future Directions in the Evaluation of Net-Zero Energy Building Retrofits: A Bibliometrics-Based Systematic Review. *Energies* **2024**, *17*, 3826. [CrossRef]
20. Bottero, M.; Dell'Anna, F.; Morgese, V. Evaluating the Transition Towards Post-Carbon Cities: A Literature Review. *Sustainability* **2021**, *13*, 567. [CrossRef]
21. Huo, Y.; Bouffard, F.; Joos, G. Spatio-Temporal Flexibility Management in Low-Carbon Power Systems. *IEEE Trans. Sustain. Energy* **2020**, *11*, 2593–2605. [CrossRef]
22. Gassar, A.A.A.; Cha, S.H. Energy prediction techniques for large-scale buildings towards a sustainable built environment: A review. *Energy Build.* **2020**, *224*, 110238. [CrossRef]
23. Gassar, A.A.A.; Koo, C.; Kim, T.W.; Cha, S.H. Performance Optimization Studies on Heating, Cooling and Lighting Energy Systems of Buildings during the Design Stage: A Review. *Sustainability* **2021**, *13*, 9815. [CrossRef]
24. Gao, Y.; Jia, R.; Yao, Y.; Xu, J. Evolutionary Game Theory and the Simulation of Green Building Development Based on Dynamic Government Subsidies. *Sustainability* **2022**, *14*, 7294. [CrossRef]
25. Yuan, M.; Li, Z.; Li, X.; Luo, X. Managing stakeholder-associated risks and their interactions in the life cycle of prefabricated building projects: A social network analysis approach. *J. Clean. Prod.* **2021**, *323*, 129102. [CrossRef]
26. Qiang, G.; Cao, D.; Wu, G.; Zhao, X.; Zuo, J. Dynamics of Collaborative Networks for Green Building Projects: Case Study of Shanghai. *J. Manag. Eng.* **2021**, *37*, 5021001.1–5021001.13. [CrossRef]
27. Liu, K.; Zhu, B.; Chen, J. Low-Carbon Design Path of Building Integrated Photovoltaics: A Comparative Study Based on Green Building Rating Systems. *Buildings* **2021**, *11*, 469. [CrossRef]
28. Zhang, X.; Zhang, X. A subproject-based quota approach for life cycle carbon assessment at the building design and construction stage in China. *Build. Environ.* **2020**, *185*, 107258. [CrossRef]
29. Akbarieh, A.; Jayasinghe, L.B.; Waldmann, D.; Teferle, F.N. BIM-Based End-of-Lifecycle Decision Making and Digital Deconstruction: Literature Review. *Sustainability* **2020**, *12*, 2670. [CrossRef]
30. Luo, D.; Chen, D.; Wang, J.; Zhu, G.; Xu, W. The smart robot crafting approach to computing materials. *Constr. Robot.* **2020**, *4*, 239–249. [CrossRef]
31. Xu, Y.; Luo, D.; Qian, Q.K.; Chan, E.H.W. Correction to: Are green buildings more liveable than conventional buildings? An examination from the perspective of occupants. *J. Hous. Built Environ.* **2023**, *38*, 1067. [CrossRef]
32. Liu, Z.; Hu, B.; Zhao, Y.; Lang, L.; Guo, H.; Florence, K.; Zhang, S. Research on Intelligent Decision of Low Carbon Supply Chain Based on Carbon Tax Constraints in Human-Driven Edge Computing. *IEEE Access* **2020**, *8*, 48264–48273. [CrossRef]
33. Schwartz, Y.; Raslan, R.; Korolija, I.; Mumovic, D. A decision support tool for building design: An integrated generative design, optimisation and life cycle performance approach. *Int. J. Archit. Comput.* **2021**, *19*, 401–430. [CrossRef]
34. Chen, Q. Application of low-carbon design concept of network energy consumption based on SDN architecture in high-rise residential building design. *Alex. Eng. J.* **2022**, *61*, 3303–3312. [CrossRef]
35. Xiong, J.; Wei, W.; Huang, S.; Zhang, Y. Association between the emission rate and temperature for chemical pollutants in building materials: General correlation and understanding. *Environ. Sci. Technol.* **2013**, *47*, 8540–8547. [CrossRef] [PubMed]

36. Tawasil, D.N.B.; Aminudin, E.; Abdul Shukor Lim, N.H.; Nik Soh, N.M.Z.; Leng, P.C.; Ling, G.H.T.; Ahmad, M.H. Coconut Fibre and Sawdust as Green Building Materials: A Laboratory Assessment on Physical and Mechanical Properties of Particleboards. *Buildings* **2021**, *11*, 256. [CrossRef]
37. Liu, Z.; Guo, A. Application of Green Building Materials and Multi-objective Energy-Saving Optimization Design. *Int. J. Heat. Technol.* **2021**, *39*, 299–308. [CrossRef]
38. Sandanayake, M.; Gunasekara, C.; Law, D.; Zhang, G.; Setunge, S.; Wanijuru, D. Sustainable criterion selection framework for green building materials—An optimisation based study of fly-ash Geopolymer concrete. *Sustain. Mater.Technol.* **2020**, *25*, e00178. [CrossRef]
39. *GB/T51231-2016*; Technical Standard for Assembled Buildings with Concrete Structure. China Architecture & Building Press: Beijing, China, 2017.
40. Lee, Y.-L.; Chang, Y.-H.; Li, J.-L.; Lin, C.-Y. Using Intelligence Green Building Materials to Evaluate Color Change Performance. *Sustainability* **2020**, *12*, 5630. [CrossRef]
41. Wang, T.; Song, Z.; Zhou, J.; Sun, H.; Liu, F. Low-Carbon Transition and Green Innovation: Evidence from Pilot Cities in China. *Sustainability* **2022**, *14*, 7264. [CrossRef]
42. *GB/T51003-2014*; Technical Code for Application of Mineral Admixture. China Architecture & Building Press: Beijing, China, 2014.
43. Agyekum, K.; Kissi, E.; Danku, J.C. Professionals' views of vernacular building materials and techniques for green building delivery in Ghana. *Sci. Afr.* **2020**, *8*, e00424. [CrossRef]
44. Suh, S.; Tomar, S.; Leighton, M.; Kneifel, J. Environmental performance of green building code and certification systems. *Environ. Sci. Technol.* **2014**, *48*, 2551–2560. [CrossRef]
45. Kalombe, R.M.; Ojumu, V.T.; Eze, C.P.; Nyale, S.M.; Kevern, J.; Petrik, L.F. Fly Ash-Based Geopolymer Building Materials for Green and Sustainable Development. *Mater* **2020**, *13*, 5699. [CrossRef] [PubMed]
46. Nowicka-Krawczyk, P.; Komar, M.; Gutarowska, B. Towards understanding the link between the deterioration of building materials and the nature of aerophytic green algae. *Sci. Total Environ.* **2022**, *802*, 149856. [CrossRef] [PubMed]
47. Rezaei, A.; Samadzadegan, B.; Rasoulian, H.; Ranjbar, S.; Samareh Abolhassani, S.; Sanei, A.; Eicker, U. A New Modeling Approach for Low-Carbon District Energy System Planning. *Energies* **2021**, *14*, 1383. [CrossRef]
48. Qian, D.; Dargusch, P.; Hill, G. Carbon Management behind the Ambitious Pledge of Net Zero Carbon Emission—A Case Study of PepsiCo. *Sustainability* **2022**, *14*, 2171. [CrossRef]
49. Parrella, V.F.; Molari, L. Building Retrofitting System Based on Bamboo-Steel Hybrid Exoskeleton Structures: A Case Study. *Sustainability* **2021**, *13*, 5984. [CrossRef]
50. Chang, C.-W. Develop a ranking algorithm for the green building project. *Qual. Quant.* **2012**, *48*, 911–921. [CrossRef]
51. Murthy, V.; Ramakrishna, S. A Review on Global E-Waste Management: Urban Mining towards a Sustainable Future and Circular Economy. *Sustainability* **2022**, *14*, 647. [CrossRef]
52. Jaffal, I.; Ouldboukhitine, S.-E.; Belarbi, R. A comprehensive study of the impact of green roofs on building energy performance. *Renew. Energy* **2012**, *43*, 157–164. [CrossRef]
53. Ikäheimo, J.; Weiss, R.; Kiviluoma, J.; Pursiheimo, E.; Lindroos, T.J. Impact of power-to-gas on the cost and design of the future low-carbon urban energy system. *Appl. Energy* **2022**, *305*, 117713. [CrossRef]
54. Lu, S.; Liu, S.; Huang, J.; Kong, X. Establishment and experimental verification of PCM room's TRNSYS heat transfer model based on latent heat utilization ratio. *Energy Build.* **2014**, *84*, 287–298. [CrossRef]
55. Zhang, Y.; Hao, J.; Ge, Z.; Zhang, F.; Du, X. Optimal clean heating mode of the integrated electricity and heat energy system considering the comprehensive energy-carbon price. *Energy* **2021**, *231*, 120919. [CrossRef]
56. Zhang, C.; Hu, M.; Laclau, B.; Garnesson, T.; Yang, X.; Tukker, A. Energy-carbon-investment payback analysis of prefabricated envelope-cladding system for building energy renovation: Cases in Spain, the Netherlands, and Sweden. *Renew. Sustain. Energy Rev.* **2021**, *145*, 111077. [CrossRef]
57. Woo, J.; Fatima, R.; Kibert, C.J.; Newman, R.E.; Tian, Y.; Srinivasan, R.S. Applying blockchain technology for building energy performance measurement, reporting, and verification (MRV) and the carbon credit market: A review of the literature. *Build. Environ.* **2021**, *205*, 108199. [CrossRef]
58. Grazieschi, G.; Asdrubali, F.; Thomas, G. Embodied energy and carbon of building insulating materials: A critical review. *Clean. Environ. Syst.* **2021**, *2*, 100032. [CrossRef]
59. Żabnieńska-Góra, A.; Khordehgah, N.; Jouhara, H. Annual performance analysis of the PV/T system for the heat demand of a low-energy single-family building. *Renew. Energy* **2021**, *163*, 1923–1931. [CrossRef]
60. Wang, Y.; Mauree, D.; Sun, Q.; Lin, H.; Scartezzini, J.L.; Wennersten, R. A review of approaches to low-carbon transition of high-rise residential buildings in China. *Renew. Sustain. Energy Rev.* **2020**, *131*, 109990. [CrossRef]

61. Stermieri, L.; Delmastro, C.; Becchio, C.; Corgnati, S.P. Linking Dynamic Building Simulation with Long-Term Energy System Planning to Improve Buildings Urban Energy Planning Strategies. *Smart Cities* **2020**, *3*, 1242–1265. [CrossRef]
62. Huo, T.; Xu, L.; Feng, W.; Cai, W.; Liu, B. Dynamic scenario simulations of carbon emission peak in China's city-scale urban residential building sector through 2050. *Energy Policy* **2021**, *159*, 112612. [CrossRef]
63. Li, H.; Wang, S. Coordinated robust optimal design of building envelope and energy systems for zero/low energy buildings considering uncertainties. *Appl. Energy* **2020**, *265*, 114779. [CrossRef]
64. Lee, J.; McCuskey Shepley, M.; Choi, J. Exploring the localization process of low energy residential buildings: A case study of Korean passive houses. *J. Build. Eng.* **2020**, *30*, 101290. [CrossRef]
65. Ding, L.; Chen, C. Thermal Environment Optimization and Energy Saving of Residential Buildings under the Demand of Low-Carbon Operation. *Int. J. Heat Technol.* **2021**, *39*, 659–668. [CrossRef]
66. Wang, J.; Huang, Y.; Teng, Y.; Yu, B.; Wang, J.; Zhang, H.; Duan, H. Can buildings sector achieve the carbon mitigation ambitious goal: Case study for a low-carbon demonstration city in China? *Environ. Impact Assess. Rev.* **2021**, *90*, 106633. [CrossRef]
67. Chai, Y.; Lin, X.; Wang, D. Industrial Structure Transformation and Layout Optimization of Beijing-Tianjin-Hebei Region under Carbon Emission Constraints. *Sustainability* **2021**, *13*, 643. [CrossRef]
68. Zhang, S.; Xiang, X.; Ma, Z.; Ma, M.; Zou, C. Carbon Neutral Roadmap of Commercial Building Operations by Mid-Century: Lessons from China. *Buildings* **2021**, *11*, 510. [CrossRef]
69. Wang, H.; Lu, X.; Xu, P.; Yuan, D. Short-term Prediction of Power Consumption for Large-scale Public Buildings based on Regression Algorithm. *Procedia Eng.* **2015**, *121*, 1318–1325. [CrossRef]
70. Liu, L.R.; Wu, J.; Wang, X.X.; Li, Z.S.; Cheng, T. Indoor Air Quality Audit and Grey Incidence Analysis of an Air-Conditioned Institutional Library in the Subtropics. *Appl. Mech. Mater.* **2012**, *260–261*, 391–396. [CrossRef]
71. Liu, X.; Zhang, T.; Liu, X.; Li, L.; Lin, L.; Jiang, Y. Energy saving potential for space heating in Chinese airport terminals: The impact of air infiltration. *Energy* **2021**, *260–261*, 119175. [CrossRef]
72. Fu, L.; Li, Y.; Wu, Y.; Wang, X.; Jiang, Y. Low carbon district heating in China in 2025- a district heating mode with low grade waste heat as heat source. *Energy* **2021**, *230*, 120765. [CrossRef]
73. Mata, É.; Wanemark, J.; Cheng, S.H.; Ó Broin, E.; Hennlock, M.; Sandvall, A. Systematic map of determinants of buildings' energy demand and CO_2 emissions shows need for decoupling. *Environ. Res. Lett.* **2021**, *16*, 055011. [CrossRef]
74. Gil, G.O.; Chowdhury, J.I.; Balta-Ozkan, N.; Hu, Y.; Varga, L.; Hart, P. Optimising renewable energy integration in new housing developments with low carbon technologies. *Renew. Energy* **2021**, *169*, 527–540. [CrossRef]
75. Cai, M.; Shi, Y.; Ren, C. Developing a high-resolution emission inventory tool for low-carbon city management using hybrid method–A pilot test in high-density Hong Kong. *Energy Build.* **2020**, *226*, 110376. [CrossRef]
76. Habert, G.; Röck, M.; Steininger, K.; Lupísek, A.; Birgisdottir, H.; Desing, H.; Chandrakumar, C.; Pittau, F.; Passer, A.; Rovers, R.; et al. Carbon budgets for buildings: Harmonising temporal, spatial and sectoral dimensions. *Build. Cities* **2020**, *1*, 429–452. [CrossRef]
77. Zhang, Y.; Xia, J.; Fang, H.; Jiang, Y.; Liang, Z. Field tests on the operational energy consumption of Chinese district heating systems and evaluation of typical associated problems. *Energy Build.* **2020**, *224*, 110269. [CrossRef]
78. Mata, É.; Korpal, A.K.; Cheng, S.H.; Jiménez Navarro, J.P.; Filippidou, F.; Reyna, J.; Wang, R. A map of roadmaps for zero and low energy and carbon buildings worldwide. *Environ. Res. Lett.* **2020**, *15*, 113003. [CrossRef]
79. Dougill, A.J.; Hermans, T.D.G.; Eze, S.; Antwi-Agyei, P.; Sallu, S.M. Evaluating Climate-Smart Agriculture as Route to Building Climate Resilience in African Food Systems. *Sustainability* **2021**, *13*, 9909. [CrossRef]
80. Weiler, L.; Vollpracht, A.; Matschei, T. Leaching of Carbon Reinforced Concrete-Part 2: Discussion of Evaluation Concepts and Modelling. *Materials* **2020**, *13*, 4937. [CrossRef] [PubMed]
81. Trofimova, P.; Cheshmehzangi, A.; Deng, W.; Hancock, C. Post-Occupancy Evaluation of Indoor Air Quality and Thermal Performance in a Zero Carbon Building. *Sustainability* **2021**, *13*, 667. [CrossRef]
82. Jain, N.; Burman, E.; Robertson, C.; Stamp, S.; Shrubsole, C.; Aletta, F.; Barrett, E.; Oberman, T.; Kang, J.; Raynham, P.; et al. Building performance evaluation: Balancing energy and indoor environmental quality in a UK school building. *Build. Serv. Eng. Res. Technol.* **2019**, *41*, 343–360. [CrossRef]
83. Liu, D.; Sun, Y.; Wilson, R.; Wu, Y. Comprehensive evaluation of window-integrated semi-transparent PV for building daylight performance. *Renew. Energy* **2020**, *145*, 1399–1411. [CrossRef]
84. Luo, X.; Ma, C.; Ge, J. Evaluation Model and Strategy for Selecting Carbon Reduction Technology for Campus Buildings in Primary and Middle Schools in the Yangtze River Delta Region, China. *Sustainability* **2020**, *12*, 534. [CrossRef]
85. Liyanage, D.R.; Hewage, K.; Karunathilake, H.; Chhipi-Shrestha, G.; Sadiq, R. Carbon Capture Systems for Building-Level Heating Systems—A Socio-Economic and Environmental Evaluation. *Sustainability* **2021**, *13*, 10681. [CrossRef]

86. Wang, H.; Zhang, H.; Hou, K.; Yao, G. Carbon emissions factor evaluation for assembled building during prefabricated component transportation phase. *Energy Explor. Exploit.* **2020**, *39*, 385–408. [CrossRef]
87. Li, K.; Ma, M.; Xiang, X.; Feng, W.; Ma, Z.; Cai, W.; Ma, X. Carbon reduction in commercial building operations: A provincial retrospection in China. *Appl. Energy* **2022**, *306*, 118098. [CrossRef]
88. Felmer, G.; Morales-Vera, R.; Astroza, R.; González, I.; Puettmann, M.; Wishnie, M. A Lifecycle Assessment of a Low-Energy Mass-Timber Building and Mainstream Concrete Alternative in Central Chile. *Sustainability* **2022**, *14*, 1249. [CrossRef]
89. Du, Q.; Pang, Q.; Bao, T.; Guo, X.; Deng, Y. Critical factors influencing carbon emissions of prefabricated building supply chains in China. *J. Clean. Prod.* **2021**, *280*, 124398. [CrossRef]
90. Trovato, M.R.; Nocera, F.; Giuffrida, S. Life-Cycle Assessment and Monetary Measurements for the Carbon Footprint Reduction of Public Buildings. *Sustainability* **2020**, *12*, 3460. [CrossRef]
91. Carcassi, O.B.; Minotti, P.; Habert, G.; Paoletti, I.; Claude, S.; Pittau, F. Carbon Footprint Assessment of a Novel Bio-Based Composite for Building Insulation. *Sustainability* **2022**, *14*, 1384. [CrossRef]
92. Huo, T.; Cao, R.; Du, H.; Zhang, J.; Cai, W.; Liu, B. Nonlinear influence of urbanization on China's urban residential building carbon emissions: New evidence from panel threshold model. *Sci. Total Environ.* **2021**, *772*, 145058. [CrossRef] [PubMed]
93. Liu, Y.; Suk, S. Constructing an Evaluation Index System for China's Low-Carbon Tourism Region—An Example from the Daxinganling Region. *Sustainability* **2021**, *13*, 12026. [CrossRef]
94. Ahmed Ali, K.; Ahmad, M.I.; Yusup, Y. Issues, Impacts, and Mitigations of Carbon Dioxide Emissions in the Building Sector. *Sustainability* **2020**, *12*, 7427. [CrossRef]
95. *GB/T 50378*; Assessment Standard for Green Building. China Architecture & Building Press: Beijing, China, 2019.

processes

MDPI

Article

A Mobile Energy Storage Configuration Method for Power Grids Considering Power Losses and Voltage Stability

Xinhui Lai [1], Haojie Du [2], Fei Long [2,*], Yi An [1] and Muliang Cai [1]

[1] State Grid Jiangxi Electric Power Research Institute, Nanchang 330001, China; laixh_jxdky@163.com (X.L.); dky_anyi@jx.sgcc.com.cn (Y.A.); dky_caiml@jx.sgcc.com.cn (M.C.)

[2] College of Electrical Engineering and New Energy, China Three Gorges University, Yichang 443002, China; haojie21037@ctgu.edu.cn

* Correspondence: feilongcug2020@163.com

Abstract: The generation output of distributed power sources and the load possess periodic changes, which cause stability problems in the operation of the power grid. To ensure stability, energy storage devices are generally installed in the power grid, but their effectiveness is also limited by their fixed installation location and capacity. In this paper, to overcome the drawback of stationary energy storage devices, mobile energy storage devices are introduced to reduce power losses and enhance voltage stability. The installation location and capacity of these mobile energy storage devices can be changed with the generation output and load demands. Firstly, the influence of the mobile energy storage devices on power losses and voltage stability is analyzed. Next, a multi-objective optimization model is established to select the optimal installation location and the ideal capacity, which has optimization objectives including the improvement of voltage stability, the minimization of power losses, and the enhancement of utilization of energy storage devices. Finally, the particle swarm optimization algorithm is applied to solve the multi-objective optimization problem. The simulation results prove that the novel method of this paper enhances the dynamics of the power grid, i.e., the voltage vulnerability and power losses are reduced by 29% and 36%, respectively. In addition, it also increases the utilization rate of energy storage devices by 33.5% in different operating scenarios.

Keywords: voltage stability; power losses; mobile energy storage devices; particle swarm optimization

1. Introduction

Because of the rapid development of renewable energy, a larger number of distributed power sources are being integrated into the power grid. However, these renewable energy sources are characterized by fluctuations. For instance, wind power generation relies on wind resources, and the instability of wind power generation results in fluctuations in power output. Similarly, photovoltaic power generation relies on the intensity and duration of sunlight, and its capacity for generating electricity decreases significantly on cloudy days or during nighttime. For example, the "Zhangbei Renewable Energy Base" is one of the largest renewable energy demonstration bases in China, which includes wind and photovoltaic power generation and achieves a high percentage of renewable energy by connecting to the grid utilizing smart grid technology [1]. Due to the randomness of distributed generation, its large-scale integration poses challenges to the stability of the power grid [2–5]. This randomness often leads to some technical problems in the distribution network, including voltage imbalance, power loss, and reversed power flow. When the demand for electricity is low, the new energy generation system continuously supplies power to the grid, which increases the system voltage. This leads to shortened equipment lifetimes, tripping, or equipment damage. In terms of load demands, there are variations in electricity consumption at different times, characterized by peaks and troughs. The rapid growth of emerging loads, such as the electric vehicle charging stations, further exacerbates the uncertainty of load demands [6–9]. The fluctuations of the distributed generation output and the load demands show the characteristic of periodic changes and bring about stability problems in the power grid [10]. To balance supply and demand in real time and reduce fluctuations, one of the most effective solutions to the stability problem is the installation of energy storage devices [5,11]. As flexible power regulation equipment, energy storage devices store excess energy when load demands are low and release it when demands are high. Therefore, the introduction of energy storage devices can balance the load demand and generation output and can maintain the voltage stability of the power grid. Therefore, the study of the application of energy storage devices to the power system is essential for the operation of power grids.

1.1. Literature Review

The suitable installation location and the allowable minimum capacity of energy storage devices in the power grid are the main concerns for the configuration, which has been researched in the past few decades with numerous studies proposed. Sanjay et al. used the hybrid gray wolf optimization algorithm to determine the optimal configuration of distributed generator sets [12]. Khan et al. used an improved grey wolf optimizer to determine the optimal distributed generation (DG) allocation in order to reduce the power losses of the power grid [13]. Onlam et al. employed an adaptive optimization algorithm to address the DG configuration problem, aiming to reduce power losses and enhance voltage stability [14]. Sangeeta et al. proposed a sensitivity analysis method for voltage stability to determine the optimal location for DG [15]. Carpinelli et al. analyzed the benefits of integrating energy storage devices into the grid but did not consider the impact of energy storage devices on the power grid [16]. Ghofrani et al. studied the economic effects of energy storage devices on the grid, but they did not consider its impact on voltage stability [17]. Ahmed et al. determined the optimal size and location of DG and PV with the goal of minimizing costs without considering the stability of the grid [18]. In order to restrain the fluctuation of wind power and reduce the economic cost of the energy storage system, Yuming et al. proposed an optimization method to choose the optimal capacity, which also did not take into account the stability of the power grid [19]. Zhang et al. used the adaptive grey wolf optimization algorithm to determine the location and capacity of energy storage devices for the purpose of improving system reliability and voltage frequency stability, but this paper did not consider economic factors [20].

Most research focuses on stationary energy storage devices whose installation location and capacity cannot be changed with varying load demands; therefore, the effectiveness of stationary energy storage devices is limited if the operating scenario changes. Fortunately, mobile energy storage technology provides a way to overcome the drawbacks of stationary energy storage devices; however, little attention has been paid to this. As stated in [21], electric vehicles are typical mobile energy storage devices that can be used to improve the stability of power grids through vehicle-to-grid (V2G) technology. Therefore, electric vehicles can be considered a practical application of mobile energy storage devices. For example, in [22,23], the optimal capacity of the mobile energy storage devices of electric vehicles and the plan of charge or discharge were investigated for the stability problem of the power grid. In [24], the connection of electric vehicles with the power grid through V2G technology was studied, showing a positive influence on the integration of photovoltaic power generation. Meanwhile, by considering the mobile energy storage devices in the power grid, the configuration problem was investigated with regard to economic benefits [25]. By considering the lowest fluctuation of power as the optimization objective, the optimal capacity of energy storage devices was investigated [26]. By taking the largest economic benefits as the objective, the optimal location and capacity were determined using mixed-integer second-order cone programming [27].

Based on the analysis of the above, it is found that most previous research has been on stationary energy storage devices. Secondly, research on the configuration of mobile energy storage devices in the power grid does not attract enough attention and there is still room for further study. As for the configuration of mobile energy storage devices, the main focus is on the economic aspects of power grid operation, but few studies address the effect of mobile energy storage devices on the voltage stability and power losses of power grid operation in depth. Therefore, this paper aims to further explore how to determine the optimal configuration of mobile energy storage devices to enhance the voltage stability of power grids and reduce power losses, which are essential for the stable operation of power grids.

1.2. Contributions

As for the fluctuations of load demands and the distributed generation output, mobile energy storage technology provides an effective way of improving the stability of the power grid by adjusting the installation location and capacity according to the operation. In this paper, a mobile energy storage configuration method is proposed to enhance the utilization rate of the mobile energy storage devices and grid stability.

The major contributions of this paper are as follows:

(1) Firstly, this study investigates the impact of energy storage devices on voltage stability and power losses with changes in the installation location and the capacity, which has not been studied in previous research;

(2) Secondly, by fully considering the properties of the load demand of the power grid within a day, the load period is divided into three load scenarios, i.e., low, flat, and peak, and then a multi-objective optimization model is established to determine the optimal position and ideal capacity of energy storage devices for three different scenarios. The optimization objectives include the minimization of power loss, the improvement of voltage stability, and the improvement of the utilization of energy storage devices;

(3) Finally, the IEEE 33-node system is used for the simulation in MATLAB R2023a. By comparing the simulation results of the stationary energy storage and the mobile energy storage, it is verified that the proposed method effectively improves the voltage stability and utilization rate of energy storage devices and reduces power losses.

1.3. Research Methodology

To make this paper readable, Table 1 shows the research problem, research objectives, research method, techniques, tools, and simulation.

Table 1. Conceptual framework of this study.

1. Research problem	The stability problem of the power grid caused by the fluctuation of new energy generation and load.
2. Research objectives	Main objectives: To determine the optimal configuration of mobile energy storage devices to enhance power grid stability.
	Specific objectives: To improve the voltage stability and utilization rate of energy storage devices and reduce power losses.
3. Research method	A mobile energy storage method by optimizing the optimal location and capacity of energy storage.
4. Techniques and tools	Multi-objective particle swarm optimization: Solving multi-objective optimization problems. Matlab R2023a: For simulation and example analysis.
5. Simulation	Simulation objects: Take the IEEE 33-node system (including wind turbines, photovoltaic generators, and energy storage devices) as an example. The stationary energy storage device and the mobile energy storage device are used as simulation objects. Simulation results: The voltage vulnerability and power losses are reduced by 29% and 36%, respectively. The utilization rate of energy storage devices increases by 33.5%.

1.4. Organization

The rest of this paper is organized as follows. In Section 2, the influence of mobile energy storage devices on the power grid and the characteristics of mobile energy storage devices are analyzed. In Section 3, a multi-objective optimization model is proposed to determine the optimal location and capacity of the energy storage device. In Section 4, particle swarm optimization (PSO) is introduced for solving multi-objective optimization problems. In Section 5, the validity of the proposed method is verified by simulation analysis. Finally, the conclusions are presented in Section 6.

2. The Influence of Mobile Energy Storage Devices on the Power Grid and the Characteristics of Mobile Energy Storage Devices

To prove the usefulness of mobile energy storage devices, the influence of introducing mobile energy storage devices on voltage stability and power losses is studied, and their characteristics are also introduced.

2.1. The Influence of Mobile Energy Storage Devices on the Power Grid

Integrating energy storage devices into the power grid enhances both its safety and economy. These devices charge or discharge according to load demands to improve voltage stability [28]. Additionally, power losses are related to the grid voltage. These losses increase or decrease in accordance with fluctuations in voltage levels [29,30]. Utilizing mobile energy storage devices to deal with fluctuations in distributed power sources and loads greatly reduces power losses and enhances voltage stability. Therefore, the impact of mobile energy storage devices on power losses and voltage stability is discussed in this section, which provides insights for the following research.

In the following example, the IEEE 33-node system is used in a simulation to study the influence of energy storage devices on voltage and power losses during the discharge state.

The influence of energy storage devices on voltage and power losses at various locations and capacities is shown in Figure 1. The nodes in various colors represent the different positions of the energy storage device. According to part (a) of Figure 1, for all nodes, the voltage increases as the capacity increases, but the location changes lead to complicated influences on the voltage. For example, when the capacity is 1000 kWh, the voltage of node 13 is significantly higher than the other nodes, but when the capacity is 3000 kWh, the voltage of node 6 becomes the highest. Part (b) of Figure 1 indicates that energy storage devices with an appropriate capacity can effectively reduce power losses. However, as the capacity continues to increase, power losses also rise. For instance, when the capacity is less than 100 kWh, the location has little influence on power losses; however, when the capacity exceeds 1000 kWh, the influence on power loss becomes more pronounced at nodes 13 and 18.

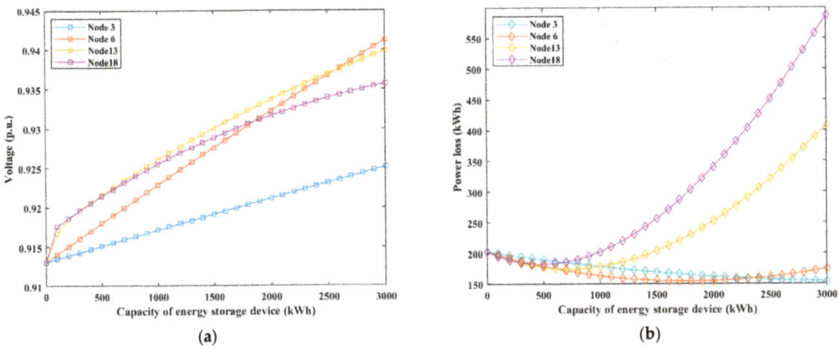

Figure 1. (**a**) The influence of energy storage devices on voltage at various locations and capacities. (**b**) The influence of energy storage devices on power loss at various cases of locations and capacities.

Although large-capacity energy storage devices are effective in improving voltage stability, excessive capacity leads to an increase in power losses and higher investment costs, as well as the movement of devices. The capacity should be reduced as little as possible while ensuring the improvement of voltage stability. Therefore, when determining the location and capacity of energy storage devices, voltage stability and power losses cannot be ignored.

2.2. Analysis of the Characteristics of Mobile Energy Storage Devices

Previous research has mainly focused on stationary energy storage devices. These devices have some limitations when installed in fixed locations, which limits their applications in power systems. The construction of stationary energy storage devices also requires careful consideration of factors such as location, land availability, and infrastructure. If there are discrepancies between the predicted demands and the actual demands, the cost of subsequent adjustments is high.

With the development of high-energy batteries and intelligent energy management systems, the installation location and the capacity of energy storage devices can be changed [22]. When mobile energy storage devices are deployed, their capacity, power, and location can be adjusted to meet the specific demands of different situations, which significantly improves the utilization rate of these devices [31].

To accurately reflect the utilization rate of energy storage devices, Formula (1) is employed to quantify this rate. A higher utilization rate indicates a longer operational time. The utilization rate is calculated as follows:

$$\beta = \frac{T_{use}}{T_{all}} \tag{1}$$

where T_{use} is the operating time of energy storage devices within a research cycle, and T_{all} is the duration of one research cycle.

The mobile energy storage devices utilized in this paper consist of a number of energy storage battery units. Its design is convenient for disassembly and installation, and the number of devices can be increased according to the total capacity required to avoid the energy waste caused by the excessive distributed generation output. Because of their high energy density, these energy storage devices have a variety of advantages in volume and can be flexibly arranged in different locations according to the needs of various scenarios.

According to the characteristics of mobile energy storage devices, a management center is established within the power grid. When the location and capacity of energy storage need to be changed to meet different demands, this center serves as a platform for the coordination and maintenance of energy storage devices. Initially, devices will be adjusted between nodes within the network. Subsequently, new devices will be dispatched from the management center, or unused devices will be returned to the management center based on load demands. The overall configuration method is illustrated in Figure 2.

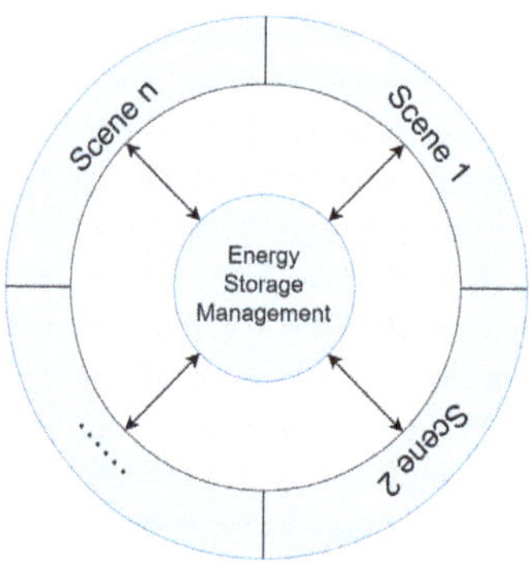

Figure 2. Configuration scheme of energy storage devices.

3. Multi-Objective Optimization Model for the Determination of the Optimal Location and Capacity of the Energy Storage Devices

The determination of the optimal location and capacity is a multi-objective optimization problem. In Section 2, the location and capacity changes of energy storage devices have been proven to have an obvious influence on the voltage stability and values of power losses. On this basis, it is important to ensure the proper configuration of mobile energy storage devices to realize voltage stability, a reduction in power losses, and an increase in the utilization rate. In this paper, voltage vulnerability is introduced to quantify the voltage stability. The smaller the voltage vulnerability, the more stable the system. The smaller the power loss and the capacity of energy storage devices, the lower the investment cost. The multi-objective optimization model for the determination of the optimal location and capacity of the energy storage devices is established in this paper and its objective function and constraints are shown below.

3.1. Objective Functions

(1) Voltage stability

During the operation of the power grid, disturbances lead to voltage instability, causing the voltage of each node to deviate from the normal range. Therefore, in order to minimize these impacts, voltage vulnerability is introduced to measure voltage stability as an objective function. Voltage vulnerability is derived from the standard voltage deviation at each node within the power grid and serves as a measure of voltage stability. A higher voltage vulnerability indicates poorer voltage quality at the node [29,32,33]. The process for constructing the voltage vulnerability is as follows:

$$\begin{cases} v(t,j) = \frac{|U_{t,j} - U_{e,j}|}{\Delta U_{max}} \\ V_{(t,j)} = \frac{v(t,j) - v(t,j_{min})}{v(t,j_{max}) - v(t,j_{min})} \end{cases} \tag{2}$$

$$B(t) = \frac{1}{N}\sum_{j=1}^{N} V(t,j) \tag{3}$$

$$f_1 = \frac{1}{T}\sum_{t=1}^{T} B(t) \tag{4}$$

where $U_{e,j}$ is the rated voltage of node j, $v(t,j)$ and $U_{t,j}$ are the voltage vulnerability and rated voltage of node j at time t, respectively, ΔU_{max} is the maximum allowable voltage deviation, $V(t,j)$ is the voltage vulnerability of node j after standardization at time t, $v(t,j_{max})$ and $v(t,j_{min})$ represent the maximum and minimum voltage vulnerability values of all nodes at time t before standardization, respectively, $B(t)$ is the integrated voltage vulnerability at time t, T is the unit cycle of the system operation with $T = 24$, and N is the number of nodes of the system.

The average voltage vulnerability f_1 ranges from 0 to 1. A higher value indicates greater vulnerability, signifying the decreased security of the power grid. A value close to 0 reflects that the power grid's operation is more stable and reliable.

(2) Power loss

Power loss is an important research objective in power system research. The minimization of power loss is essential for improving system efficiency [32,34]. According to the load demands, energy storage devices either inject or absorb power at various nodes. This process redistributes power flow from overloaded lines to those with lighter loads, thereby reducing the current in the overloaded lines and minimizing power losses. The power loss is calculated as follows:

$$P_{loss} = \sum_{l=1}^{L} k_{t,l} \cdot r_l \tag{5}$$

$$f_2 = \sum_{t=1}^{T} P_{loss} \tag{6}$$

where P_{loss} represents power losses within one hour, L represents the total count of lines within the power grid, $k_{t,l}$ is the square of the current in the line l after the energy storage device is connected, and r_l is the resistance of the line l.

(3) Capacity of energy storage devices

The total capacity of energy storage devices must be carefully determined. If the capacity is too small, these devices will be unable to effectively enhance voltage stability, particularly during significant load changes. Conversely, if the capacity is too large, it will lead to an increase in investment costs. Therefore, the appropriate choice of capacity is crucial. The designated capacities of energy storage devices are defined by the maximum charge or discharge energy within the operational cycle. The capacity is determined by the following formula:

$$f_3 = \sum_{m=1}^{M} \sum_{t=t_0}^{t_0+\Delta t} P_i(t) \cdot \Delta t \tag{7}$$

where t_0 and $t_0 + \Delta t$ are the start and end times of charging or discharging, respectively, $P_i(t)$ is the power of the storage device i at time t, Δt is the time interval between the start and end moments, and M is the number of energy storage devices.

The multi-objective optimization functions, encompassing voltage stability, power losses, and the capacity of energy storage devices, are formulated as follows:

$$min\ f = [f_1, f_2, f_3] \tag{8}$$

3.2. Constraint Conditions

The constraints of the multi-objective optimization model are divided into two components: operation constraints of the power grid and the energy storage devices. The power grid operation constraints include both power flow balance constraints and power balance constraints. The energy storage devices' operation constraints include both capacity and power constraints, as well as state of charge (SOC) constraints.

(1) Power flow balance constraint

$$\begin{cases} P_i(t) = V_i(t) \sum\limits_{j \in i} V_i(t) \left(G_{i,j} cos\theta_{i,j} + B_{i,j} sin\theta_{i,j} \right) \\ Q_i(t) = V_i(t) \sum\limits_{j \in i} V_i(t) \left(G_{i,j} sin\theta_{i,j} - B_{i,j} cos\theta_{i,j} \right) \end{cases} \tag{9}$$

where $P_i(t)$ and $Q_i(t)$ are the active power and reactive power injected into node i at time t, respectively, $G_{i,j}$ and $B_{i,j}$ are the conductance and susceptance of the line connecting node i to node j, respectively, $V_i(t)$ and $V_j(t)$ are the voltages at node i and node j at time t, respectively, and $\theta_{i,j}$ is the phase angle difference between node i and node j.

(2) Power balance constraint

$$P_{grid}(t) + \sum_{i=1}^{N} P_{DG,i}(t) + \sum_{i=1}^{N} P_{ESS,i}(t) = P_{Load}(t) + P_{Loss}(t) \tag{10}$$

where $P_{grid}(t)$ is the active power flowing from the main grid to the distribution network at time t, $P_{DG,i}(t)$ is the active power injected into node i by the distributed generation source at time t, $P_{ESS,i}(t)$ is the active power supplied by the energy storage devices at node i at time t, $P_{Load}(t)$ is the total load at time t, and $P_{Loss}(t)$ is the power losses at time t.

(3) Capacity and power constraints

$$P_{ESS,min} \le P_{ESS}(t) \le P_{ESS,max} \tag{11}$$

$$E_{ESS,min} \le E_{ESS}(t) \le E_{ESS,max} \tag{12}$$

where $P_{ESS}(t)$ and $E_{ESS}(t)$ are the power and energy stored in the energy storage devices at time t, respectively, $P_{ESS,min}$ and $P_{ESS,max}$ are the minimum and maximum power thresholds, respectively, and $E_{ESS,min}$ and $P_{ESS,max}$ are the minimum and maximum rated capacities, respectively.

(4) SOC constraint

The energy storage devices are assumed to operate in a cyclic mode during the research period, meaning that the amount of charging is equal to the amount of discharging points in a period, and the SOC remains unchanged. The specific constraints are as follows:

$$S_{n,s} = S_{n,e} \tag{13}$$

$$S_{n,t} + \alpha_{ch} \cdot P_{ESS}^{n,t+1} = S_{n,t+1},\ t \in H_o \tag{14}$$

$$S_{n,t} + \frac{P_{ESS}^{n,t+1}}{\alpha_{dis}} = S_{n,t+1},\ t \in H_p \tag{15}$$

$$S_{n,t} = S_{n,t+1},\ t \in H_a \tag{16}$$

$$S_{min} \leq S_{n,t} \leq S_{max} \tag{17}$$

$$\sum_{t \in H_p} P_{ESS}^{n,t} = \alpha_{ch} \cdot \alpha_{dis} \cdot \sum_{t \in H_o} P_{ESS}^{n,t} \tag{18}$$

where $S_{n,s}$ and $S_{n,e}$ are the SOC at the beginning and end of the energy storage device n during a research cycle, respectively, $S_{n,t}$ and $S_{n,t+1}$ are the SOC of device n at time t and time $t+1$, respectively, α_{ch} and α_{dis} are the charging and discharging efficiencies, respectively, $P_{ESS}^{n,t+1}$ is the power of device n at time t and S_{min} and S_{max} are the minimum and maximum values of the charged state of the battery, respectively.

4. Multi-Objective Particle Swarm Optimization (MOPSO) Algorithm

As shown in Section 3, it is observed that the determination of the optimal location and capacity includes multiple optimization objectives. To deal with the multi-objective optimization problem, various optimization algorithms have been used in previous research, such as the genetic algorithm, fireworks algorithm, cuckoo search algorithm, ant colony search algorithm, runner-root algorithm, and shuffled frog leaping algorithm [14]. To clearly show the advantages of the mobile energy storage device over the stationary one, MOPSO, as an effective method of solving the multi-objective optimization problem, is applied to the configurations of the two devices. Note that other advanced optimization algorithms may lead to better results, which will be studied in the near future. In the particle swarm optimization (PSO) algorithm, the problem space is regarded as a multidimensional, continuous space, where each particle is a candidate solution. A particle swarm is made up of multiple particles, each of which has a position and a speed that are constantly updated during the iterative process of the algorithm. Each particle has an adaptation value determined by the objective functions, which reflects the current position of the particle. Each particle also records two positions: the best position it has ever been and the best position found by all particles in the entire population. Each particle then updates its speed and position based on its own best position and the group's best position. This process continues until an end condition is met, such as the maximum number of iterations [35,36]. In this algorithm, the number of particles and iterations can be set freely. However, these two parameters are related to the computational complexity and execution time of the simulation because each additional particle requires one more calculation. In addition, the execution time increases with the number of iterations. Therefore, the number of particles and iterations should be chosen reasonably. Among them, the updated formulas for speed and position are as follows:

$$V_{id}^{(k+1)} = wV_{id}^{(k)} + c_1 r_1 \left(P_{id}^{(k)} - X_{id}^{(k)}\right) + c_2 r_2 \left(g_d^{(k)} - X_{id}^{(k)}\right) \tag{19}$$

$$x_{id}^{(k+1)} = x_{id}^{(k)} + V_{id}^{(k+1)} \tag{20}$$

where w is the inertia weight, c_1 and c_2 are learning factors, and random numbers r_1 and r_2 vary between 0 and 1. $P_{id}^{(k)}$ and $g_d^{(k)}$ are the components of the optimal position vectors of the particle i in the k iteration and the population after the k iteration, respectively.

MOPSO is an extension of PSO that is specifically designed to address optimization problems with multiple objectives. It introduces the external archives, which maintain a collection of solutions that are not dominated. The individuals in these external archives are considered elite individuals. The algorithm directs the evolution of the population through these elite individuals, guiding the swarm toward the true Pareto front. Once the algorithm converges, the particles in the external archive are considered the approximated Pareto optimal solutions [37,38].

The problem of the capacity and location selection of energy storage is modeled as a multi-objective optimization problem in this paper, and a set of Pareto optimal solutions is obtained by solving the multi-objective problem. In order to select the optimal solution, that is, to make the optimal balance between different objectives, this paper adopts the technique for order preference by similarity to ideal solution (TOPSIS) [39]. By introducing positive and negative ideal solutions (the maximum and minimum values of each objective function), the method calculates the relative distance of each scheme to the ideal scheme (the distance between the positive ideal solution and the negative ideal solution) and then sorts the schemes to select the optimal scheme. This is based on the principle that among all the candidate solutions, the optimal solution should have the property of being the closest to the ideal solution and the furthest from the negative ideal solution. In this paper, the three objective functions (voltage vulnerability, power loss, and the total capacity of energy storage devices) are first normalized. The optimal and worst solutions (positive and negative ideal solutions) for each objective function are then determined based on the solution set. Finally, the relative distance between each solution and the positive and negative ideal solutions is calculated separately, and the relative distance is calculated using Formula (21). The larger the value of C is, the better the solution is.

$$C = \frac{D^-}{D^+ + D^-} \tag{21}$$

where C is the relative distance and D^+ and D^- are the distances from each solution to the positive and negative ideal solutions, respectively.

A set of Pareto-optimal solutions sets is obtained by solving multi-objective problems, from which decision-makers can choose the best option according to their preference criteria. This paper employs the technique for order preference by similarity to ideal solution (TOPSIS) method to identify the optimal solution [39]. In the TOPSIS method, the first step is to eliminate the influence of dimensions by standardizing the indicators into a common scale. Next, the score is calculated by comparing the distance between each assessed object and the best and worst values. The distance values are then used to compute the final evaluation score. The solution with the highest score is considered the optimal choice.

In the optimization model introduced in this paper, the optimization variables include location, capacity, and power. Consequently, these variables are encoded during the optimization process, and the encoding format is as follows:

$$z = [x_1, x_2, \cdots, x_n, y_1, y_2, \cdots, y_n, P_{1,1}, P_{1,2}, \cdots, P_{n,t}] \quad (22)$$

where x_n is the location where the energy storage device n is connected, it is an integer, and thus it must be rounded up when initializing and updating the location. y_n is the capacity of the energy storage device n, and $P_{n,t}$ is the power of the energy storage device n at time t.

Figure 3 shows the solution process with MOPSO. The specific process is to search for the optimal location and capacity of the energy storage device as follows:

- Enter the basic parameters of the system and initialize the basic quantities in the algorithm;
- Calculate the fitness using Equations (4), (6) and (7). Then, input the non-inferior set of solutions. Finally, determine the individual and group extremes.
- Update particle velocity and position using Equations (19) and (20).
- Recalculate the fitness using Equations (4), (6) and (7) and form a new non-dominated set of solutions.
- Update the non-dominated set and select the global optimum solution.
- Check the termination condition. If the stopping condition is met, output the Pareto solution set and use TOPSIS to select the optimal configuration. Otherwise, return to Step 3 and continue the iterative process.

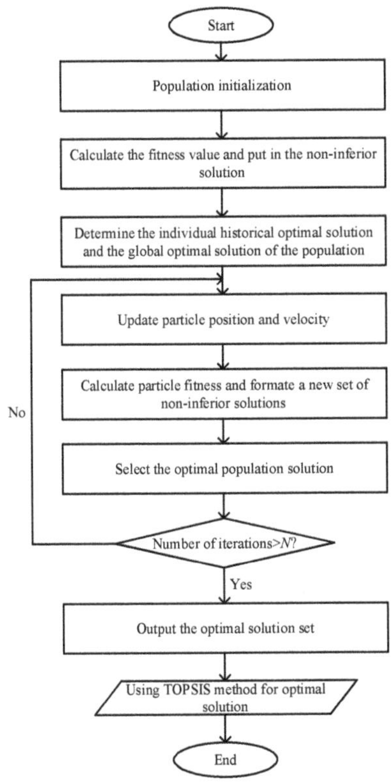

Figure 3. Flow chart of MOPSO.

5. Simulation Analysis

In this section, based on the optimization model established in Section 3 and MOPSO proposed in Section 4, the feasibility of the method proposed in this paper is verified using the IEEE 33-node system as an example in MATLAB R2023a. Based on the wind power, photovoltaic, and load data over a 24 h period, the optimal location and ideal capacity of the energy storage devices can be determined using the method in this paper. This section contains two simulations. In both simulations, the data and the solution methods used are identical, but the types of energy storage devices differ. Specifically, the first simulation (simulation 5.1) features stationary energy storage devices that maintain their location and capacity throughout the day. In contrast, the second simulation (simulation 5.2) introduces mobile energy storage devices. According to the net load, the day is divided into three scenarios, and the energy storage devices are configured separately in each scenario.

5.1. Stationary Energy Storage Devices

For the simulation analysis, this study uses the IEEE 33-node distribution network system, as illustrated in Figure 4. The total network load is 3715 kW + j2300 kvar. The allowable node voltage range is between 0.9 and 1.05 per unit. Detailed system parameters are shown in Table A1 in Appendix A [40].

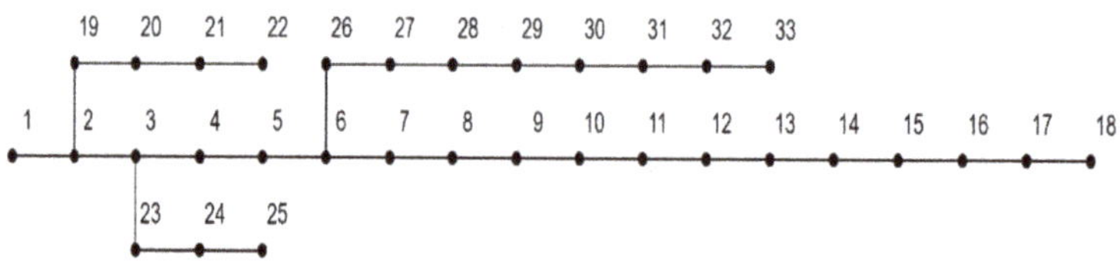

Figure 4. IEEE 33-node distribution network node location diagram.

The typical daily load curve is illustrated in part (a) of Figure 5. The system connects two energy storage devices, with nodes 2 to 33 serving as their access points. The power range of these devices varies from 30 kW to 200 kW, and their capacity ranges from 80 kWh to 300 kWh. The SOC_{min} and SOC_{max} are set as 0.2 and 0.9, respectively. In MOPSO, the population size is set to 50, the number of iterations is 100, the inertia weight is 0.8, and the learning factor is 0.5.

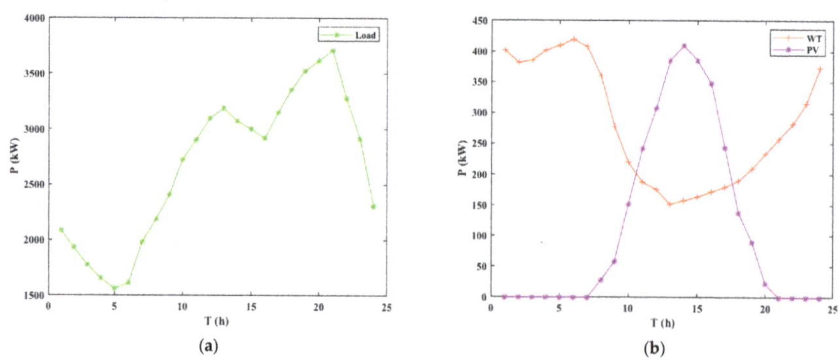

Figure 5. (**a**) Typical load daily curve. (**b**) Typical daily curve for PV and WT.

In this simulation, wind turbine generators (WT) are connected at nodes 14 and 20, and photovoltaic (PV) systems are connected at nodes 9 and 30. The typical daily output curves for both the WT and PV systems are shown in part (b) of Figure 5.

Using the above data, the Pareto solution set obtained by MOPSO is shown in part (a) of Figure 6. Each yellow asterisk represents a solution, and the three axes represent the values of the three objective functions (average voltage vulnerability, power loss and capacity of the energy storage device). The TOPSIS is used to determine the optimal solution, which determines the final locations at nodes 20 and 32 (see part (b) of Figure 6). The black dot represents the node of the IEEE33-node system, and the red star represents the location of the energy storage device. The corresponding capacities for these installations are 204 kWh and 257 kWh, respectively.

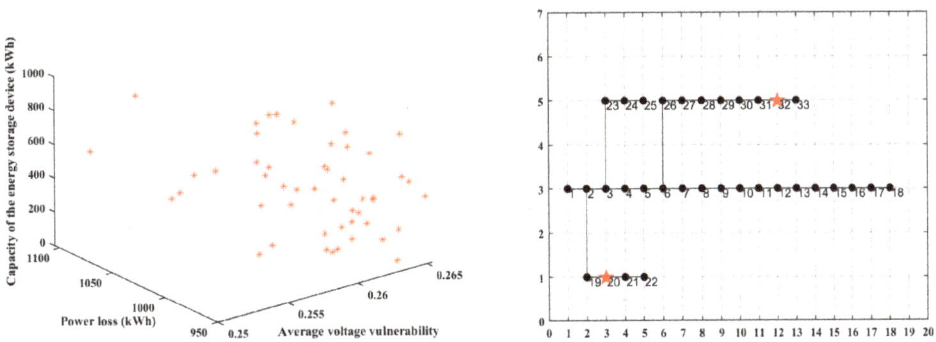

(a) (b)

Figure 6. (**a**) Pareto solution set for the location and capacity selection problem of stationary energy storage devices. (**b**) Optimal location of stationary energy storage devices.

Figure 7 presents the 24 h power and SOC for two energy storage devices. Focusing on the first energy storage device, part (a) of Figure 7 shows the power over 24 h, represented by the blue bar chart. If the power exceeds zero, the energy storage system is in the charging state. Conversely, if it falls below zero, the device is in the discharging state. If the power is zero, neither charging nor discharging occurs. The yellow asterisk represents the SOC of the energy storage device. It is assumed that the energy storage device maintains a balance of charge and discharge during a study cycle, that is, the SOC remains unchanged at the beginning and end of the study. The initial SOC of the energy storage device is set to 0.6. The SOC changes with the charging and discharging behavior of the device.

By observing the load curve and the voltage vulnerability curve, a correlation between them becomes evident. When the load is high, the voltage vulnerability at the nodes increases. If the energy storage devices are connected, it is clear that the voltage vulnerability decreases, indicating an improvement in voltage stability. Figure 8 displays the voltage vulnerability and power loss before and after energy storage devices are integrated into the power grid, respectively. After the energy storage devices are connected, according to Formula (4), the average voltage vulnerability is reduced from 0.3639 to 0.2583, it is significantly reduced by 29%, and the voltage stability is greatly improved. Power loss decreased from 1650 kwh to 1050 kwh, reducing the total power loss by 36%. This shows that more electricity can be utilized efficiently without increasing the total amount of electricity generated, thereby improving the economy of the power grid.

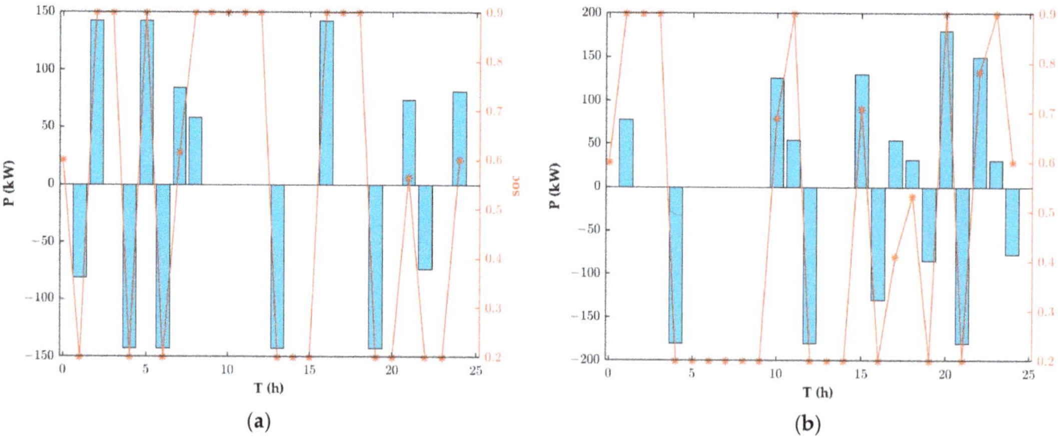

Figure 7. (**a**) The power and SOC of the first stationary energy storage device. (**b**) The power and SOC of the second stationary energy storage device.

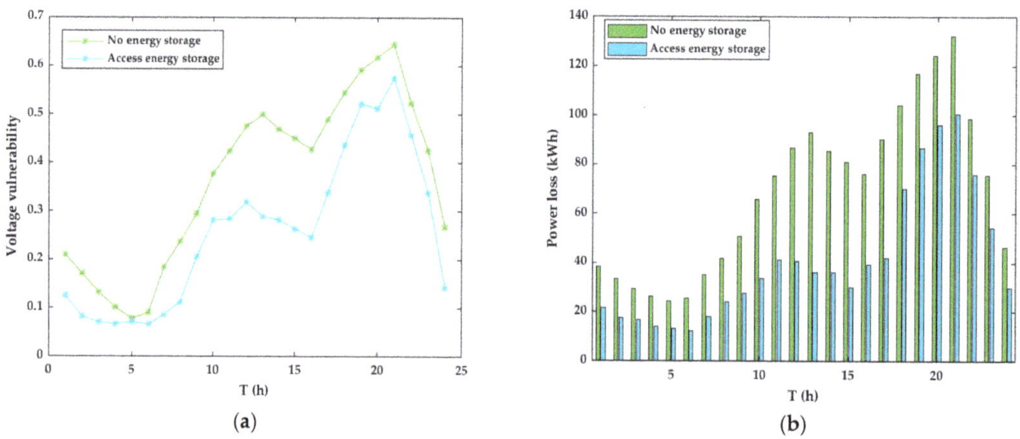

Figure 8. (**a**) The voltage vulnerability before and after energy storage devices are integrated into the power grid. (**b**) The power loss before and after energy storage devices are integrated into the power grid.

5.2. Mobile Energy Storage Devices

Through the simulation of the stationary energy storage devices, the accuracy and feasibility of the proposed model and method are validated. However, the location and capacity of the energy storage devices remained stationary throughout the day. As a result, there is a period of time when the energy storage devices neither charge nor discharge, resulting in low utilization. By utilizing the mobile energy storage method, their location and capacity can be dynamically adjusted throughout the day. Therefore, according to net load demands, the day is divided into various scenarios, and the mobile energy storage devices are dynamically reconfigured to meet load demands in each scenario.

The net load curve is obtained from Figure 5 and shown in Figure 9. The 24 h day is segmented into three time periods based on the average net load over eight consecutive hours. Table 2 lists the time and average load value of each scenario in detail.

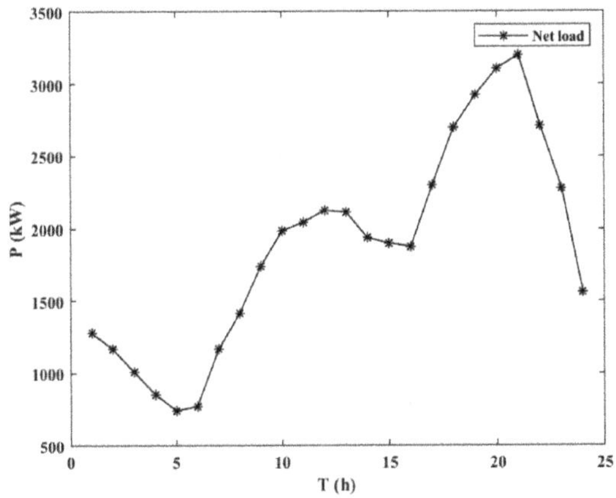

Figure 9. Net load curve.

Table 2. Time and average load for different scenarios.

Scenario	T (h)	Average Net Load (kW)	Net Load Segments
1	01:00–08:00	1050.87	Low
2	09:00–16:00	1965.62	Flat
3	17:00–24:00	2598.12	Peak

In the three scenarios, the configuration schemes of the mobile energy storage devices are shown in Figure 10. During periods of low load, two energy storage devices are connected at nodes 22 and 26, with a capacity of 167 kWh and 123.6 kWh, respectively. As the load increases, two energy storage devices are connected at nodes 2 and 33, with a capacity of 200 kWh and 172 kWh. In this scenario, the existing devices are insufficient to meet the load demands. Therefore, new energy storage devices are dispatched from the energy storage management center. In the highest load scenario, a larger energy storage capacity is required. Consequently, additional energy storage devices must be deployed to the corresponding locations.

The configuration schemes of energy storage devices in each scenario are summarized in Table 3. The utilization rate of these devices is calculated using Formula (1). Without altering the location and capacity of the energy storage devices, their utilization rates are 54% and 63%. However, when the devices are dynamically adjusted to adapt to different scenarios, their utilization rates increase to 87.5% and 92%, respectively. This calculation demonstrates that dynamically adjusting the location and capacity of energy storage devices can significantly improve their overall utilization.

In Figure 11, the comparisons of the voltage vulnerability and the power loss between the stationary and mobile energy storage strategies in the three scenarios are proposed. As for voltage vulnerability, it is clear that the mobile energy storage strategy brings about smaller values of voltage vulnerability in the three scenarios in comparison to the stationary energy storage strategy. As for power loss, except for a few cases at 6 h and 13 h, the power loss of the mobile energy storage strategy is slightly larger than that of the stationary energy storage strategy. The mobile energy storage strategy also reduces the power loss produced by the stationary energy storage strategy in general.

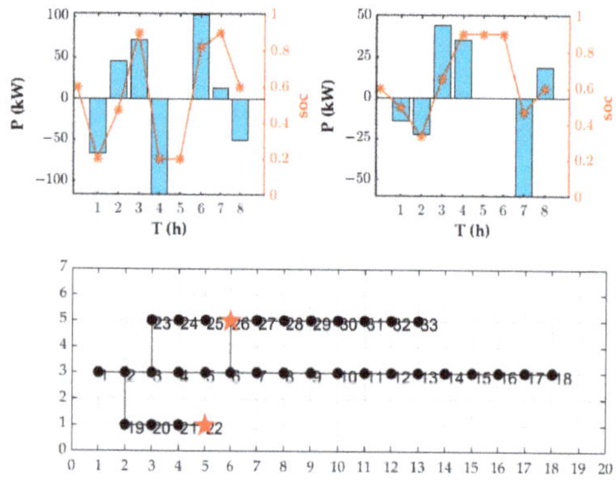

(**a**)

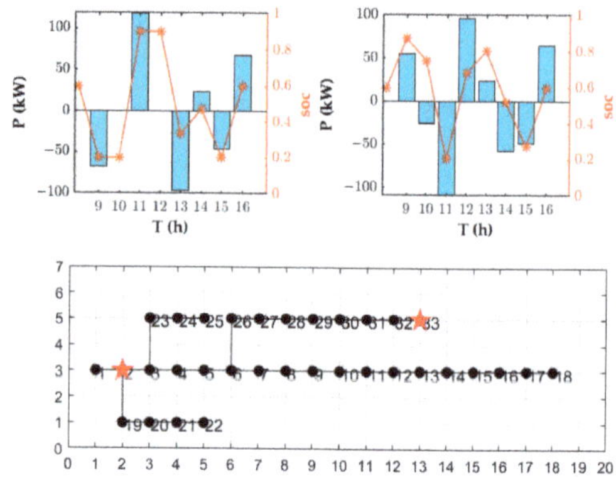

(**b**)

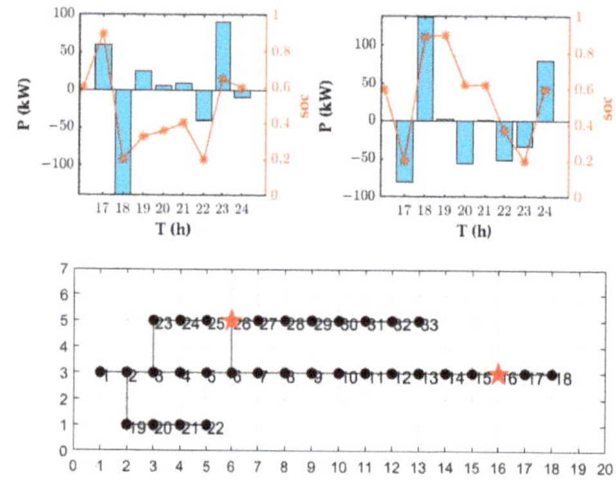

(**c**)

Figure 10. (**a**) Configuration scheme of the mobile energy storage devices in scenario 1. (**b**) Configuration scheme of the mobile energy storage devices in scenario 2. (**c**) Configuration scheme of the mobile energy storage devices in scenario 3.

Table 3. The configuration schemes of energy storage devices in different scenarios.

Scenario	Node	Capacity (kWh)	T_{use} (h)
1	22 26	167 123.6	7 6
2	2 33	200 172	6 8
3	16 26	200 200	8 8
Stationary energy storage device	30 32	204 257	13 15

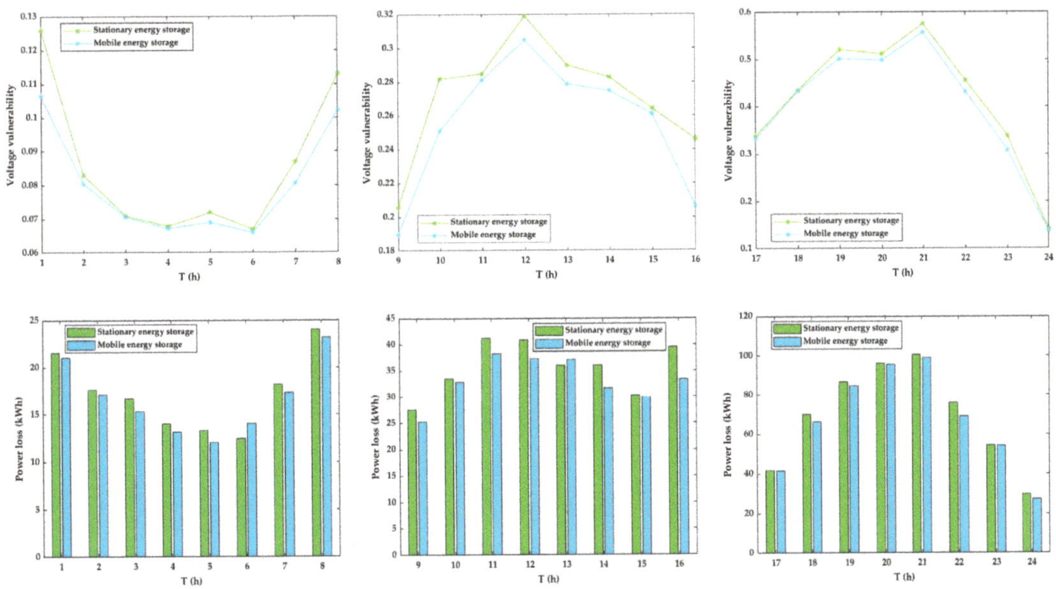

Figure 11. The voltage vulnerability and power loss of the stationary or mobile energy storage strategies in the three scenarios.

To sum up, the application of mobile energy storage devices is effective in enhancing voltage stability, reducing power loss, and increasing utilization. Moreover, the IEEE 33-node system can be changed with actual operational situations to enhance the reliability of the simulation study, and real-world cases will be studied in future research.

6. Conclusions

For the purposes of enhancing the voltage stability and utilization of energy storage devices and reducing power loss, mobile energy storage devices and a configuration method were proposed in this paper as renewable energy generation is connected to the power grid. Firstly, it has been proven that the changes to the location and capacity of the energy storage devices have significant influences on voltage stability and power loss. Secondly, by considering the minimization of power loss, the enhancement of voltage stability, and the improvement of the utilization of energy storage devices as the optimization objectives, a multi-objective optimization model has been constructed, which has been solved by the multi-objective particle swarm optimization (MOPSO) algorithm. By using this method, a simulation analysis has been conducted to examine the advantages of mobile energy storage devices within the power grid over stationary ones. The results have demonstrated that the optimal location and ideal capacity of energy storage devices can be effectively determined while enhancing voltage stability, minimizing power losses, and improving the utilization of these devices.

However, it should be noted that the electricity price during the charging or discharging of the energy storage device is not taken into account in this study. Further optimization of the charging or discharging of the energy storage device, combined with peak and valley electricity prices to improve the economy of the grid, is considered to be a future work. In addition, the deployment of energy storage devices is considered to be an ideal state and not a reality; therefore, according to the power grid and the combination of the road network, the fast and efficient deployment of energy storage is a future research direction. In addition, the typical load, wind power, and photovoltaic curves used as examples in this paper are not universal, and their combination with AI-based forecasting will also be the focus of future research.

Author Contributions: Conceptualization, X.L. and H.D.; methodology, X.L., H.D. and F.L.; validation, Y.A. and M.C.; writing—original draft, X.L. and H.D.; writing—review and editing, F.L. All authors have read and agreed to the published version of the manuscript.

Funding: This research was funded by the State Grid Jiangxi Electric Power Co., LTD, Science and technology project under Grant 521820240004, and the Key R&D Program of Jiangxi Province under Grant 20223BBE51013.

Data Availability Statement: The original contributions presented in this study are included in the article. Further inquiries can be directed to the corresponding author.

Conflicts of Interest: The authors declare no conflicts of interest.

Abbreviations

The following abbreviations are used in this manuscript:

SOC	State of charge
PSO	Particle Swarm Optimization
MOPSO	Multi-Objective Particle Swarm Optimization
TOPSIS	Technique for Order Preference by Similarity to Ideal Solution
WT	Wind Turbine
PV	Photovoltaic
DG	Distributed Generation
V2G	Vehicle-to-grid

Appendix A

Table A1. Branch parameters of the IEEE 33-node distribution network.

Node *i*	Node *j*	Node Impedance (Ω)	Load of Node j (kW)	Node *i*	Node *j*	Node Impedance (Ω)	Load of Node j (kW)
1	2	0.0239 + j0.0350	100 + j60	17	18	0.0348 + j0.0509	90 + j40
2	3	0.0195 + j0.0285	90 + j40	2	19	0.0309 + j0.0452	90 + j40
3	4	0.0344 + j0.0503	120 + j80	19	20	0.0279 + j0.0409	90 + j40
4	5	0.0168 + j0.0246	60 + j30	20	21	0.0175 + j0.0257	90 + j40
5	6	0.0243 + j0.0355	60 + j20	21	22	0.0223 + j0.0327	90 + j40
6	7	0.0260 + j0.0381	200 + j100	3	23	0.0253 + j0.0370	90 + j50
7	8	0.0204 + j0.0299	200 + j100	23	24	0.0150 + j0.0220	420 + j200
8	9	0.0127 + j0.0186	60 + j20	24	25	0.0066 + j0.0097	420 + j200
9	10	0.0168 + j0.0246	60 + j20	6	26	0.0213 + j0.0312	60 + j25
10	11	0.0259 + j0.0380	45 + j30	26	27	0.0149 + j0.0219	60 + j25
11	12	0.0230 + j0.0337	60 + j35	27	28	0.0200 + j0.0293	60 + j20
12	13	0.0298 + j0.0436	60 + j35	28	29	0.0119 + j0.0174	120 + j70
13	14	0.0218 + j0.0319	120 + j80	29	30	0.0063 + j0.0093	200 + j600
14	15	0.0239 + j0.0349	60 + j10	30	31	0.0119 + j0.0174	150 + j70
15	16	0.0360 + j0.0526	60 + j20	31	32	0.0168 + j0.0246	210 + j70
16	17	0.0321 + j0.0470	60 + j20	32	33	0.0079 + j0.0116	60 + j40

References

1. Fang, Z.; Lyu, J.; Jing, J.; Shao, H.; Li, X.; Cai, X. Research on real-time simulation modelling of Zhangbei renewable energy base based on RTDS. In Proceedings of the 8th Renewable Power Generation Conference (RPG 2019), Shanghai, China, 24–25 October 2019; pp. 1–6.
2. Al-Shetwi, A.Q. Sustainable development of renewable energy integrated power sector: Trends, environmental impacts, and recent challenges. *Sci. Total Environ.* **2022**, *822*, 153645. [CrossRef] [PubMed]
3. Husin, H.; Zaki, M. A critical review of the integration of renewable energy sources with various technologies. *Prot. Control Mod. Power Syst.* **2021**, *6*, 3.
4. Impram, S.; Nese, S.V.; Oral, B. Challenges of renewable energy penetration on power system flexibility: A survey. *Energy Strategy Rev.* **2020**, *31*, 100539.
5. Zheng, F.; Meng, X.; Xu, T.; Sun, Y.; Zhang, N. Voltage Zoning Regulation Method of Distribution Network with High Proportion of Photovoltaic Considering Energy Storage Configuration. *Sustainability* **2023**, *15*, 10732. [CrossRef]
6. Elghitani, F.; El-Saadany, E.F. Efficient assignment of electric vehicles to charging stations. *IEEE Trans. Smart Grid* **2020**, *12*, 761–773. [CrossRef]
7. Narasipuram, R.P.; Mopidevi, S. A technological overview & design considerations for developing electric vehicle charging stations. *J. Energy Storage* **2021**, *43*, 103225.
8. Sayed, M.A.; Atallah, R.; Assi, C.; Debbabi, M. Electric vehicle attack impact on power grid operation. *Int. J. Electr. Power Energy Syst.* **2022**, *137*, 107784. [CrossRef]
9. Tavakoli, A.; Saha, S.; Arif, M.T.; Haque, M.E.; Mendis, N.; Oo, A.M. Impacts of grid integration of solar PV and electric vehicle on grid stability, power quality and energy economics: A review. *IET Energy Syst. Integr.* **2020**, *2*, 243–260. [CrossRef]
10. Zhang, S.; Cheng, H.; Xing, H.; Yao, L.; Zhang, Y. Review of DG planning considering uncertainties for distribution network. *Electr. Power Autom. Equip.* **2016**, *36*, 1–9.
11. Zhang, T.; Ma, Y.; Wu, Y.; Yi, L. Optimization configuration and application value assessment modeling of hybrid energy storage in the new power system with multi-flexible resources coupling. *J. Energy Storage* **2023**, *62*, 106876.
12. Sanjay, R.; Jayabarathi, T.; Raghunathan, T.; Ramesh, V.; Mithulananthan, N. Optimal allocation of distributed generation using hybrid grey wolf optimizer. *IEEE Access* **2017**, *5*, 14807–14818. [CrossRef]
13. Khan, M.S.; Waheed, D.; Shahzad, S.; Afzal, S.; Killic, H. Optimal allocation of distributed generation in meshed power networks: A metaheuristic approach. *IET Renew. Power Gener.* **2024**, *18*, 1570–1595. [CrossRef]
14. Onlam, A.; Yodphet, D.; Chatthaworn, R.; Surawanitkun, C.; Siritaratiwat, A.; Khunkitti, P. Power loss minimization and voltage stability improvement in electrical distribution system via network reconfiguration and distributed generation placement using novel adaptive shuffled frogs leaping algorithm. *Energies* **2019**, *12*, 553. [CrossRef]
15. Das, S.; Das, D.; Patra, A. Operation of distribution network with optimal placement and sizing of dispatchable DGs and shunt capacitors. *Renew. Sustain. Energy Rev.* **2019**, *113*, 109219. [CrossRef]
16. Carpinelli, G.; Celli, G.; Mocci, S.; Mottola, F.; Pilo, F.; Proto, D. Optimal integration of distributed energy storage devices in smart grids. *IEEE Trans. Smart Grid* **2013**, *4*, 985–995. [CrossRef]
17. Ghofrani, M.; Arabali, A.; Etezadi-Amoli, M.; Fadali, M.S. A framework for optimal placement of energy storage units within a power system with high wind penetration. *IEEE Trans. Sustain. Energy* **2013**, *4*, 434–442. [CrossRef]
18. Ahmed, H.M.; Sindi, H.F.; Azzouz, M.A.; Awad, A.S. Optimal sizing and scheduling of mobile energy storage toward high penetration levels of renewable energy and fast charging stations. *IEEE Trans. Energy Convers.* **2021**, *37*, 1075–1086. [CrossRef]
19. Yuming, C.; Peng, J.; Gaojun, M.; Yao, W.; Tiantian, L. Optimal capacity configuration of hybrid energy storage system considering smoothing wind power fluctuations and economy. *IEEE Access* **2022**, *10*, 101229–101236. [CrossRef]
20. Zhang, D.; Shafiullah, G.; Das, C.K.; Wong, K.W. Optimal allocation of battery energy storage systems to improve system reliability and voltage and frequency stability in weak grids. *Appl. Energy* **2025**, *377*, 124541. [CrossRef]
21. Ai, B. Study on the SCUC Model and Algorithm Considering Mobile Energy Storage Technology. Master's Thesis, North China Electric Power University, Beijing, China, 2017.
22. Khodayar, M.; Wu, L.; Shahidehpour, M. Hourly coordination of electric vehicle operation and volatile wind power generation in SCUC. *IEEE Trans. Smart Grid* **2012**, *3*, 1271–1279. [CrossRef]
23. Khodayar, M.; Wu, L.; Li, Z. Electric vehicle mobility in transmission- constrained hourly power generation scheduling. *IEEE Trans. Smart Grid* **2013**, *4*, 779–788.
24. Huang, G.; Lei, X.; Yang, Y.; Wang, Y.; Chen, X. Two-layer smart charge-discharge strategies for electric vehicles considering wing generation and users' satisfaction. *Trans. China Electrotech. Soc.* **2015**, *30*, 85–97.
25. Wu, J.; Wen, C.; Li, S.; Shi, Q. Optimal allocation of PV and accumulator system based on time-of-use price. *Adv. Technol. Electr. Eng. Energy* **2018**, *37*, 23–30.

26. Li, J.; Guo, B.; Niu, M.; Xiu, X.; Tian, L. Optimal configuration strategy of energy storage capacity in wind/PV storage hybrid system. *Trans. China Electrotech. Soc.* **2018**, *33*, 1189–1196.
27. Du, X.; Xiong, X.; Wang, J.; Cheng, J. Energy storage planning method of active distribution network based on load ordered clustering. *J. Electr. Power Sci. Technol.* **2023**, *38*, 187–197.
28. Yang, J.; Peng, L.; Luo, L.; Yang, T. Control strategy for energy-storage systems to smooth wind power fluctuation based on interval and fuzzy control. *IEEE Access* **2023**, *11*, 20979–20993.
29. Sivakumar, S.; Ramasamy Gunaseelan, S.B.; Reddy Krishnakumar, M.V.; Krishnan, N.; Sharma, A.; Aguila Téllez, A. Analysis of distribution systems in the presence of electric vehicles and optimal allocation of distributed generations considering power loss and voltage stability index. *IET Gener. Transm. Distrib.* **2024**, *18*, 1114–1132.
30. Nguyen, T.T.; Nguyen, T.T.; Nguyen, N.A. Optimal Network Reconfiguration to Reduce Power Loss Using an Initial Searching Point for Continuous Genetic Algorithm. *Complexity* **2020**, *2020*, 2420171.
31. Hosseini, S.S.; Badri, A.; Parvania, M. A survey on mobile energy storage systems (MESS): Applications, challenges and solutions. *Renew. Sustain. Energy Rev.* **2014**, *40*, 161–170.
32. Ntombela, M.; Kabeya, M. Reduction of Power Losses and Voltage Profile Improvement in a Smart Grid Incorporated with Electric Vehicles. *Sustainability* **2023**, *15*, 10132. [CrossRef]
33. Ettehadi, M.; Ghasemi, H.; Vaez-Zadeh, S. Voltage Stability-Based DG Placement in Distribution Networks. *IEEE Trans. Power Deliv.* **2012**, *28*, 171–178.
34. Mahmoud, K.; Yorino, N.; Ahmed, A. Optimal Distributed Generation Allocation in Distribution Systems for Loss Minimization. *IEEE Trans. Power Syst.* **2016**, *31*, 960–969.
35. Shami, T.M.; El-Saleh, A.A.; Alswaitti, M.; Al-Tashi, Q.; Summakieh, M.A.; Mirjalili, S. Particle swarm optimization: A comprehensive survey. *IEEE Access* **2022**, *10*, 10031–10061.
36. Wang, F.; Zhang, H.; Zhou, A. A particle swarm optimization algorithm for mixed-variable optimization problems. *Swarm Evol. Comput.* **2021**, *60*, 100808.
37. Cui, Y.; Meng, X.; Qiao, J. A multi-objective particle swarm optimization algorithm based on two-archive mechanism. *Appl. Soft Comput.* **2022**, *119*, 108532.
38. Han, F.; Chen, W.T.; Ling, Q.H.; Han, H. Multi-objective Particle Swarm Optimization with Adaptive Strategies for Feature Selection. *Swarm Evol. Comput.* **2021**, *62*, 100847.
39. Elikbilek, Y.; Tüysüz, F. An in-depth review of theory of the TOPSIS method: An experimental analysis. *J. Manag. Anal.* **2020**, *7*, 281–300.
40. Goswami, S.K.; Basu, S.K. A new algorithm for the reconfiguration of distribution feeders for loss minimization. *IEEE Trans. Power Deliv.* **1992**, *7*, 1484–1491.

MDPI AG
Grosspeteranlage 5
4052 Basel
Switzerland
Tel.: +41 61 683 77 34

Processes Editorial Office
E-mail: processes@mdpi.com
www.mdpi.com/journal/processes

www.ingramcontent.com/pod-product-compliance
Lightning Source LLC
Chambersburg PA
CBHW040820250926
51850CB00007BA/1225
* 9 7 8 3 7 2 5 8 4 9 6 7 3 *